The Total Least Squares Problem
Computational Aspects and Analysis

Frontiers in Applied Mathematics

Frontiers in Applied Mathematics is a series that presents new mathematical or computational approaches to significant scientific problems. Beginning with Volume 4, this series reflects a change in both philosophy and format. Each volume focuses on a broad application of general interest to applied mathematicians as well as engineers and other scientists.

This unique series will advance the development of applied mathematics through the rapid publication of short, inexpensive books that lie on the cutting edge of research.

Frontiers in Applied Mathematics

Vol. 1 Ewing, Richard E., The Mathematics of Reservoir Simulation
Vol. 2 Buckmaster, John D., The Mathematics of Combustion
Vol. 3 McCormick, Stephen F., Multigrid Methods
Vol. 4 Coleman, Thomas F. and Van Loan, Charles, Handbook for Matrix Applications
Vol. 5 Grossman, Robert, Symbolic Computation: Applications to Scientific Computing
Vol. 6 McCormick, Stephen F., Multilevel Adaptive Methods for Partial Differential Equations
Vol. 7 Bank, Randolph E., PLTMG: A Software Package for Solving Differential Equations
Vol. 8 Castillo, José E., Mathematical Aspects of Numerical Grid Generation
Vol. 9 Van Huffel, Sabine and Vandewalle, Joos, The Total Least Squares Problem: Computational Aspects and Analysis

The Total Least Squares Problem
Computational Aspects and Analysis

Sabine Van Huffel
Joos Vandewalle
Katholieke Universiteit Leuven

Society for Industrial and Applied Mathematics
Philadelphia, 1991

Library of Congress Cataloging-in-Publication Data

Huffel, Sabine van
The total least squares problem : computational aspects and analysis / Sabine Van Huffel and Joos Vandewalle.
p. cm. -- (Frontiers in applied mathematics ; 9)
Includes bibliographical references and index.
ISBN 0-89871-275-0
1. Least squares. I. Vandewalle, J. (Joos), 1948-
II. Title III. Series
QA275.H84 1991
511'.42--dc20 91-18739

10 9 8 7 6 5 4 3 2

This book is dedicated with love to Adriaan
April 4, 1988–April 28, 1990

Contents

Foreword

Total least squares (TLS) is a concept that leads to a procedure that has been independently developed in various literatures. It has been known by various names; for example, it is known as the "errors-in-variables" model in statistical literature. For the linear case, the technique requires computationally intensive methods for obtaining the numerical solution. With the advent of modern computer technology, it is now possible to compute the solution for problems involving large numbers of variables and observations. A fundamental tool has been the singular value decomposition (SVD), which yields the solution in a transparent manner. Other matrix decompositions are required for solving multivariate problems and for problems with constraints. These numerical procedures have reached a very sophisticated level in the last 25 years.

Thus, TLS represents a technique that synthesizes statistical and numerical methodologies for solving problems arising in many application areas. The authors of this monograph have been the leaders in showing how to use TLS for solving a variety of problems, especially those arising in a signal processing context. They give an elegant presentation of the various aspects of the TLS problem. Their survey encompasses the many elements required to understand the problem. It is a pleasure to read such a clear account, which is presented using standard mathematical ideas and nomenclature.

We owe the authors a debt of gratitude for their complete and careful discussion of a tool that can be of great value in many situations.

Gene H. Golub
Stanford University

Preface

Actual computing power of mainframes, minicomputers, personal computers, workstations, and integrated circuits is increasing substantially each year and is expected to continue to increase well into the nineties. On the other hand, measurements, data capture, sensing, and storage is greatly facilitated for many application areas of engineering, manufacturing, processing, quality control, economics, physics, and health care.

Hence, there is a considerable practical incentive for new approaches, algorithms, and software that can take advantage of the computing power to extract more precise information out of the usually inaccurate measured data. The total least squares (TLS) method is precisely such an approach that has already shown its effectiveness in many diverse applications. It can be applied to all sets of linear equations $AX \approx B$, where both matrices A and B are inaccurately known. Then we typically do not think of the standard numerical analysis environment of computations, where roundoff errors are the only source of inaccuracies; instead, we think of computations on inherently inaccurate measurement data (a few decimal digits accuracy).

In typical applications, gains of 10–15 percent in accuracy can be obtained using TLS with respect to the standard least squares methods, almost at no extra computational cost. Moreover, it becomes more effective when more measurements can be made. Although a systematic investigation of TLS in numerical analysis was only started in 1980 with the SIAM paper of Golub and Van Loan, the subject is now sufficiently worked out to justify an entire book on TLS. Since the concept of TLS, its computation and its properties have been rediscovered many times in statistics, identification, signal processing, and numerical analysis, so the subject is characterized by many scattered results in different domains. In fact, it is expected that a unified presentation of the TLS method—its statistical and numerical properties, the algorithms and software, and its applications—can greatly facilitate its use in an even wider range of applications.

This book is an outgrowth of the doctoral thesis of Sabine Van Huffel in 1987 at the ESAT division of the Electrical Engineering Department of the Katholieke Universiteit Leuven. It contains a comprehensive description of the state of the art on TLS, from its conception up to the summer of 1990.

Using the Book

This book has grown out of our joint research on the TLS problem during the last seven years. Our attempt has not only been to summarize our findings in a unified form and extend these to some degree, but also to include some practical hints for the user. We have chosen a mathematical level of text that we hope will appeal to both those interested in the technical details and those merely interested in the applications of TLS. To understand the book, the reader is assumed to have some basic knowledge of linear algebra and matrix computations and to have some notion of elementary statistics. We have included some background material in §§ 1.4 and 2.2.2 to make the book reasonably self-contained.

We anticipate three categories of readers: (1) researchers in technical areas, specifically in numerical analysis, econometrics, and regression analysis; (2) practicing engineers and scientists; and (3) graduated students in the aforementioned areas.

For researchers, we have analyzed the different types of TLS problems separately and in increasing order of difficulty. Considerable attention is devoted to the computational aspects of each problem. Various algorithms are discussed and their merits are mutually extensively compared. A large amount of sensitivity analysis is presented, and many properties of the TLS problem are proved. For ease of understanding, the TLS properties are compared in detail, wherever possible, with the properties of other methods that are more familiar to the reader; e.g., least squares, principal components, latent root regression, etc. This book encompasses different scientific domains: Chapters 2–7 are devoted to numerical analysis, while Chapters 8 and 9 are mainly statistically oriented. Needless to say, the presentation of both analyses in the same book should be particularly fruitful. On one hand, numerical analysts can appreciate the statistical merits of the TLS problem that is not numerically better conditioned than the least squares problem. On the other hand, statisticians can learn how to improve the numerical properties and efficiency of their solution methods.

For readers who are interested only in practical applications, we recommend a careful reading of Chapters 1 and 2 and of §§ 7.4, 8.6, and 10.2, thus skipping the more theoretical and algorithmic parts of the analysis. In particular, the extensive set of examples presented in § 1.2 should convince the practicing engineer of the merits of TLS in diverse fields of applications. If the basic TLS algorithm outlined in § 2.3.3 is not satisfactory, then Chapter 3 should be read also. Users interested in more efficient algorithms should read Chapters 4 and 5.

The book can also be used by graduate students (and their instructors) in several types of courses; for example, a course on linear equations and least squares problems (Chapters 2, 3, 6, and 7) or a course on errors-in-variables regression (Chapters 2, 3, 8, and 9). To acquire a sufficient level of understanding and insight, many homework problems can be generated. The algorithms can be programmed, for example, in Matlab, and applied to simulated or real life problems, using a small set of measurements. Also, some proofs can be further elaborated.

It is possible to obtain the Fortran 77 code of the TLS algorithms outlined in this book through netlib, a system for the distribution of mathematical software through electronic mail. Simply send a message containing the line "help" to the Internet address netlib@research.att.com or the uucp address uunet!research!netlib. You will then receive information on how to use netlib and how to retrieve our routines, collected in the VANHUFFEL library, from netlib.

Acknowledgments

It is a pleasure to acknowledge the assistance of many friends and colleagues in writing this book.

First, we express our sincere thanks to our friend Professor Gene Golub. He introduced total least squares in numerical analysis and is due credit for his primary role in making the TLS method popular, not only in numerical analysis, but also in a variety of other disciplines such as signal processing, medicine, econometrics, system identification, acoustics, vibration analysis, harmonic retrieval, beamforming and direction finding, parameter estimation, inverse scattering, geology, etc. Gene enthusiastically shared his knowledge with us and encouraged us to study the TLS problem. His diligence and excitement, provocative questions, and valuable suggestions led us to a more careful and complete analysis of the TLS problem.

It was the creative and fertile doctoral research of Dr. Jan Staar that motivated our initial investigations.

We are also very grateful to Dr. Bart De Moor and Hongyuan Zha, not only for their valuable guidance in our research but also for providing constructive criticism and detailed comments on specific chapters in this book.

Furthermore, we express our feelings of thanks to our dear colleagues of our research group ESAT-SISTA for their assistance in our research and for the cheerful and stimulating research environment they have created. In particular, we are indebted to Ingrid Tokka and Hilde Devoghel for typing the manuscript.

It is a pleasure to publicly acknowledge the research support of the Katholieke Universiteit Leuven, of the National Fund for Scientific Research, and of the European Communities. Their generous financial support and the academic freedom made our deep involvement in fundamental research possible. We hope that this book is a modest contribution.

We also express our appreciation to our publisher, the Society for Industrial and Applied Mathematics, and to the editing staff for their consistent cooperation in the publication of this book. In particular, we are indebted to Eric Grosse for inviting us to write this book within the scope of the series and for making many valuable suggestions.

Last but not least, we express our special feelings of thanks to our families, especially Johan, Eva, Liesbeth, Rita, Patrick, Johan, and Ellen for their immense patience and heartfelt encouragement during the realization of this work.

Chapter 1

Introduction

1.1. A simple example

In this book a thorough analysis is made of the method of *total least squares* (TLS), which is one of several linear parameter estimation techniques that have been devised to compensate for data errors.

The problem of *linear parameter estimation* arises in a broad class of scientific disciplines such as signal processing, automatic control, system theory, and in general engineering, statistics, physics, economics, biology, medicine, etc. It starts from a model described by a linear equation:

$$\alpha_1 x_1 + \cdots + \alpha_n x_n = \beta, \tag{1.1}$$

where $\alpha_1, \cdots, \alpha_n$ and β denote the variables and $x = [x_1, \cdots, x_n]^T \in \mathcal{R}^n$ plays the role of a parameter vector that characterizes the special system. A basic problem of applied mathematics is to determine an estimate of the *true* but *unknown* parameters from certain measurements of the variables. This gives rise to an *overdetermined* set of m linear equations ($m > n$):

$$Ax \approx b, \tag{1.2}$$

where the ith row of the data matrix $A \in \mathcal{R}^{m \times n}$ and the vector of observations $b \in \mathcal{R}^m$ contain the measurements of the variables $\alpha_1, \cdots, \alpha_n$ and β, respectively.

In the classical least squares (LS) approach the measurements A of the variables α_i (the left-hand side of (1.2)) are assumed to be free of error; hence, all errors are confined to the observation vector b (the right-hand side of (1.2)). However, this assumption is frequently unrealistic: sampling errors, human errors, modeling errors, and instrument errors may imply inaccuracies of the data matrix A as well. TLS is one method of fitting that is appropriate when there are errors in *both* the observation vector b and the data matrix A. It amounts to fitting a "best" subspace to the measurement data (a_i^T, b_i), $i = 1, \cdots, m$, where a_i^T is the ith row of A.

To illustrate the effect of the use of TLS as opposed to LS, we consider here the *simplest example of parameter estimation*, i.e., only one parameter ($n = 1$) must be estimated. Hence, (1.1) reduces to the following:

$$\alpha x = \beta. \tag{1.3}$$

An estimate for parameter x is to be found from m measurements of the variables α and β:

$$\begin{aligned} a_i &= a_i^0 + \Delta a_i, \\ b_i &= b_i^0 + \Delta b_i, \qquad i = 1, \cdots, m \end{aligned} \tag{1.4}$$

by solving the set (1.2) with $A = [a_1, \cdots, a_m]^T$ and $b = [b_1, \cdots, b_m]^T$. Δa_i and Δb_i represent the random errors added to the true values a_i^0 and b_i^0 of the variables α and β.

If α can be observed exactly, i.e., $\Delta a_i = 0$, errors only occur in the measurements of β contained in the right-hand side vector b. Hence, the use of LS for solving (1.2) is appropriate. This method perturbs the observation vector b by a minimum amount r so that $b - r$ can be predicted by Ax. This is done by minimizing the sum of squared differences $\sum_{i=1}^m (b_i - a_i x)^2$. The best estimate x' of x then follows immediately:

$$x' = \frac{\sum_{i=1}^m a_i b_i}{\sum_{i=1}^m a_i^2}. \tag{1.5}$$

This LS estimation has a nice geometric interpretation in Fig. 1.1(a).

If β can be measured without error, i.e., $\Delta b_i = 0$, the use of LS is again appropriate. Indeed we can write (1.3) as

$$\frac{\beta}{x} = \alpha \tag{1.6}$$

and confine all errors to the measurements of α contained in the right-hand side vector A of the corresponding set $bx^{-1} \approx A$. By minimizing the sum of squared differences between the measured values a_i and the predicted values b_i/x, the best estimate x'' of x follows immediately (see Fig. 1.1(b)):

$$x'' = \frac{\sum_{i=1}^m b_i^2}{\sum_{i=1}^m a_i b_i}. \tag{1.7}$$

In many problems of physics, engineering, biology, etc., however, *both* variables are measured with errors, i.e., $\Delta a_i \neq 0$ and $\Delta b_i \neq 0$. If the errors are independently and identically distributed with zero mean and common variance σ_ν^2, the best estimate $\widehat{x}$ is obtained by minimizing the sum of squared distances of the observed points from the fitted line, i.e., $\sum_{i=1}^m (b_i - a_i x)^2/(1 + x^2)$. This

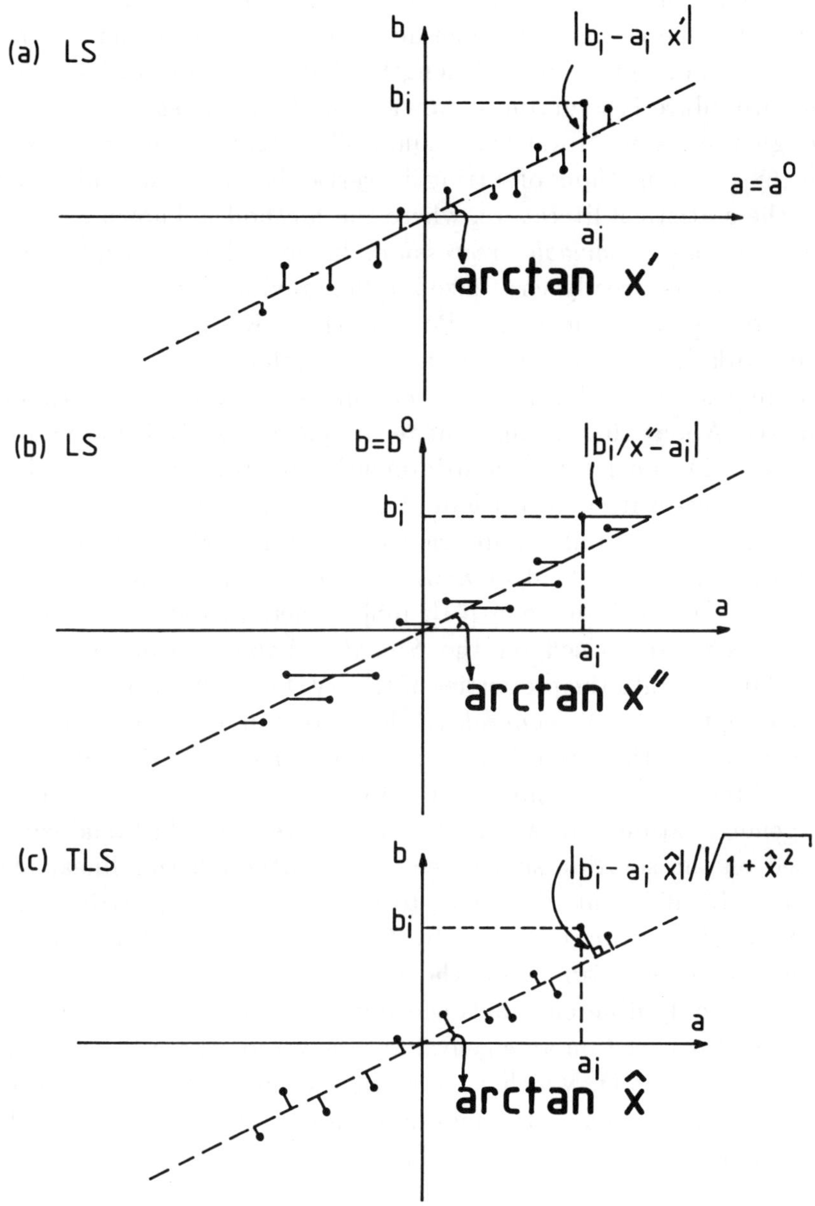

FIG. 1.1. *Geometric interpretation of one-parameter estimation* $\alpha x = \beta$ *with errors in* (a) *the measurements* b_i *of* β *only* (LS *solution*), (b) *the measurements* a_i *of* α *only* (LS *solution*), *and* (c) *both the measurements* a_i *of* α *and* b_i *of* β (TLS *solution*).

is in fact the solution $\widehat{x}$ we obtain by solving (1.2) with TLS for $n = 1$. Figure 1.1(c) illustrates the estimation. The deviations are orthogonal to the fitted line: it is the sum of squares of their lengths that is minimized. Therefore, this estimation procedure is sometimes known as *orthogonal regression.*

Although the name "total least squares" appeared only recently in the literature [68], this method of fitting is certainly not new and has a long history in the statistical literature, where the method is known as *orthogonal regression* or *errors-in-variables regression*. Indeed, the univariate line fitting problem ($n = 1$) was already scrutinized in the previous century [6]. Some well-known contributors are Adcock [6], Pearson [121], Koopmans [92], Madansky [101], and York [210] (see [10] for a list of references). The method of orthogonal regression has been discovered and rediscovered many times, often independently. About 20 years ago, the technique was extended to multivariate problems ($n > 1$) and later to multidimensional problems, which deal with more than one observation vector b in (1.2), e.g., [155], [59].

More recently, the TLS approach to fitting has also attracted interest outside statistics. In the field of *numerical analysis*, this problem was first studied by Golub and Van Loan [61], [68]. Their analysis, as well as their algorithm, is strongly based on the *Singular Value Decomposition* (SVD). Geometrical insight into the properties of the SVD brought us independently to the same concept [156]. We *generalized* the algorithm of Golub and Van Loan to all cases in which their algorithm fails to produce a solution, described the properties of these so-called nongeneric TLS problems, and proved that the proposed generalization still satisfies the TLS criteria if additional constraints are imposed on the solution space (see §3.4). Although this linear algebraic approach is quite different, it is easy to see that the multivariate errors-in-variables regression estimate, given by Gleser [59], coincides with the TLS solution of Golub and Van Loan whenever the TLS problem has a unique minimizer. The only difference between both methods is the algorithm used: Gleser's method is based on an eigenvalue-eigenvector analysis, while the TLS algorithm uses the SVD, which is numerically more robust. Furthermore, the TLS algorithm computes the minimum norm solution whenever the TLS problem lacks a unique minimizer. These extensions are not considered by Gleser.

Also, in the field of *experimental modal analysis*, the TLS technique (more commonly known as the H_v technique), was studied recently [97], [129]. Finally, in the field of *system identification*, Levin [98] first studied the same problem. His method, called the *eigenvector method* or *Koopmans–Levin method* [49], computes the same estimate as the TLS algorithm whenever the TLS problem has a unique solution.

Mixed LS-TLS problems, in which *some* of the columns of A in (1.2) are *error free*, are much less considered. It is quite easy to generalize the classical

TLS algorithm to solve these problems (see §§3.5 and 3.6.3). In particular, this mixed LS-TLS algorithm computes the same estimate as the *Compensated Least Squares* (CLS) method [70], [168] used in system identification when the inputs fed to the system are error free. Hence, the results described in this book also hold for all methods mentioned above and their respective applications.

Remember that TLS is *only one* possible fitting technique for estimating the parameters of a linear multivariate problem. It gives the "best" estimates (in a statistical sense) when all variables are subject to independently and identically distributed errors with zero mean and common covariance matrix equal to the identity matrix, up to a scaling factor. Several other and more general approaches to this problem have led to as many other fitting techniques for the linear as well as for the nonlinear case, see, e.g., [53], [43], [86], [127], [141], [155].

1.2. Total least squares applications: an overview

The renewed interest in the TLS method is mainly due to the development of computationally efficient and numerically reliable TLS algorithms, e.g., [68]. Therefore, much attention in this book is paid to the computational aspects of TLS and new algorithms are presented. The improved results obtained so far in TLS applications also confirm its practical use and enhance the widespread use of this method. Indeed, until now TLS has been successfully applied in very diverse fields, as reviewed below.

There are basically three situations in which TLS is most useful. First, TLS has been proven useful in models with *only measurement error*. These models, referred to as classical *errors-in-variables* (EIV) models, are characterized by the fact that the true values of the observed variables satisfy one or more unknown but exact linear relations of the form (1.1). If the errors in the observations are independent random variables with zero mean and equal variance, TLS gives better estimates than does LS, as confirmed by simulations [190], [8], [7] (see also Chapter 8). This situation may occur far more often in practice than is recognized. It is very common in agricultural, medical, and economic science, in humanities, business, and many other data analysis situations. Hence TLS should prove to be a quite useful tool to data analysts. An important caveat should be noted. The EIV model is useful when the primary goal is model parameter estimation rather than prediction. If we wish to predict new values b_i^0 of β given additional measurements a_i of α in model (1.3)–(1.4), ordinary LS should normally be used. Also, if the data significantly violate the model assumptions, e.g., when outliers are present, the accuracy of the TLS estimates deteriorates considerably and may be quite inferior to that of the LS estimates [8], [7], [197].

A second application of TLS stems from the immediate connection between TLS and *orthogonal least squares fitting*. TLS fits a linear manifold of dimension s, $1 \leq s \leq n$, to a given set of points in $\mathcal{R}^{n+1}$ (the rows of $[A, b]$ in (1.2)) such that the sum of orthogonal squared distances from these points to the manifold attains a minimum [150].

Third, TLS is particularly useful in modeling situations in which the variables α_i and β in (1.1) should be treated *symmetrically*. These situations frequently occur in scientific and technical measurement when we are interested in the parameters of a model only, and not in predicting one variable from other variables. For example, in models with only measurement error, the variables enter the model in a symmetric manner from a distributional point of view but we generally choose to identify one variable as the β variable.

We now point out more specific TLS applications in different fields to illustrate the improvements obtained so far with TLS. This set of examples is certainly not exhaustive and should motivate the reader for the diverse fields of applications. Hence, readers who are not familiar with some of these fields can easily skip these examples without loss of continuity.

First, the use of TLS for handling *multicollinearities*, i.e., nearly exact linear relations among the independent variables α_i of model (1.1), is discussed. In multivariate regression—linear or nonlinear—multicollinearities among the independent variables sometimes cause severe problems. The estimated coefficients can be very unstable and therefore far from their target values. In particular, this makes predictions by the regression model poor. In many chemical applications of multivariate regression, such as the example presented in [209] of relationships between chemical structure and biological activity, the predictive properties of the model are of prime importance and the regression estimates therefore often need to be stabilized. The example [209] shows that the minimum norm TLS solution to the multivariate regression problem is stabilized, as opposed to the LS solution, and has (at least in the example investigated) minimal prediction error compared to all other biased regression estimators studied here. Stabilization is performed by reducing the matrix of observations $[A; b]$ to a matrix of smaller rank. All variables remain in the model.

In applied work, the multicollinearity problem is often handled by selecting a subset of variables such that the size of the subset equals the estimated rank r of the matrix $[A; b]$. How to choose these r variables α_i, which are to be used in approximating the response β in (1.1), is the problem of *subset selection*. If all variables now are observed with zero-mean random errors of equal variance, the TLS method can be successfully applied. A subset selection algorithm SAB-TLS, based on TLS, has been developed [182], [177]. Its selection properties and accuracy in parameter estimation and prediction have been evaluated with respect to the subset selection algorithm SA-LS of Golub, Klema, and Stewart [69, §12.2]. Only the need of predicting future responses β in the true system is considered here, e.g., in hypotheses and

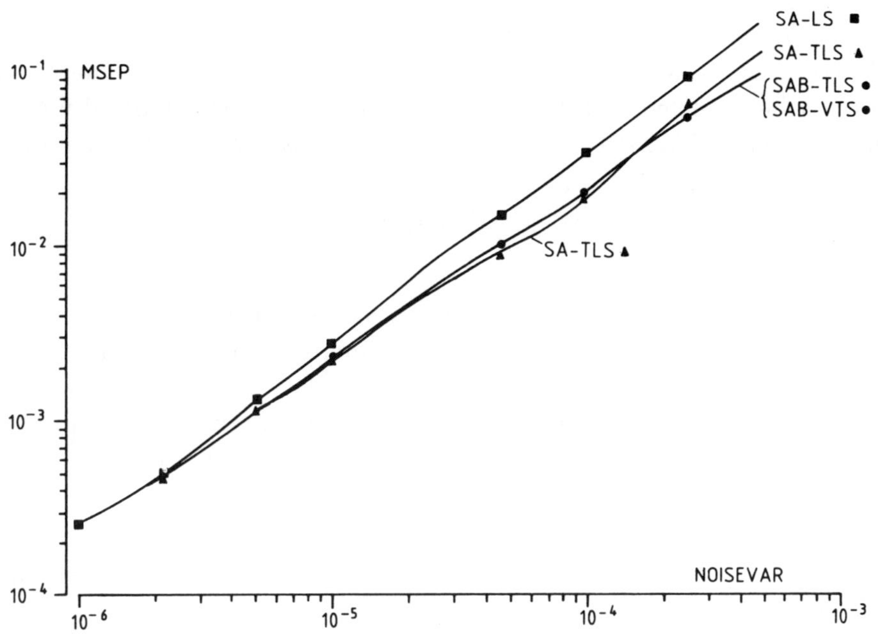

FIG. 1.2. *Comparison of* SA-LS, SA-TLS, SAB-TLS, *and* SAB-VTS: *mean squared error of prediction* (MSEP), *computed over* 100 *noisy sets of equations, versus the variance of the noise added to the exact original model* $(A_0)_{30\times 10}x = b_0$. *The singular values of* A_0 *are* 10, 5, 1, .5, .1, 0, 0, 0, 0, 0. *Hence, five variables must be deleted.* $b_0 = \sum_{i=1}^{5}(1/\sqrt{5})u_i$, *where* u_i *is the ith left singular vector of* A_0.

simulation studies. The TLS concept typically applies to those cases. The SA-LS algorithm is recommended if the variables in the data matrix A are assumed to be free of error. As soon as the variables are unobservable and only perturbed data are available, the TLS technique should be used to obtain the most accurate estimates for the parameters in any selected subset model. This is shown in Fig. 1.2. If the perturbations are sufficiently small, subset selection based on the rth right singular subspace of A (SA-TLS) enables a slightly better prediction accuracy in the case of bad subset condition. In all other cases, namely, in the case of well-conditioned or highly perturbed datasets, subset selection based on the rth right singular subspace of $[A; b]$ (SAB-TLS) yields a better *prediction accuracy.* By taking into account the extra information provided by the observation vector b, SAB-TLS computes the best subset and attains a better and longer stability in its subset choice. A computationally attractive variant SAB-VTS of the subset selection algorithm SAB-TLS, which enables the same prediction accuracy (see Fig. 1.2), was also derived.

In the following we explore the use of TLS in the field of *time domain system identification* and *parameter estimation*. A wide variety of different models are used to describe the system behavior in the time domain. If the process can be modeled as a linear, time-invariant, causal, finite-dimensional system with zero-initial state, then an *impulse response model* may be used:

$$y(t) = \sum_{k=0}^{n} h(k)u(t-k). \tag{1.8}$$

$h(k)$ is the impulse response of the system at time instant k. The system is identified if its impulse response can be estimated from observations of the inputs $u(t)$ and outputs $y(t)$ over a certain interval of time $t = -n, \cdots, m$. This so-called *deconvolution* problem is essentially reduced to a problem of solving a set of linear equations by writing (1.8) for $t = 0, \cdots, m$:

(1.9)

$$Y = UH, \text{ i.e., } \begin{bmatrix} y(0) \\ \vdots \\ y(m) \end{bmatrix} = \begin{bmatrix} u(0) & u(-1) & \cdots & u(-n) \\ u(1) & u(0) & \cdots & u(1-n) \\ \vdots & \vdots & \ddots & \vdots \\ u(m) & u(m-1) & \cdots & u(m-n) \end{bmatrix} \begin{bmatrix} h(0) \\ \vdots \\ h(n) \end{bmatrix}.$$

U is the data matrix obtained from (1.8) and H contains the unknown model parameters $h(k)$. By taking more data samples such that $m > n$, the accuracy can be improved. Assume now that all observed inputs and outputs are perturbed by independent, time-invariant, zero-mean white noise processes of equal variance; then one obtains a more accurate impulse response by solving (1.9) with TLS instead of LS, as confirmed by simulations [177], [181]. Taking into account the Toeplitz structure of the data matrix U further improves the accuracy but these so-called "constrained" TLS problems are hard to solve [2]–[5] (see also §10.3).

The TLS method has been applied as a deconvolution technique in a practical real-life problem, encountered in *renography* [192]. An intravenous injection of a radioactive tracer that is cleared by the kidneys is administered, and the radioactivity over each kidney, as well as over the renal artery, is measured during a certain time period and visualized in a diagram (see Fig. 1.3). Using these samples, the desired renal retention function, which yields useful clinical data, is obtained by deconvolution. Results from simulations, as well as from clinical data, show the advantages of the TLS technique with respect to the matrix algorithm [89], [172]. This last method is the most current deconvolution technique for calculating the renal retention function from (1.9) and is based on back-substitution or Gaussian elimination. As is well known in linear algebra, this method is very sensitive to errors on the data. To counteract this problem, the original curves are then usually smoothed.

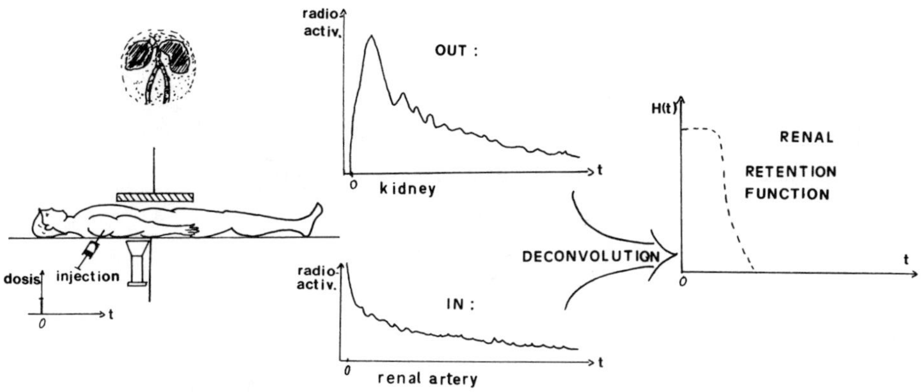

FIG. 1.3. *After the injection of a radioactive tracer into an antecubital vein of the patient, the radioactive counts in each kidney, as well as in the renal artery, are enregistered during a certain time interval* (20-30 *minutes*) *and visualized in an activity-time curve. Using these samples, the desired renal retention function is computed by deconvolution.*

It is concluded that the TLS method is more accurate, more powerful, and more reliable than the matrix algorithm (MA), even if the curves are smoothed. As shown in Fig. 1.4, the accuracy of MA depends strongly on the number of smoothings carried out on the curves. Using TLS, the curves need not be smoothed. Indeed, smoothing does not improve the accuracy of the retention function computed with TLS. Moreover, MA can only solve full rank problems. This implies that the method fails to compute the retention function when the data result in a (nearly) rank-deficient set (1.9) of convolution equations. However, TLS can easily reduce the rank of the matrix involved and can still compute a reliable retention function.

Another frequently used model to describe a system behavior in the time domain is the *transfer function* model, which essentially expresses an *autoregressive moving average* (ARMA) relation between the inputs and outputs fed to the system. In polynomial form, we have the following:

$$A(q^{-1})y(t) = B(q^{-1})u(t), \tag{1.10}$$

where $A(q^{-1}) = 1 + a_1 q^{-1} + \cdots + a_{na} q^{-na}$ and $B(q^{-1}) = b_1 q^{-1} + \cdots + b_{nb} q^{-nb}$ are polynomials of order na and nb, respectively, in the backward shift operator q^{-1}. Since $q^{-1}y(t) = y(t-1)$, (1.10) reduces to:

$$y(t) + a_1 y(t-1) + \cdots + a_{n_a} y(t-n_a) = b_1 u(t-1) + \cdots + b_{n_b} u(t-n_b). \tag{1.11}$$

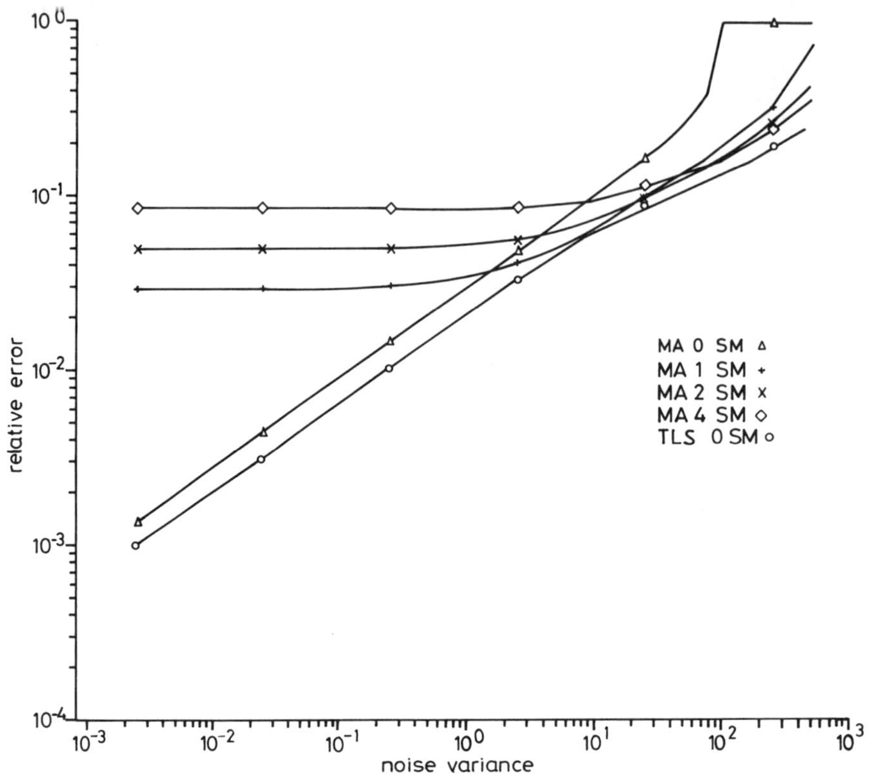

FIG. 1.4. *Comparison of the accuracy of the impulse response computed by* MA *with smoothing and* TLS *without smoothing: a* log-log *diagram shows the average relative error (over* 100 *noisy sets) versus the variance of the noise added to an exact set of convolution equations, derived from the ideal curves. Errors larger than one are set equal to one.* $0, 1, 2,$ *and* 4 *correspond to the number of smoothings carried out on the curves.*

The scalars or vectors $\{u(t)\}$ and $\{y(t)\}$ are the input and output sequences, respectively, and $\{a_j\}$ and $\{b_j\}$ are the unknown constant parameters of the system. If sufficient observations are taken, (1.11) gives rise to an overdetermined Toeplitz-like set of equations. If the observed inputs and outputs are perturbed by mutually independent time-invariant zero-mean white noise sequences of equal variance, the TLS solution of this set, coinciding with the Koopmans–Levin estimate, can be computed here. Aoki and Yue [12] studied the statistical properties of this estimate. The TLS method, although theoretically inferior to true maximum likelihood estimation methods, has all the attractive asymptotic properties of these methods (e.g., strong consistency and mean-square convergence) if certain input conditions and system properties are satisfied, such as equality of the input and output noise variances. Using two published examples of models of the form (1.11), Fernando and Nicholson [49] demonstrated that the accuracy of the TLS estimates is comparable to that of the joint output method proposed by Söderström [147] and superior to all other methods described in [147]. Moreover, the TLS method based on the SVD is numerically more robust and requires less computational effort than the joint output method.

Now assume that only the outputs are disturbed. Then, the mixed LS-TLS solution that takes into account the error-free columns corresponding to the noise-free inputs must be computed and coincides with the compensated LS estimate presented by Stoica and Söderström [168]. However, as shown in §3.6.3, our TLS algorithm based on the SVD is computationally more efficient. Stoica and Söderström proved that the covariance matrix of this estimate is bounded from below by the covariance matrix of a commonly used instrumental variable (IV) estimate and also showed that the TLS method is asymptotically not necessarily more accurate than the ordinary IV method with delayed inputs as instruments. However, these results are only valid for large sample lengths for which asymptotic theory holds quite well and do not necessarily hold for small sample lengths, as shown by Van Huffel and Vandewalle [191]. By means of simulations, the latter compared the accuracy of the TLS parameter estimates with those obtained by applying some commonly used (extended) IV methods [148] in models of the form (1.10) for short sample lengths. They concluded that TLS outperforms the IV methods when *short* sample lengths are used and when the outputs and possibly the inputs are perturbed by independent time-invariant, zero-mean *white noise* processes with equal variance. Mostly, the TLS method is superior in terms of both bias and variance and gives estimates with smaller mean squared error (MSE) than the ordinary IV methods. The MSE becomes comparable by extending the IV sufficiently. The better accuracy of TLS is particularly clear in cases where the zeros of the polynomial $A(z)$ in (1.10) are *close* to the unit circle or where *both* the inputs and outputs are noisy. In other cases, TLS and IV methods give quite similar results.

TLS has also been proved useful in estimating the *autoregressive* (AR) parameters of an ARMA *model* from noisy measurements [166]. These models are described by (1.10), where $\{u(t)\}$ is a white noise process with zero mean and unit variance. Here, we assume that $A(q^{-1}) = a_0 + a_1q^{-1} + \cdots + a_{n_a-1}q^{-n_a+1} + q^{-n_a}$, $(a_0 \neq 0)$. $B(q^{-1})$ is a similarly defined polynomial. A modification of the High-Order Yule–Walker (HOYW) method was proposed in [167] to estimate the AR parameters $\{a_i\}$ of this ARMA model and it is applicable to AR-plus-noise models, as well as to general ARMA processes. This method provides estimates of the coefficients of a polynomial that contains among its roots the poles of the ARMA model. To obtain these coefficients, a linear system of the form $Ax \approx b$ must be solved. Since both A and b

are noisy, the use of TLS instead of LS for solving the HOYW equations is recommended. Simulation results show that, in general, the HOYW-TLS method outperforms the HOYW-LS technique described in [167]. The better accuracy of TLS with respect to LS is particularly clear in cases where the zeros of the ARMA model approach the unit circle, no matter what the pole configuration is. The accuracy increases significantly with increasing dimension of the HOYW system.

Another application in the field of system identification to which TLS has been applied is *structural identification*; i.e., based on some criterion and using noise-corrupted data, a suitable structure for a linear multivariable system is selected out of a sequence of increasing-complexity models. The approach of Beghelli, Guidorzi, and Soverini [16] is based on the use of canonical models that are described by a minimal number of parameters and directly exhibit the system structure. A criterion based on the predicted percentage of reconstruction errors (PPCRE) associated with this sequence of models, whose parameters are estimated with the TLS method, is proposed to select the best model. The properties of the PPCRE for a canonical model computed with reference to the TLS estimate are similar to those for a canonical model computed with reference to the LS estimate [71]. Two main differences can be noted: while in a LS environment the PPCRE is a monotone (nonincreasing) function, the PPCRE associated with the TLS method can increase when the order of the associated subsystem becomes larger than the real one. The second difference regards the larger values of the TLS PPCRE when compared with the LS PPCRE; this is because the reconstruction error (on the noisy data) associated with a LS model is lower than the error associated with a maximum likelihood model. This last model, however, gives a more accurate description of the real process than the LS model does. Like LS PPCRE, the TLS PPCRE can also be used not only with reference to canonical state-space and input-output models but also with reference to multistructural (overlapping) models, which can be used advantageously in on-line identification problems.

Furthermore, the TLS approach has been applied to the *identification of state-space models* from *noisy* input-output measurements [38], [105], [106]. A linear discrete-time, time-invariant system with m inputs and l outputs is considered with state-space representation:

$$\begin{aligned} x(t+1) &= Ax(t) + Bu(t), \\ y(t) &= Cx(t) + Du(t). \end{aligned} \tag{1.12}$$

Vectors $u(t)$, $y(t)$, and $x(t)$ denote the input, output, and state at time t, the dimension of $x(t)$ being the minimal system order n. A, B, C, and D are the unknown system matrices to be identified, making use only of measured input-output sequences. A fundamental structural matrix input-output equation is derived from the state-space equations (1.12) and provides a much more elegant framework for the formulation and solution of the multivariable identification problem. Indeed, as opposed to other, mostly stochastic, identification schemes, no variance-covariance information whatsoever is involved, and only a limited number of input-output data are required for the determination of A, B, C, D. The algorithm consists of two steps. First, a state vector sequence is realized as the intersection of the row spaces of two block Hankel matrices, constructed with input-output data. This corresponds to a TLS approach that applies when both input and output are corrupted by the same amount of noise. Next, the system matrices are obtained by solving a set of linear equations. The algorithm is easily converted into an adaptive version for slowly time-varying systems and can be extended to the case where the input and output data are corrupted by colored noise. Examples, including the identification of an industrial plant, demonstrate the robustness of the identification scheme with respect to noise and over- and underestimation of the system order [38], [105].

In the field of modern *signal processing*, the TLS approach is also under study and is very promising.

First, we consider the classical problem of estimating the *frequencies* and *amplitudes* or powers of multiple *complex sinusoids in white noise.* This problem arises in numerous applications encountered in such diverse fields as radar, sonar, exploration seismology, and radio astronomy. A variety of algorithms for this so-called *harmonic retrieval* problem [123] has been proposed and analyzed over the past few decades. We only cite here the TLS-based contributions (although not always explicitly stated), such as the algebraic approach of Pisarenko [123], the linear-prediction-based work of Rahman and Yu [125], the rotational invariance technique ESPRIT of Roy and Kailath [131], as well as the Procrustes-rotations-based ESPRIT algorithm proposed recently by Zoltowski and Stavrinides [220].

Mathematically, the problem can be described as follows:

$$(1.13) \qquad x_t = \sum_{k=1}^{M} a_k e^{j2\pi f_k t} + e_t \qquad t = 0, 1, \cdots, N-1,$$

where $\{x_t\}$ and $\{e_t\}$ are the measured samples and noise samples, respectively, of the received signal; f_k and a_k are the frequency and amplitude of the kth sinusoid, $k = 1, \cdots, M$, which are to be estimated from the given set of N data samples.

To solve this problem, the following set of linear prediction equations is formed:

$$(1.14) \quad Ac \approx b \text{ or } \begin{bmatrix} x_0 & x_1 & \cdots & x_{L-1} \\ x_1 & x_2 & \cdots & x_L \\ \vdots & \vdots & & \vdots \\ x_{N-L-1} & x_{N-L} & & x_{N-2} \end{bmatrix} \begin{bmatrix} c_L \\ c_{L-1} \\ \vdots \\ c_1 \end{bmatrix} \approx \begin{bmatrix} x_L \\ x_{L+1} \\ \vdots \\ x_{N-1} \end{bmatrix},$$

where L is the prediction order satisfying $M \leq L \leq N - M$, A is the linear prediction (LP) data matrix using the data in the forward direction, c is the LP vector, and b is the observation vector. The matrix A and the vector b will have different elements if we use the data matrix in the backward or forward-backward direction.

The different possible choices for the prediction order L, the number of data samples N, and the ways of solving the set (1.14) (e.g., in a LS or TLS sense) give rise to a wide range of existing estimation methods. For example, if $L = M$ and $N = 2M$, Prony's method is obtained. In this case, (1.14) can be solved exactly but the method can only be applied to noise-free signals. If $L = M$ and the prediction vector c is extracted from the appropriately normalized eigenvector $[c^T; -1]^T$, associated with the minimal eigenvalue of the $(2M+1)$th-order covariance matrix R_{2M+1} of the noisy data, the *Pisarenko* harmonic decomposition method is obtained [123]. Its close relationship to the TLS problem is evident from the following possible formulation of the Pisarenko harmonic decomposition:

(1.15)
$$\min_{c} \left\{ [c^T; -1] R_{2M+1} \begin{bmatrix} c \\ -1 \end{bmatrix} / (c^T c + 1) \right\} \quad \text{with} \quad R_{2M+1} = [A; b]^T [A; b].$$

The solution c to this problem is clearly given by the minimal eigenvector (appropriately normalized) of R_{2M+1} or, in other words, c equals the TLS solution of $[A; b][c^T; -1]^T \approx 0$.

To improve the spectral resolution for short data length and low signal-to-noise ratio, a prediction order L much higher than the number of sinusoids but lower than the number of samples N is used [171], [125]. Tufts and Kumaresan [171] formulated the principal eigenvector method; i.e., using the SVD, the LP data matrix A is first reduced to a matrix A_r of lower rank r; c is then given by the minimum norm LS solution of $A_r c \approx b$. Rahman and Yu [125] applied a TLS approach for solving the LP equations (1.14) to combat the noise effects from both the data matrix A and observation vector b simultaneously. Here, c is given by the minimum norm TLS solution of $[\widehat{A}; \widehat{b}][c^T; -1]^T \approx 0$, where $[\widehat{A}; \widehat{b}]$ is the best rank r approximation of $[A; b]$. The best accuracy is obtained for $r = M$.

Once the prediction vector c is determined, the frequencies can be estimated from the zeros of the prediction-error filter polynomial:

$$C(z) = 1 - c_1 z^{-1} - c_2 z^{-2} - \cdots - c_L z^{-L}. \tag{1.16}$$

The order L of the polynomial may lie between M and $N - M$. So, there will be $L - M$ extraneous zeros, and M signal-related zeros in $C(z)$. The frequencies $\{f_k\}$ in (1.13) can be derived from the angular positions of the latter zeros.

Finally, the amplitudes can be solved from the set (1.13) of observed samples that are linear in amplitudes.

The resolution of the estimated closely spaced frequencies of the multiple sinusoids degrades as the signal-to-noise ratio of the received signal becomes low. This resolution can be improved by using the TLS method in solving (1.14), as shown in [125]. The TLS method performs better than the principal eigenvector method in resolving *closely spaced* frequencies for both damped and undamped sinusoids in terms of both mean squared error and bias. The improvement is especially significant at *low* prediction order. Moreover, it decreases the *threshold signal-to-noise ratio* below the value that can be achieved by the principal eigenvector method. This improvement in spectral estimation provided by the TLS method [125] is apparent in the quantitative analysis of multidimensional *Nuclear Magnetic Resonance* (NMR) spectra, in particular, when the data are truncated or the signal-to-noise ratio is low. This is the case when the NMR data are obtained from dilute or low gyromagnetic ratio nuclei [170].

The TLS approach has also been applied to the more general class of practical signal processing problems whose objective is to estimate from measurements a set of constant parameters upon which the received signals depend. For example, high resolution *direction-of-arrival* (DOA) estimation is important in many sensor systems such as radar, sonar, and electronic surveillance. In such problems, the functional form of the underlying signals can be assumed to be known, e.g., narrowband plane waves, complex sinusoids, etc. The quantities to be estimated are parameters (e.g., frequencies and DOAs of plane waves, sinusoid frequencies, etc.) upon which the sensor outputs depend. Several approaches to such problems have been proposed, as surveyed in [152]. Although often successful and widely used, these methods have certain fundamental limitations (especially bias and sensitivity in parameter estimates), largely because they do not explicitly use the underlying data model. Schmidt and, independently, Bienvenu were the first to correctly exploit the measurement model in the case of sensor arrays of arbitrary form. Schmidt's algorithm, called MUSIC (MUltiple Signal Classification), has been widely used [139]. Although the performance advantages of MUSIC are substantial, they are achieved at a considerable cost in computation (searching over parameter space) and storage (of array calibration data). Recently, a new algorithm, called ESPRIT (Estimation of Signal Parameters via Rotational Invariance Techniques), which considerably reduces the aforementioned computation and storage costs, has been developed [130], [134]. In the context of DOA estimation, the reductions are achieved by requiring that the sensor array possesses a displacement invariance; i.e., the sensors must occur in matched pairs, the two sensors in each pair must be identical to each other in response, and the displacement between both sensors (separation and angular orientation) must be identical for each pair. In many problems, these conditions are satisfied. For example, linear arrays of equispaced identical sensors are commonplace in sonar applications, as are regular rectangular arrays of identical elements in radar applications. Although the preliminary version of ESPRIT is manifestly more robust, i.e., less sensitive, with respect to array imperfections than previous techniques, including MUSIC, the LS criterion inherently employed leads to parameter estimate bias at low signal-to-noise ratios. By applying the TLS approach, Roy and Kailath [131]–[133] improved their early LS version of ESPRIT and called their algorithm TLS-ESPRIT. By means of simulations, it is shown that the application of the TLS criterion eliminates the small parameter bias at low signal-to-noise ratios, resulting in an improved performance. In the other cases, the differences between the LS and TLS parameter estimates are insignificant. More details can be found in [130], [131], [132], [112], [113], [126].

Recently, Zoltowski and Stavrinides [220] proposed the *Procrustes-rotations*-based ESPRIT algorithm, called PRO-ESPRIT, as an alternative to the TLS-ESPRIT algorithm of Roy and Kailath. Given data derived from two identical sensor arrays X and Y that are displaced relative to one another, PRO-ESPRIT exploits the inherent redundancy built into the ESPRIT geometry to compensate, in part at least, for imperfections in the array system. This redundancy is manifested by two fundamental properties of the noiseless X and Y data matrices comprising the ESPRIT data pencil: they have the same row space, the signal subspace, and the same column space, the source subspace. These properties are advantageously exploited in PRO-ESPRIT by invoking the solution of the Procrustes problem [69, §12.4.1] for optimally approximating an invariant subspace rotation.

By invoking the TLS concept in PRO-ESPRIT, this redundancy is further exploited. Note, however, that the way in which the TLS concept is applied in PRO-ESPRIT differs from the way it is applied in TLS-ESPRIT, where it is used to solve a set of linear equations. In contrast, PRO-ESPRIT utilizes TLS as a means for perturbing each of the two estimates of the signal subspace in some minimal fashion until they are equal. The common subspace after perturbation is then taken as a better estimate of the signal subspace. A better estimate of the source subspace is obtained in the same way. This TLS-based variant of PRO-ESPRIT, called TLS-PRO-ESPRIT, offers the potential for better performance than either PRO-ESPRIT or TLS-ESPRIT, at the expense of increased computation.

Furthermore, Zoltowski also demonstrated the improvements in performance obtained by applying the minimum norm TLS solution method to the *minimum variance distortionless response* (MVDR) *beamforming* problem [217] and to the covariance level ESPRIT problem [134] encountered in the field of sensor array signal processing and high-resolution frequency estimation [218],

[219].

In the field of *experimental modal analysis* the TLS method has been successfully applied in estimating *frequency response functions* from measured input forces and response signals applied to mechanical structures [97], [129]. An application example [129] shows that, without special measurement considerations, the estimates obtained with the TLS method are superior to the ones obtained with the LS method. In particular, around the resonance frequencies the estimate using TLS is better than the estimate using LS. The latter estimate is known to be strongly influenced by leakage, causing an underestimate of the true value. The estimate obtained with the TLS technique is less influenced by errors in the data, such as leakage. However, if special attention is given to reducing errors in the measurement process, e.g., by the use of special excitation signals, then both techniques give comparable results.

In *acoustic radiation* problems the use of TLS can also improve the accuracy. Formulated with boundary element techniques, these problems require the solution of a system of linear equations $Ax \approx b$ [75]. The matrix A is dense, the vector x contains the unknowns on the boundary (acoustic pressure, normal motion, impedance, or some combination of the three), and the vector b is calculated from information known or measured on the boundary. A, x, and b are all complex. The elements of both A and b include errors due to numerical inaccuracy and data uncertainty. It is shown [75] that the TLS method is more robust and computes the acoustic surface pressure more accurately (around 3 percent) than the LS method at or near the characteristic frequencies where the nonuniqueness problem of the Helmholtz integral equation occurs. Apart from these frequencies, the TLS and LS solutions are comparable in accuracy.

Even in the field of *geology* the use of TLS has been investigated. When interpreting metamorphic mineral assemblages, it is often necessary to identify assemblages that may represent equilibrium states to determine whether differences between assemblages reflect changes in metamorphic grade or variations in bulk decomposition or to characterize isograd reactions. In multicomponent assemblages these questions can be best approached by investigating the rank, composition space (range), and reaction space (null space) of a matrix representing the compositions of the phases involved. The TLS method based on the SVD of the data matrix is a useful tool for computing these quantities [50] and, moreover, enables us to find a model matrix of specified rank that is closest (in Frobenius norm) to an observed assemblage. These models permit quantitative testing of the role played by minor components in an assemblage and accurate determination of isograd reactions. Moreover, this TLS approach allows for errors in all observations, is computationally simpler, more direct, and more stable than the currently used nonlinear least squares algorithms for finding linear dependencies, and it enables us to treat large composite assemblages as single entities, instead of requiring examination of numerous subsets of the full assemblage.

Inverse scattering is another class of problems in which TLS has been successfully applied. Succinctly, the inverse scattering problem is to infer the shape, size, and constitutive properties of an object from scattering measurements that result from one of the following: seismic, acoustic, or electromagnetic probes. Under ideal conditions, theoretical results exist. However, when the scattering measurements are noisy, as is the case in practical scattering experiments, direct application of the classical inverse scattering solutions results in numerically unstable algorithms. Applying the TLS method, Silvia and Tacker [144] were able to provide a regularization to the one-dimensional inverse scattering problem, which arises, for example, in the exploration for oil and natural gas. Specifically, they show how to use multiple data sets in a Marchenko-type inversion scheme and how the theory of TLS introduces an adaptive spectral balancing parameter that explicitly depends on the scattering data. This is a clear advantage of the use of TLS, as opposed to LS techniques, which utilize nonexplicit and nonadaptive spectral balancing parameters, generally derived by ad hoc considerations.

In *theoretical* studies the analysis of the TLS problem can also be useful. In [19], Bloch shows how the TLS problem for a countable sequence of data points in a separable Hilbert space gives rise to an infinite-dimensional *Hamiltonian system* that can be explicitly integrated. Moreover, the Hamiltonian flow mirrors statistical information associated with the given data. In particular, it reflects asymmetries in the principal components of the data. These results shed light on the analysis of quite general problems in nonlinear estimation and control.

Finally, the TLS concept has been applied to *nonlinear* models in which the true variables are nonlinearly related to each other. The independent variables, as well as the observations, may have errors. Golub and LeVeque [65] presented two algorithms for solving separable nonlinear TLS problems. Other algorithms are discussed in [20], [100], [141], [142].

This list of TLS applications is certainly not exhaustive. Currently, the use of TLS in geophysical tomographic imaging [85] and oceanographic data analysis, in three-dimensional motion registration, and in parameter estimation for partial differential equations, as well as in nonlinear regression, is being investigated. However, there still remain many *unexplored* problems, where TLS could be successfully applied to improve the quality of the results. It is hoped that this book will aid and stimulate the reader to apply TLS in his or her own applications and problems.

1.3. Outline of the book

In this book, the main emphasis is put on the *linear algebraic* approach of the TLS technique. Therefore, it is assumed that the reader is somewhat familiar with this domain (see [69] for a comprehensive survey of most tools used in this book). Less attention is given to the viewpoint of the statistician.

This book is organized into ten chapters, each treating a different aspect of the TLS problem. In the following two chapters, the TLS problem is fully analyzed.

Chapter 2 surveys the main principles of the basic TLS problem $Ax \approx b$ and shows how to compute its solution in a reliable way by means of the SVD. By "basic" we mean that only one right-hand side vector b is considered and that a solution of the TLS problem exists and is unique. For ease of understanding, the SVD and LS problems, as well as their main properties, are surveyed. A geometric comparison between TLS and LS problems enlightens the main differences between both principles.

Extensions of the basic TLS problem are investigated in Chapter 3. We discuss consecutively multidimensional TLS problems $AX \approx B$ having more than one right-hand side vector, problems in which the TLS solution is no longer unique, TLS problems that fail to have a solution altogether, and mixed LS-TLS problems that assume some of the columns of the data matrix A to be error free. At the chapter's end, the TLS computations are summarized in one algorithm that takes into account all the extensions given above.

The next two chapters show how to improve the *efficiency* of the TLS computations.

In Chapter 4, the TLS computations are sped up in a *direct* way by modifying the SVD computations appropriately. These modifications are summarized in a computationally improved algorithm PTLS. An analysis of its operation counts, as well as computational results, shows its relative efficiency with respect to the classical TLS computations.

Chapter 5 describes how the TLS computations can be sped up in an *iterative* way if a priori information about the TLS solution is available. In particular, inverse iteration and (inverse) Chebyshev iteration methods for solving slowly varying TLS problems are investigated. Different algorithms are presented and their convergence properties are analyzed. This analysis allows us to compare the efficiency of these algorithms with that of the direct computation methods and shows for which class of problems each method is computationally most efficient. Simulated examples confirm the theoretical analysis. Additionally, the efficiency of the inverse iteration method is illustrated in a practical real-life problem encountered in experimental modal analysis. Furthermore, Rayleigh quotient iteration and the Lanczos methods are briefly discussed, in particular, their applicability in solving slowly varying TLS problems.

In the next four chapters, the *properties* of the TLS method are analyzed to delineate its domain of applicability and to evaluate its practical significance.

Chapter 6 proves interesting *algebraic* connections between the TLS and LS problem with respect to their solutions, their residuals, their corrections applied to data fitting, and their approximate subspaces.

In Chapter 7 the *sensitivity* of the TLS problem is compared with that of the LS problem in the presence of errors in all data. The perturbation effects on their solutions and on the SVD of their associated matrices are analyzed and confirmed by simulations.

Chapter 8 presents the *statistical* properties of the TLS technique based on knowledge of the distribution of the errors in the data and evaluates their practical relevance.

In Chapter 9, *algebraic* connections, interrelations, and differences in approach between the *classical linear regression* estimators (least squares, principal component, ridge regression, and latent root regression) and (non) generic TLS estimation in the presence of multicollinearities are outlined, using the SVD and geometric concepts.

Finally, Chapter 10 summarizes the conclusions of this book and surveys some recent extensions of the classical TLS problem currently under investigation. Suggestions for further research are also made.

1.4. Notation and preliminaries

Before starting, we introduce some notation, list the assumptions, and define the elementary statistical concepts used throughout this book:

- A *matrix* is always denoted by a *capital* letter, e.g., A. The corresponding lowercase letter with the subscripts i and ij refers to the ith column and (i,j)th entry, respectively, e.g., a_i, a_{ij}.

- A *vector* is represented by a *lowercase* letter, e.g., b. The individual components are denoted with single subscripts, e.g., b_i.

- The superscript T denotes the *transpose* of a vector or matrix.

- $R(S)$ (respectively, $R_r(S)$) denotes the column (respectively, row) space of matrix S, and $N(S)$ denotes the null space or kernel of S.

- A special notation is convenient for *diagonal* matrices. If A is an $m \times n$ matrix and we write

$$A = \mathrm{diag}(\alpha_1, \cdots, \alpha_p), \quad p = \min\{m, n\},$$

  then $a_{ij} = 0$ whenever $i \neq j$ and $a_{ii} = \alpha_i$ for $i = 1, \cdots, p$.

- The $m \times m$ *identity* matrix is denoted by I_m or, more simply, by I.

- The *set of m linear equations* in $n \times d$ unknowns X is represented in matrix form by

$$AX \approx B. \tag{1.17}$$

  A is the $m \times n$ data matrix and B is the $m \times d$ matrix of observations. Unless stated otherwise, we *assume* that the set of equations $AX \approx B$ is *overdetermined*, i.e., $m > n$, and that all *preprocessing* measures on the data (such as scaling, whitening, centering, standardizing, etc.) have been performed in advance. This avoids the need to burden the equations with left and right multiplications by diagonal matrices as in [68].

- X' is the $n \times d$ *minimum norm* least squares (LS) *solution* and $\widehat{X}$ is the $n \times d$ minimum norm total least squares (TLS) solution of (1.17).

- For the *one-dimensional* problem, i.e., $d = 1$, the matrices are replaced by their corresponding vector notations, e.g., the vectors b and x are used instead of the matrices B and X in (1.17). If $d > 1$, the problem is called *multidimensional.*

- The *Frobenius* norm of an $m \times n$ matrix M is defined by ("tr" denotes trace)

$$\|M\|_F = \sqrt{\sum_{i=1}^{m}\sum_{j=1}^{n} m_{ij}^2} = \sqrt{\mathrm{tr}(M^T M)}. \tag{1.18}$$

- The 2-*norm* or *Euclidean norm* of an n-dimensional *vector* y is defined by

$$\|y\|_2 = \sqrt{\sum_{i=1}^{n} y_i^2}. \tag{1.19}$$

- Based on (1.19), the 2-*norm* of an $m \times n$ *matrix* M is defined by

$$\|M\|_2 = \sup_{y \neq 0} \frac{\|My\|_2}{\|y\|_2} \tag{1.20}$$

  and equals the largest singular value of M.

- $B//u$ means that all column vectors of matrix B are *proportional* to the vector u.

- Denote the *singular value decomposition* (SVD) of the $m \times n$ matrix A, $m > n$, in (1.17) by

$$A = U'\Sigma'V'^T \tag{1.21}$$

with

$$\begin{aligned} U' &= [U_1'; U_2'], U_1' = [u_1', \cdots, u_n'], U_2' = [u_{n+1}', \cdots, u_m'], \\ &\qquad u_i' \in \mathcal{R}^m, U'^T U' = I_m, \\ V' &= [v_1', \cdots, v_n'], v_i' \in \mathcal{R}^n, V'^T V' = I_n, \\ \Sigma' &= \mathrm{diag}(\sigma_1', \cdots, \sigma_n') \in \mathcal{R}^{m \times n}, \quad \sigma_1' \geq \cdots \geq \sigma_n' \geq 0, \end{aligned}$$

and denote the SVD of the $m \times (n+d)$ matrix $[A; B]$, $m > n$, in (1.17) by

(1.22) $$[A; B] = U \Sigma V^T$$

with

$$\begin{aligned} U &= [U_1; U_2], U_1 = [u_1, \cdots, u_n], U_2 = [u_{n+1}, \cdots, u_m], \\ &\qquad u_i \in \mathcal{R}^m, U^T U = I_m, \\ V &= \underset{\quad n \qquad\quad d}{\begin{bmatrix} V_{11} & V_{12} \\ V_{21} & V_{22} \end{bmatrix}} \begin{matrix} n \\ d \end{matrix} = [v_1, \cdots, v_{n+d}], \; v_i \in \mathcal{R}^{n+d}, V^T V = I_{n+d}, \\ \Sigma &= \begin{bmatrix} \Sigma_1 & 0 \\ 0 & \Sigma_2 \end{bmatrix} = \mathrm{diag}(\sigma_1, \cdots, \sigma_{n+t}) \in \mathcal{R}^{m \times (n+d)}, \\ &\qquad t = \min\{m-n, d\}, \\ &\Sigma_1 = \mathrm{diag}(\sigma_1, \cdots, \sigma_n) \in \mathcal{R}^{n \times n}, \\ &\Sigma_2 = \mathrm{diag}(\sigma_{n+1}, \cdots, \sigma_{n+t}) \in \mathcal{R}^{(m-n) \times d}, \\ \text{and} \quad &\sigma_1 \geq \cdots \geq \sigma_{n+t} \geq 0. \end{aligned}$$

For convenience of notation, we define $\sigma_i = 0$ if $m < i \leq n+d$. The σ_i' and σ_i are the *singular values* of A and $[A; B]$, respectively. The vectors u_i' and u_i are the ith *left singular vector*, and the vectors v_i' and v_i are the ith *right* singular vector of A and $[A; B]$, respectively.

- rank(S) denotes the *rank* of the matrix S, defined by the number of its nonzero singular values. The *numerical* rank of S is defined by the number of its singular values larger than a given error-dependent bound. The *corank* of S, denoted by corank(S), equals the dimension of $N(S)$.
- The superscript $\dagger$ denotes the *pseudo-inverse* of a matrix. If (1.21) is the SVD of a rank r matrix A then $A^\dagger = V'\mathrm{diag}(1/\sigma_1', \cdots, 1/\sigma_r', 0, \cdots, 0)U'^T$.
- Let $V(\sigma_j)$ (respectively, $U(\sigma_j)$) be the right (respectively, left) *singular subspace* of $[A; B]$ spanned by the right (respectively, left) singular vectors associated with the singular value σ_j, and let $V'(\sigma_j)$ (respectively, $U'(\sigma_j)$) be the right (respectively, left) singular subspace of A spanned by the right (respectively, left) singular vectors associated with the singular value σ_j.

- $[A; B']$ is the LS *approximation* of $[A; B]$ with B' the orthogonal projection of B onto the range $R(A)$ of A. The TLS approximation of $[A; B]$, which solves the TLS problem (see Definitions 2.3 and 3.1–3.3), is denoted by $[\widehat{A}; \widehat{B}]$.

- The difference $\Delta B' = B - B'$ is called the LS *correction* and the difference $[\Delta\widehat{A}; \Delta\widehat{B}] = [A - \widehat{A}; B - \widehat{B}]$ is the TLS correction applied to the data $[A; B]$ in (1.17) to obtain a compatible set.

- We easily observe that the LS correction $\Delta B'$ equals the LS *residual* R', defined by $B - AX'$. The TLS residual $\widehat{R}$, however, is defined by $B - A\widehat{X}$ and differs from the TLS correction $[\Delta\widehat{A}; \Delta\widehat{B}]$.

- Denote by $[A_0; B_0]$ the exact but unobservable data matrix, perturbed by errors $[\Delta A; \Delta B]$ such that $[A; B] = [A_0 + \Delta A; B_0 + \Delta B]$. Then the LS and TLS solution of the corresponding unperturbed problem $A_0 X \approx B_0$ are denoted by X_0' and $\widehat{X}_0$, respectively. If $A_0 X = B_0$ is compatible, its solution is simply given by X_0.

- We call the TLS problem *generic* if for $\sigma_p > \sigma_{p+1} = \cdots = \sigma_{n+1}$ with $p \leq n$, the submatrix

  (1.23)
$$V_\gamma = \begin{bmatrix} v_{n+1,p+1} & \cdots & v_{n+1,n+d} \\ \vdots & \ddots & \vdots \\ v_{n+d,p+1} & \cdots & v_{n+d,n+d} \end{bmatrix} \quad \text{with } v_{j,i} \text{ the } j\text{th component of } v_i$$

  of V in (1.22) has full row rank d. If $\sigma_n > \sigma_{n+1}$ (i.e., $p = n$), this means that V_{22} in (1.22) is nonsingular (or $v_{n+1,n+1} \neq 0$ if $d = 1$). The TLS solution of generic problems, called the generic TLS solution, can be computed with the algorithm of Golub and Van Loan [68].

- We call the solution of the TLS problem *unique* if the problem (1.17) has only one solution satisfying the TLS criteria (see Definitions 2.3 and 3.1). This is the case for generic TLS problems where $\sigma_n > \sigma_{n+1}$.

Finally, the elementary *statistical* concepts used in this book are defined below.

It is well known that the *mean* and the *variance* are important parameters of the *distribution* of a *random variable* x. These are defined by

$$\text{mean}(x) = \mathcal{E}(x) \quad \text{and} \quad \text{var}(x) = \mathcal{E}(x - \mathcal{E}(x))^2, \tag{1.24}$$

where $\mathcal{E}$ denotes the expectation or the expected value of the estimator between brackets. The square root of the variance, called the *standard deviation*, describes the extent to which values of the variable x are scattered around their mean. Taking enough samples $x^{(k)}$ of $x, k = 1, \cdots, m$, the expected values can be estimated by the average value $\bar{x}$, called the *sample* mean and sample variance:

(1.25)

$$\text{sample mean} = \bar{x} = \frac{1}{m}\sum_{k=1}^{m} x^{(k)} \quad \text{and sample variance} \quad = \frac{1}{m}\sum_{k=1}^{m}(x^{(k)} - \bar{x})^2.$$

More generally, if $x = [x_1, \cdots, x_n]^T$ is a random *vector* of n components x_i, then its expectation or mean is itself a vector containing the expectations of each of the elements x_i of x. In this case, the distribution of x is also characterized by the covariance between its elements. The *covariance* between the elements x_i and x_j of x is defined by

$$\text{cov}(x_i, x_j) = \mathcal{E}\{(x_i - \mathcal{E}(x_i))(x_j - \mathcal{E}(x_j))\} = \text{cov}(x_j, x_i). \tag{1.26}$$

If m vector samples $x^{(k)}$ of $x, k = 1, \cdots, m$, are available, $\text{cov}(x_i, x_j)$ can be estimated by the

$$\text{sample covariance} = \frac{1}{m}\sum_{k=1}^{m}(x_i^{(k)} - \bar{x}_i)(x_j^{(k)} - \bar{x}_j). \tag{1.27}$$

Variances and covariances between the elements of x are characterized in its covariance matrix $\text{cov}(x)$, defined by

$$\text{cov}(x) = \begin{bmatrix} \text{var}(x_1) & \cdots & \text{cov}(x_1, x_n) \\ \text{cov}(x_2, x_1) & & \text{cov}(x_2, x_n) \\ \vdots & & \vdots \\ \text{cov}(x_n, x_1) & \cdots & \text{var}(x_n) \end{bmatrix}. \tag{1.28}$$

Using these concepts, the following *properties* of an estimator x can be defined [88].

- *Unbiasedness*: If, on the average, an estimator x neither tends to over- nor underestimate the true parameter value x_0, it is said to be an unbiased estimator. Notationally, we require that for all sample lengths m and x_0

  $$\mathcal{E}(x) = x_0. \tag{1.29}$$

- *Consistency*: An estimator $x^{(m)}$, computed from a sample of m values, is said to be a consistent estimator of the true value x_0 if $x^{(m)}$ converges to x_0 as m tends to infinity. Several convergence concepts are possible. When dealing with consistency we use "*convergence with probability one*" (w.p.1), also called *strong* consistency. This means that the probability for the event ($x^{(m)} \to x_0$ as $m \to \infty$) is 1. Notationally, we have

$$\lim_{m\to\infty} x^{(m)} = x_0 \qquad \text{w.p.1.} \tag{1.30}$$

In the statistical literature, the normal choice is "*convergence in probability*," which is a weaker concept (implied by, but not equivalent to, convergence w.p.1). Then it is required that the probability P

$$P(|x^{(m)} - x_0| > \varepsilon) \to 0 \text{ as } m \to \infty \text{ for every } \varepsilon > 0 \tag{1.31}$$

or, notationally equivalent,

$$\operatorname*{plim}_{m\to\infty} x^{(m)} = x_0, \tag{1.32}$$

where "plim" denotes probability limit.

- *Minimum mean squared error*: What we are really demanding of an estimator x is that it should be "close" to the true value x_0. Therefore, we consider its mean squared error (MSE) around that true value, instead of its own expected value. We have at once that

$$\text{MSE}(x) = \mathcal{E}(x - x_0)^2 = \mathcal{E}(x - \mathcal{E}(x))^2 + (\mathcal{E}(x) - x_0)^2, \tag{1.33}$$

the cross-product term on the right being equal to zero. The last term on the right is simply the square of the bias of x in estimating x_0. If x is unbiased, this last term is zero, and MSE becomes variance.

Chapter 2

Basic Principles of the Total Least Squares Problem

2.1. Introduction

The Total Least Squares (TLS) problem has been introduced in recent years in the numerical analysis literature [68] as an alternative for Least Squares (LS) problems in the case that all data are affected by errors. This chapter reviews the main principles of the basic TLS and LS problem in solving overdetermined sets of linear equations $Ax \approx b$. The word "basic" implies three conditions: (1) that only one right-hand side vector b is considered, (2) that the problem is solvable, and (3) that it has a unique solution.

First, the LS problem is reviewed in §2.2.1 and it is shown how its solution can be determined from the Singular Value Decomposition (SVD) of the matrix A (§2.2.3). For ease of understanding, the SVD and its main properties are surveyed in §2.2.2. In §2.3, the basic TLS problem is presented: its definition is formulated (§2.3.1) and it is shown how to solve this problem by making heavy use of the SVD (§2.3.2). The algorithm, outlined in §2.3.3, indicates how this can be accomplished.

Furthermore, the differences between both problems are clarified by a geometric representation of each problem in its column and row space (§§2.4.1 and 2.4.2). An extreme example illustrates the property that the data modifications applied by TLS are always less than those applied by LS to make the set of equations compatible. Finally, other TLS problem formulations are presented in §2.5. Depending on the linear relation between the columns of $[A; b]$ in which we are interested, these formulations may be more appropriate than those given in §2.3.1. Nevertheless, although formulations differ, the solution of the basic TLS problem is essentially the same.

2.2. Least squares problems

2.2.1. Problem formulation and solution. Consider the problem of finding a vector $x \in \mathcal{R}^n$ such that $Ax = b$ when the data matrix $A \in \mathcal{R}^{m \times n}$ and the observation vector $b \in \mathcal{R}^m$ are given. When there are more equations than unknowns, i.e., $m > n$, the set is *overdetermined*. Unless b belongs to

$R(A)$, the overdetermined set has no exact solution and is therefore denoted by $Ax \approx b$.

There are many possible ways of defining the "best" solution. A choice that can often be motivated for statistical reasons (see below) and also leads to a simple computational problem is outlined in the following definition [69, §5.3], [18].

DEFINITION 2.1 (Ordinary least squares problem).

$$\text{(2.1)} \qquad \underset{x \in \mathcal{R}^n}{\text{minimize}} \; \| Ax - b \|_2 \qquad\qquad A \in \mathcal{R}^{m \times n}, \quad b \in \mathcal{R}^m.$$

Any minimizing x' is called a (linear) least squares (LS) solution of the set $Ax \approx b$.

One important application of LS problems is parameter estimation in linear statistical models. Here we assume that the m observations $b = [b_1, \cdots, b_m]^T$ are related to the n unknown parameters $x = [x_1, \cdots, x_n]^T$ by

$$Ax = b_0 \qquad\qquad b = b_0 + \Delta b,$$

where $\Delta b = [\Delta b_1, \cdots, \Delta b_m]^T$ and $\Delta b_i, i = 1, \cdots, m$ are random errors. Let us assume that A has rank n and Δb has zero mean and covariance matrix $\sigma_\nu^2 I$ (i.e., Δb_i are uncorrelated random variables with equal variance). Then Gauss [57] showed that the LS estimate x' has the smallest variance in the class of estimation methods, which fulfills the following two conditions:

1. There are no systematic errors (no bias) in the estimates.
2. The estimates are linear functions of b.

Note that this property of LS estimates does *not* depend on any assumed distributional form of the error.

LS solutions are characterized by the following theorem.

THEOREM 2.1 (LS solution property).

$$x' \textit{ solves the } \text{LS } \textit{problem } (2.1) \iff A^T(b - Ax') = 0.$$

Proof. For the proof, see [18, Thm. 1.2].

This theorem asserts that the residual $r' = b - Ax'$ of a LS solution x' is orthogonal to the range $R(A)$ of A. Thus, the right-hand side b is decomposed into two orthogonal components:

$$b = b' + r' = Ax' + r' \qquad\qquad r' \perp Ax',$$

where b' is the orthogonal projection of b onto $R(A)$. This geometric interpretation is illustrated in Fig. 2.1(a) for $n = 2$. Note that this decomposition is always unique, even when the LS solution x' is not unique.

From Theorem 2.1 it follows that an LS solution satisfies the *normal* equations

$$A^T Ax = A^T b. \tag{2.2}$$

The matrix $A^T A$ is symmetric and nonnegative definite, and the normal equations are always consistent. Furthermore, we have the following corollary.

COROLLARY 2.1 (LS solution and residual). *If* $\text{rank}(A) = n$ *then* (2.1) *has a unique* LS *solution, given by*

$$x' = (A^T A)^{-1} A^T b. \tag{2.3}$$

The corresponding LS *correction is given by the residual:*

$$r' = b - Ax' = b - b', \qquad b' = P_A b, \tag{2.4}$$

where $P_A = A(A^T A)^{-1} A^T$ *is the orthogonal projector onto* $R(A)$.

If $\text{rank}(A) = r < n$, the LS problem (2.1) is *rank-deficient* and has an infinite number of solutions, for if x is a minimizer and $z \in N(A)$ then $x + z$ is also a minimizer. For reasons of stability and minimal sensitivity, a unique solution having *minimal* 2-*norm* is singled out from the set of all minimizers

$$\mathcal{X} = \{x \in \mathcal{R}^n : \|Ax - b\|_2 = \min\}.$$

We denote this solution by x' (note that in the full rank case, there is only one LS solution and so it must have minimal 2-norm).

Neat expressions for x' and $\|b - Ax'\|_2$ are provided by the SVD defined below.

2.2.2. Singular value decomposition.

The Singular Value Decomposition (SVD) is of great theoretical and practical importance for LS and TLS problems. The history of this matrix decomposition goes back more than one century but is only recently receiving more attention for conceptual, numerical, algebraic, and computational reasons [69].

THEOREM 2.2 (Singular Value Decomposition (SVD)). *If* $C \in \mathcal{R}^{m \times n}$ *then there exist orthonormal matrices* $U = [u_1, \cdots, u_m] \in \mathcal{R}^{m \times m}$ *and* $V = [v_1, \cdots, v_n] \in \mathcal{R}^{n \times n}$ *such that*

(2.5)

$$U^T C V = \Sigma = \text{diag}(\sigma_1, \cdots, \sigma_p), \quad \sigma_1 \geq \cdots \geq \sigma_p \geq 0$$
$$\textit{and} \quad p = \min\{m, n\}.$$

Proof. For the proof, see [69, Thm. 2.5.1].

The σ_i are the *singular values* of C and their set is called the *singular value spectrum*. The vectors u_i and v_i are the ith *left singular vector* and the ith *right singular vector*, respectively. The triplet (u_i, σ_i, v_i) is called a *singular triplet.* It is easy to verify by comparing columns in the equations $CV = U\Sigma$ and $C^TU = \Sigma^TV$ that

$$Cv_i = \sigma_i u_i \quad \text{and} \quad C^T u_i = \sigma_i v_i \quad i = 1, \cdots, p. \tag{2.6}$$

The SVD reveals a great deal about the structure of a matrix. If the SVD of C is given by Theorem 2.2, and we define r by

$$\sigma_1 \geq \cdots \geq \sigma_r > \sigma_{r+1} = \cdots = \sigma_p = 0,$$

then

$$\begin{aligned} \text{rank}(C) &= r, \\ R(C) &= R([u_1, \cdots, u_r]), \\ N(C) &= R([v_{r+1}, \cdots, v_n]), \\ R_r(C) = R(C^T) &= R([v_1, \cdots, v_r]), \\ N_r(C) = N(C^T) &= R([u_{r+1}, \cdots, u_m]). \end{aligned}$$

Moreover, if $U_r = [u_1, \cdots, u_r]$, $\Sigma_r = \text{diag}(\sigma_1, \cdots, \sigma_r)$, and $V_r = [v_1, \cdots, v_r]$, then we have the SVD expansion

$$C = U_r \Sigma_r V_r^T = \sum_{i=1}^{r} \sigma_i u_i v_i^T. \tag{2.7}$$

Equation (2.7), also called the *dyadic decomposition*, decomposes the matrix C of rank r in a sum of r matrices of rank one. Also, the 2-*norm* and the *Frobenius norm* are neatly characterized in terms of the SVD:

$$\begin{aligned} \|C\|_F^2 = \sum_{i=1}^{m}\sum_{j=1}^{n} c_{ij}^2 &= \sigma_1^2 + \cdots + \sigma_p^2, \qquad p = \min\{m, n\}, \\ \|C\|_2 = \sup_{y \neq 0} \frac{\|Cy\|_2}{\|y\|_2} &= \sigma_1. \end{aligned}$$

From (2.5) it follows that

$$C^TC = V\Sigma^T\Sigma V^T \quad \text{and} \quad CC^T = U\Sigma\Sigma^TU^T.$$

Thus $\sigma_i^2, i = 1, \cdots, p$ are eigenvalues of the symmetric and nonnegative definite matrices C^TC and CC^T, and v_i and u_i are the corresponding eigenvectors. Hence, in principle, the SVD can be reduced to the eigenvalue problem for symmetric matrices. This is *not* a numerically suitable way to compute the SVD, as shown below [18].

Example 2.1. Consider the case $n = 2$ and take

$$C = [c_1, c_2], \quad c_1^Tc_2 = \cos\gamma, \quad \text{and} \quad \|c_1\|_2 = \|c_2\|_2 = 1,$$

where γ is the angle between the vectors c_1 and c_2. The matrix

$$C^TC = \begin{bmatrix} 1 & \cos\gamma \\ \cos\gamma & 1 \end{bmatrix}$$

has eigenvalues $\lambda_1 = 2\cos^2(\gamma/2)$, $\lambda_2 = 2\sin^2(\gamma/2)$ and so

$$\sigma_1 = \sqrt{2}\cos\frac{\gamma}{2}, \quad \sigma_2 = \sqrt{2}\sin\frac{\gamma}{2}.$$

The eigenvectors of C^TC

$$v_1 = \frac{1}{\sqrt{2}}\begin{bmatrix} 1 \\ 1 \end{bmatrix}, \quad v_2 = \frac{1}{\sqrt{2}}\begin{bmatrix} 1 \\ -1 \end{bmatrix}$$

are the right singular vectors of C. The left singular vectors can be determined from (2.6). If $\gamma << 1$, then $\sigma_1 \approx \sqrt{2}$ and $\sigma_2 \approx \gamma/\sqrt{2}$ and we get

$$u_1 \approx \frac{1}{2}(c_1 + c_2), \quad u_2 \approx \frac{1}{\gamma}(c_1 - c_2).$$

If we assume that γ is *less* than the square root of the machine precision, the computed values of the entries $\cos\gamma$ in C^TC equal 1. If this is the case, the computed matrix C^TC is singular with eigenvalues 2 and 0, and it is not possible to retrieve the small singular value $\gamma/\sqrt{2}$. This illustrates that information may be lost in computing C^TC unless sufficient precision is used.

The SVD plays an important role in a number of matrix approximation problems. In the theorem below we consider the approximation of one matrix by another of lower rank. Several other results can be found in [69].

THEOREM 2.3 (Eckart–Young–Mirsky matrix approximation theorem). *Let the* SVD *of* $C \in \mathcal{R}^{m\times n}$ *be given by* $C = \sum_{i=1}^{r} \sigma_i u_i v_i^T$ *with* $r = \text{rank}(C)$. *If* $k < r$ *and* $C_k = \sum_{i=1}^{k} \sigma_i u_i v_i^T$, *then*

$$\min_{\text{rank}(D)=k} \|C - D\|_2 = \|C - C_k\|_2 = \sigma_{k+1}, \tag{2.8}$$

and

$$\min_{\text{rank}(D)=k} \|C - D\|_F = \|C - C_k\|_F = \sqrt{\sum_{i=k+1}^{p} \sigma_i^2}, \quad p = \min\{m,n\}. \tag{2.9}$$

Proof. For the proof, see [103] and [48]. □

The theorem was originally proved for the Frobenius norm in 1936 by Eckart and Young [48]. In 1960, Mirsky [103] proved the theorem for the 2-norm. Therefore, Theorem 2.3 is called the *Eckart–Young–Mirsky Theorem.* A generalization of this theorem is given in [62].

The following theorem presents the *interlacing property for singular values* and gives bounds on the perturbation of the singular values due to the removal of a column or row of a matrix.

THEOREM 2.4 (Interlacing theorem for singular values). *Let C be an $m \times n$ matrix with singular values $\gamma_1 \geq \cdots \geq \gamma_{\min\{m,n\}}$. Let D be a $p \times q$ submatrix of C, with singular values $\delta_1 \geq \cdots \geq \delta_{\min\{p,q\}}$, and define, for convenience, $\gamma_t = 0$ for $\min\{m,n\} < t \leq \max\{m,n\}$ and $\delta_t = 0$ for $\min\{p,q\} < t \leq \max\{p,q\}$. Then*

$$\begin{aligned} &(2.10) \qquad \gamma_i \geq \delta_i \quad \text{for } i = 1, \cdots, \min\{p,q\}, \\ &(2.11) \qquad \delta_i \geq \gamma_{i+(m-p)+(n-q)} \quad \text{for } i \leq \min\{p+q-m, p+q-n\}. \end{aligned}$$

Proof. For the proof, see [169]. □

If D results from C by deleting one column of C, then Theorem 2.4 yields

$$\begin{aligned} &(2.12) \qquad \text{if } m \geq n: \ \gamma_1 \geq \delta_1 \geq \gamma_2 \geq \delta_2 \geq \cdots \geq \delta_{n-1} \geq \gamma_n \geq 0; \\ &(2.13) \qquad \text{if } m < n: \ \gamma_1 \geq \delta_1 \geq \gamma_2 \geq \delta_2 \geq \cdots \geq \gamma_m \geq \delta_m \geq 0. \end{aligned}$$

2.2.3. Least squares solutions and the SVD. The SVD is a powerful computational tool for solving LS problems. The reason is that the orthogonal matrices that transform C to diagonal form (2.5) do not change the 2-norm of vectors. We have the following fundamental result.

THEOREM 2.5 (Minimum norm LS solution of $Ax \approx b$). *Let the* SVD *of $A \in \mathcal{R}^{m \times n}$ be given by (1.21), i.e., $A = \sum_{i=1}^{n} \sigma_i' u_i' v_i'^T$, and assume that* $\operatorname{rank}(A) = r$. *If $b \in \mathcal{R}^m$, then*

$$x' = \sum_{i=1}^{r} \sigma_i'^{-1} v_i' u_i'^T b \tag{2.14}$$

minimizes $\|Ax - b\|_2$ and has the smallest 2*-norm of all minimizers. Moreover,*

$$\rho'^2 = \|Ax' - b\|_2^2 = \sum_{i=r+1}^{m} (u_i'^T b)^2. \tag{2.15}$$

Proof. For the proof, see [69, Thm. 5.5.1]. □

We can write (2.14) and (2.15) as

$$x' = A^{\dagger} b \quad \text{and} \quad \rho'^2 = \|(I - AA^{\dagger})b\|_2^2,$$

where $A^{\dagger} = V'\Sigma'^{\dagger}U'^T$, $\Sigma'^{\dagger} = \operatorname{diag}(\sigma_1'^{-1}, \cdots, \sigma_r'^{-1}, 0, \cdots, 0) \in \mathcal{R}^{n \times m}$, defines the *pseudo-inverse* of A.

2.3. Total least squares problems

2.3.1. Motivation and problem formulation. The term "Total Least Squares" (TLS) was coined in [68], although its solution, using the SVD, had already been introduced in [67] and [61]. In this section we formulate the *main principle* of the TLS problem. A good way to introduce and motivate the method is to recast the *ordinary least squares* (LS) problem [69, §5.3],[18].

DEFINITION 2.2 (Ordinary least squares problem). Given an overdetermined set of m linear equations $Ax \approx b$ in n unknowns x, the least squares (LS) problem seeks to

$$\underset{b' \in \mathcal{R}^m}{\text{minimize}} \quad \| b - b' \|_2 \tag{2.16}$$

$$\text{subject to} \quad b' \in R(A). \tag{2.17}$$

Once a minimizing b' is found, then any x satisfying

$$Ax = b' \tag{2.18}$$

is called an LS *solution* and $\Delta b' = b - b'$ the corresponding LS correction.

Equations (2.16) and (2.17) are satisfied if b' is the orthogonal projection of b onto $R(A)$. Thus, the LS problem amounts to perturbing the observation vector b by a minimum amount $\Delta b'$ so that $b' = b - \Delta b'$ can be "predicted" by the columns of A.

The underlying assumption here is that errors *only* occur in the vector b and that the matrix A is *exactly* known. Often this assumption is not realistic because sampling or modeling or measurement errors also affect the matrix A.

One way to take errors in A into account is to introduce *perturbations in* A also and to consider the following TLS problem.

DEFINITION 2.3 (Basic total least squares problem). Given an overdetermined set of m linear equations $Ax \approx b$ in n unknowns x, the total least squares problem seeks to

$$\underset{[\widehat{A};\widehat{b}] \in \mathcal{R}^{m \times (n+1)}}{\text{minimize}} \quad \| [A;b] - [\widehat{A};\widehat{b}] \|_F \tag{2.19}$$

$$\text{subject to} \quad \widehat{b} \in R(\widehat{A}). \tag{2.20}$$

Once a minimizing $[\widehat{A};\widehat{b}]$ is found, then any x satisfying

$$\widehat{A}x = \widehat{b} \tag{2.21}$$

is called a TLS *solution* and $[\Delta\widehat{A};\Delta\widehat{b}] = [A;b] - [\widehat{A};\widehat{b}]$ the corresponding TLS correction.

The TLS solution is denoted by $\widehat{x}$.

One important application of TLS problems is parameter estimation in errors-in-variables models. Here, we assume that the m measurements in A, b are related to n unknown parameters x by

$$A_0x = b_0 \qquad\qquad A = A_0 + \Delta A \text{ and } b = b_0 + \Delta b,$$

where $\Delta A, \Delta b$ represent the measurement errors. Let us assume that A_0 has full rank and all rows of $[\Delta A; \Delta b]$ are independently and identically distributed with zero mean and covariance matrix $\sigma_\nu^2 I$ (i.e., Δa_{ij} and Δb_i are uncorrelated random variables with equal variance). Then it can be proved [59], [55] that the TLS solution $\widehat{x}$ of $Ax \approx b$ estimates the *true* parameter values x_0 , given by $A_0^\dagger b_0$, *consistently*; i.e., $\widehat{x}$ converges to x_0 as m tends to infinity. This property of TLS estimates does not depend on any assumed distributional form of the errors. Note also that the LS estimates are inconsistent in this case.

2.3.2. Basic solution. We now analyze the TLS problem by making substantial use of the SVD. As pointed out in [67] and more fully in [61], [68], this decomposition can be used to solve the TLS problem. We bring $Ax \approx b$ into the following form:

$$[A; b][x^T; -1]^T \approx 0. \tag{2.22}$$

Let (1.22) be the SVD of $[A; b]$. If $\sigma_{n+1} \neq 0$, $[A; b]$ is of rank $n + 1$ and the space S generated by the rows of $[A; b]$ coincides with $\mathcal{R}^{n+1}$. There is no nonzero vector in the orthogonal complement of S, hence the set of equations (2.22) is incompatible. To obtain a solution, the rank of $[A; b]$ must be reduced to n. Using the Eckart–Young–Mirsky Theorem 2.3, the best rank n TLS approximation $[\widehat{A}; \widehat{b}]$ of $[A; b]$, which minimizes the deviations in variance, is given by

$$[\widehat{A}; \widehat{b}] = U\widehat{\Sigma}V^T \text{ with } \widehat{\Sigma} = \operatorname{diag}(\sigma_1, \cdots, \sigma_n, 0).$$

The minimal TLS correction is then

$$\sigma_{n+1} = \min_{\operatorname{rank}([\widehat{A};\widehat{b}])=n} \|[A; b] - [\widehat{A}; \widehat{b}]\|_F$$

and is attained for

$$[A; b] - [\widehat{A}; \widehat{b}] = [\Delta\widehat{A}; \Delta\widehat{b}] = \sigma_{n+1}u_{n+1}v_{n+1}^T.$$

Note that this TLS correction matrix has rank *one*. It is clear that the approximate set

$$[\widehat{A}; \widehat{b}][x^T; -1]^T = 0$$

is now compatible and its solution is given by the only vector v_{n+1} (i.e., the last column of V) that belongs to $N([\widehat{A};\widehat{b}])$. The TLS solution is then obtained by *scaling* v_{n+1} until its last component is -1 or

$$(2.23)\qquad [\widehat{x}^T;-1]^T = \frac{-1}{v_{n+1,n+1}} v_{n+1}.$$

If $v_{n+1,n+1} \neq 0$, then $\widehat{b} = \widehat{A}\widehat{x} = -1/(v_{n+1,n+1})\widehat{A}[v_{1,n+1},\cdots,v_{n,n+1}]^T \in R(\widehat{A})$, i.e., (2.20) is satisfied, and hence $\widehat{x}$ solves the basic TLS problem.

Observe that, if σ_{n+1} is zero, $[A;b]$ is of rank n and hence $v_{n+1} \in N([A;b])$. In this case, the set of equations (2.22) is compatible and no approximation is needed to obtain the exact solution, given by (2.23).

The following theorem gives conditions for the *uniqueness* and *existence* of a TLS solution.

THEOREM 2.6 (Solution of the basic TLS problem $Ax \approx b$). *Suppose* (1.21) (*respectively,* (1.22)) *is the* SVD *of* A (*respectively,* $[A;b]$). *If* $\sigma'_n > \sigma_{n+1}$, *then*

$$(2.24)\qquad [\widehat{A};\widehat{b}] = U\widehat{\Sigma}V^T \quad \textit{and,}\ \widehat{\Sigma} = \operatorname{diag}(\sigma_1,\cdots,\sigma_n,0),$$

with corresponding TLS *correction matrix*

$$(2.25)\qquad [\Delta\widehat{A};\Delta\widehat{b}] = [A;b] - [\widehat{A};\widehat{b}] = \sigma_{n+1}u_{n+1}v_{n+1}^T,$$

solves the TLS problem (2.19) *and* (2.20) *and*

$$(2.26)\qquad \widehat{x} = -\frac{1}{v_{n+1,n+1}}[v_{1,n+1},\cdots,v_{n,n+1}]^T$$

exists and is the unique solution to $\widehat{A}x = \widehat{b}$.

Proof. The interlacing Theorem 2.4 for singular values implies that

$$\sigma_1 \geq \sigma'_1 \geq \cdots \geq \sigma_n \geq \sigma'_n \geq \sigma_{n+1}.$$

The assumption $\sigma'_n > \sigma_{n+1}$ thus ensures that σ_{n+1} is not a repeated singular value of $[A;b]$. If $[A;b]^T[A;b][\begin{smallmatrix}y\\0\end{smallmatrix}] = \sigma_{n+1}^2[\begin{smallmatrix}y\\0\end{smallmatrix}]$ and $0 \neq y \in \mathcal{R}^n$, then it clearly follows that $A^TAy = \sigma_{n+1}^2 y$, a contradiction since $\sigma_n'^2$ is the smallest eigenvalue of A^TA. Thus $N([\widehat{A};\widehat{b}])$ must contain a vector whose $(n+1)$th component is nonzero. This implies that the TLS problem has a solution. Since $N([\widehat{A};\widehat{b}])$ has dimension one, this solution must be unique. Formulas (2.24)–(2.26) follow straightforwardly from the application of the Eckart–Young–Mirsky Theorem 2.3, as proved above. □

It is interesting to note here that the conditions

$$\sigma'_n > \sigma_{n+1} \iff \sigma_n > \sigma_{n+1} \text{ and } v_{n+1,n+1} \neq 0$$

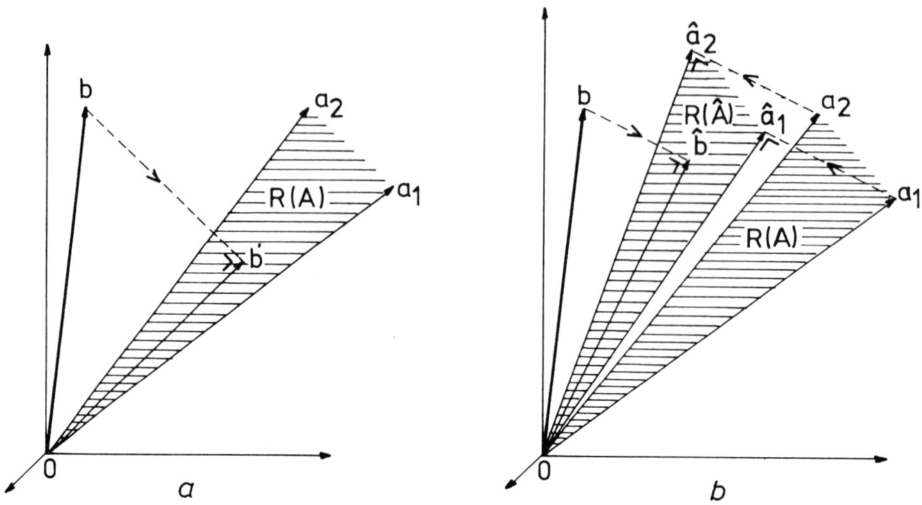

FIG. 2.1. (a) *The* LS *solution is obtained by projecting b orthogonally onto $R(A)$ and solving $Ax = b'$.* (b) *The* TLS *solution is obtained by approximating the columns a_i of A and b by $\widehat{a}_i$ and $\widehat{b}$ until $\widehat{b}$ is in the space $R(\widehat{A})$, generated by the columns $\widehat{a}_i$, and $\widehat{A}x = \widehat{b}$.*

are equivalent, as will be proved in Corollary 3.4 (set $p = n$ and $d = 1$).

An illustration of the *geometry* of the TLS solution in the column space of A is depicted in Fig. 2.1(b) for $n = 2$. The TLS problem is tantamount to finding a "closest" subspace $R([\widehat{A}; \widehat{b}])$ to the $n + 1$ columns of $[A; b]$. The sum of squared *perpendicular* distances from each column $a_i, b_i \in \mathcal{R}^m$ to $R([\widehat{A}; \widehat{b}])$ is minimized, and each column a_i, b_i is approximated by its orthogonal projection $\widehat{a}_i, \widehat{b}_i$ onto that subspace.

A well-known and useful characterization of the TLS solution $\widehat{x}$ and the minimal TLS correction σ_{n+1} is proved in the next theorem.

THEOREM 2.7 (Closed-form expression of the basic TLS solution). *Let* (1.21) (*respectively,* (1.22)) *be the* SVD *of A* (*respectively,* $[A; b]$). *If $\sigma'_n > \sigma_{n+1}$, then*

$$\widehat{x} = (A^T A - \sigma_{n+1}^2 I)^{-1} A^T b \tag{2.27}$$

and

$$\sigma_{n+1}^2 \left[1 + \sum_{i=1}^{n} \frac{(u_i'^T b)^2}{\sigma_i'^2 - \sigma_{n+1}^2} \right] = \rho'^2 = \min_{x \in \mathcal{R}^n} \|Ax - b\|_2^2. \tag{2.28}$$

Proof. The condition $\sigma'_n > \sigma_{n+1}$ ensures that $\widehat{x}$ exists and is unique, as given by (2.23). Since the singular vectors v_i are eigenvectors of $[A; b]^T [A; b]$,

$\widehat{x}$ also satisfies the following eigenvector equations:

$$(2.29)\quad [A;b]^T[A;b]\begin{bmatrix}\widehat{x}\\-1\end{bmatrix}=\begin{bmatrix}A^TA & A^Tb\\ b^TA & b^Tb\end{bmatrix}\begin{bmatrix}\widehat{x}\\-1\end{bmatrix}=\sigma_{n+1}^2\begin{bmatrix}\widehat{x}\\-1\end{bmatrix}.$$

The formula (2.27) follows directly from the top part of (2.29).

To establish (2.28), transform (2.29) to

$$\begin{bmatrix}\Sigma'^T\Sigma' & g\\ g^T & \|b\|_2^2\end{bmatrix}\begin{bmatrix}z\\-1\end{bmatrix}=\sigma_{n+1}^2\begin{bmatrix}z\\-1\end{bmatrix}\qquad\text{with } g=\Sigma'^TU'^Tb,\ z=V'^T\widehat{x}.$$

From this equation we see that

$$(\Sigma'^T\Sigma'-\sigma_{n+1}^2I)z=g\quad\text{and}\quad\sigma_{n+1}^2+g^Tz=\|b\|_2^2.$$

Substituting z in the latter expression by the former yields

$$\sigma_{n+1}^2+g^T(\Sigma'^T\Sigma'-\sigma_{n+1}^2I)^{-1}g=\|b\|_2^2.$$

This can be rewritten as

$$\sigma_{n+1}^2+\sum_{i=1}^{n}\frac{\sigma_i'^2(u_i'^Tb)^2}{\sigma_i'^2-\sigma_{n+1}^2}=\sum_{i=1}^{m}(u_i'^Tb)^2$$

or

$$\sigma_{n+1}^2\left[1+\sum_{i=1}^{n}\frac{(u_i'^Tb)^2}{\sigma_i'^2-\sigma_{n+1}^2}\right]=\sum_{i=n+1}^{m}(u_i'^Tb)^2.$$

Equation (2.28) now follows since

$$\min_x\|b-Ax\|_2^2=\min_w\|U'^Tb-\Sigma'w\|_2^2=\sum_{i=n+1}^{m}(u_i'^Tb)^2.\qquad\square$$

2.3.3. Basic algorithm. The following algorithm computes (if possible) a TLS solution $\widehat{x}$ of $Ax\approx b$ such that $(A-\Delta\widehat{A})\widehat{x}=b-\Delta\widehat{b}$ and $\|[\Delta\widehat{A};\Delta\widehat{b}]\|_F$ is minimal.

ALGORITHM 2.1. Basic TLS solution of $Ax\approx b$.

Given: $A\in\mathcal{R}^{m\times n}$ and $b\in\mathcal{R}^m$.

Step 1. Compute the SVD (1.22), i.e., $[A;b]=U\Sigma V^T$.

Step 2. If $v_{n+1,n+1}\neq 0$ then $\widehat{x}=-1/(v_{n+1,n+1})[v_{1,n+1},\cdots,v_{n,n+1}]^T$.

END

The conditions $\sigma'_n > \sigma_{n+1}$ or, equivalently, $\sigma_n > \sigma_{n+1}$ and $v_{n+1,n+1} \neq 0$, ensure that Algorithm 2.1 computes the *unique* TLS solution of $Ax \approx b$. These conditions are generically satisfied provided A is of full rank and the set $Ax \approx b$ is not too conflicting. Hence, most TLS problems that arise in practice can be solved by means of Algorithm 2.1.

Now suppose that $\sigma_n = \sigma_{n+1}$. Then the TLS problem (2.19) and (2.20) may still have a solution, although it may not be unique. Indeed, any linear combination of singular vectors $v_{p+1}, \cdots, v_{n+1}$ associated with the multiple singular value $\sigma_{p+1} = \cdots = \sigma_{n+1}$, $p < n$ and $\sigma_p > \sigma_{p+1}$, solves the TLS problem provided it results in a vector of the form $[\hat{x}^T; -1]^T$. The same conclusions apply to the case $m < n$; i.e., the set $Ax \approx b$ is *underdetermined*, since then the conditions $\sigma_{m+1} = \cdots = \sigma_{n+1} = 0$ hold. Whenever $\sigma_n = \sigma_{n+1}$, Algorithm 2.1 still computes a TLS solution that may be appropriate for the user. However, we often want to single out the TLS solution with *minimal norm* $||\hat{x}||_2$ since this solution has minimal sensitivity. If this is the case, one should apply the procedure outlined in §3.3.1.

Whenever $v_{n+1,n+1} \neq 0$, the TLS problem (2.19) and (2.20) is *solvable* and is therefore called *generic*. The TLS solution can, for instance, be computed by Algorithm 2.1. Problems may occur whenever $v_{n+1,n+1} = 0$. In particular, if $\sigma_n > \sigma_{n+1}$ and $v_{n+1,n+1} = 0$, the TLS problem *fails* to have a solution altogether, as shown below. More generally, if $\sigma_p > \sigma_{p+1} = \cdots = \sigma_{n+1}$, $p \leq n$, the TLS problem (2.19) and (2.20) fails to have a solution if all $v_{n+1,i} = 0$ $i = p+1, \cdots, n+1$. These problems are called *nongeneric*. Let us consider an example [18].

Example 2.2.(a) Let $n = 1$ and

$$[A; b] = \begin{bmatrix} 1 & 0 \\ 0 & 2 \end{bmatrix}.$$

Here $[A; b]$ is already in diagonal form and has a unique smallest singular value $\sigma_2 = 1$. Clearly, $v_2 = [1 \;, \; 0]^T$ and hence $v_{2,2} = 0$. The unique minimizing $[\Delta\hat{A}; \Delta\hat{b}]$ gives

$$[\hat{A}; \hat{b}] = [A - \Delta\hat{A}; b - \Delta\hat{b}] = \begin{bmatrix} 0 & 0 \\ 0 & 2 \end{bmatrix},$$

so the approximated set $\hat{A}x \approx \hat{b}$ is not compatible. However, an arbitrary small perturbation ε in the (2,1) entry will give a compatible set with solution $x = 2/\varepsilon$. Hence, there is no solution to the TLS problem (2.19) and (2.20).

(b) If $\text{rank}(A) < n$ then the TLS problem has a solution only in the trivial case that $b \in R(A)$. Consider the example

$$A = \begin{bmatrix} 1 & 0 \\ 0 & 0 \end{bmatrix}, \quad b = \begin{bmatrix} 1 \\ 1 \end{bmatrix}.$$

Here we have $b \in R(A - \Delta\widehat{A})$ for $\Delta\widehat{A} = \text{diag}(0, \varepsilon)$ for any $\varepsilon \neq 0$, so there does not exist a smallest value of $\|[\Delta\widehat{A}; \Delta\widehat{b}]\|_F$. In the case when $\text{rank}(A) < n$ we must add the constraint

$$\text{rank}(A - \Delta\widehat{A}) \leq \text{rank}(A)$$

to the problem (2.19) and (2.20) to find a solution of the form $[\Delta\widehat{A}; \Delta\widehat{b}] = [A; b]vv^T$.

Nongeneric TLS problems occur whenever A is rank-deficient ($\sigma_n' \approx 0$), or when the set of equations is highly conflicting ($\sigma_n' \approx \sigma_{n+1}$ large). The latter situation is easily detected by inspecting the size of the smallest σ_i for which $v_{n+1,i} \neq 0$. If this singular value is large, the user can simply conclude that the data are not appropriate for linear modeling and as such reject the problem as irrelevant from a linear modeling point of view. In the first situation ($\sigma_n' \approx 0$), two options are possible. First, the user can remove the dependency between the columns of A by removing appropriate columns in A such that the remaining submatrix A_p of dimension $m \times p$, $p < n$, has full rank and then apply TLS to the reduced problem $A_p x \approx b$ to obtain an estimate of the remaining parameters. The position of the nonzero entries in v_{n+1} determines which columns of A, i.e., which variables in the model, are dependent. How to pick these columns is a problem of *subset selection*. Procedures for discarding or selecting columns in A are available in a variety of settings, e.g. see [79], [80], [182], [69, §12.2]. Another way is to solve a nongeneric TLS problem; i.e., additional constraints are imposed on the TLS problem (2.19) and (2.20) to make the problem solvable. The properties of these problems and their solutions are studied in detail in §3.4.

It should be noted that nongeneric TLS problems do not occur if we are only interested in estimating a linear relationship between the columns of $[A; b]$, no matter which column is used as right-hand side. If then $v_{n+1,n+1} = 0$, we simply replace b by any column a_i of A provided $v_{i,n+1} \neq 0$.

Finally, only TLS problems $Ax \approx b$ with *one* right-hand side vector b are considered here. These problems are called *one-dimensional* in contrast to multidimensional problems $AX \approx B_{m\times d}$ characterized by d (> 1) right-hand side vectors. The latter problems can be solved by computing the TLS solution of each subproblem $Ax_i \approx b_i$, $i = 1, \cdots, d$ separately. Here, Algorithm 2.1 can be used. A better way of computation, at least when all data are equally perturbed and all subproblems $Ax_i \approx b_i$ have the same degree of incompatibility, is to solve the multidimensional TLS problem $AX \approx B$. These problems are studied in detail in §3.2. As illustrated in Fig. 7.6, the solution $\widehat{X}$ of these multidimensional TLS problems exhibits an improved accuracy compared to the accuracy of the solution of each subproblem $Ax_i \approx b_i$. The accuracy of LS solutions, on the contrary, is independent of the number of right-hand side vectors considered in the computation.

2.4. Geometric comparison between the TLS and the LS problem

2.4.1. Viewed from the column space. The basic principle of the TLS problem can be best understood by comparing it with ordinary least squares from a geometric point of view.

The classical least squares (LS) problem of $Ax \approx b$ aims at finding a best approximation b' to b such that $Ax = b'$ is compatible. Geometrically, the fact that $Ax \approx b$ has no solution implies that the n-dimensional subspace $R(A)$ of $\mathcal{R}^m$, generated by the columns of A, does not contain the vector b. The best approximation b', satisfying (2.16) and (2.17), is then the orthogonal projection of b into that subspace (Fig. 2.1(a)). In many applications, both A and b are inaccurate. It seems then unfair to "correct" only b. To overcome this deficiency, the total least squares problem was proposed.

By looking at Fig. 2.1(b), clearly TLS bends b *and* the set of columns a_i of A toward each other until the new set of equations is compatible (see (2.20)) and such that all correction vectors $\Delta\widehat{a}_i$ and $\Delta\widehat{b}$ applied to the columns of A and b are minimized in the sense of (2.19). Observe that *all* $\Delta\widehat{a}_i$ and $\Delta\widehat{b}$ are orthogonal to $R(\widehat{A})$ and *parallel* with each other. Indeed, both LS and TLS decompose the data matrix $C = [A;b]$ of rank $n+1$ in two orthogonal parts: $C = \widehat{C} + \Delta\widehat{C}$ with the additional properties that $\widehat{C}^T\Delta\widehat{C} = 0$ and $\text{rank}(\Delta\widehat{C}) = 1$. Geometrically, this means that $R(\widehat{C}) \perp R(\Delta\widehat{C})$ and all columns of $\Delta\widehat{C}$ are parallel with each other. In TLS problems, $\widehat{C}$ and $\Delta\widehat{C}$ are the TLS approximation $[\widehat{A};\widehat{b}]$ and TLS correction matrix $[\Delta\widehat{A};\Delta\widehat{b}]$, respectively, defined by (2.24) and (2.25). In LS problems, $\widehat{C}$ and $\Delta\widehat{C}$ are given by the LS approximation $[A;b']$ and LS correction matrix $[0;r']$, respectively, and defined by the residual $r' = b - b' = b - Ax'$.

2.4.2. Viewed from the row space. Here we present an interesting geometric interpretation of the difference between TLS and LS in the row space $R_r([A;b])$. If no errors are present in the data, the set $Ax \approx b$ is compatible. Hence the rank of $[A;b]$ equals n and the space $R_r([A;b])$, generated by the rows $r_1, \cdots, r_m$ of $[A;b]$, is n-dimensional. Any n linearly independent rows determine a hyperplane $R_r([A;b])$. As shown in Fig. 2.2, the normal of this hyperplane characterizes the exact solution x. If *errors* occur, then the observed row vectors $r_1, \cdots, r_m$ of $[A;b]$ are *scattered* around the hyperplane and the space spanned by $r_1, \cdots, r_m$ coincides with the ambient space $\mathcal{R}^{n+1}$. Because $[A;b]$ is of rank $n+1$ now, the set $Ax \approx b$ is no longer compatible: a method of fitting a lower-dimensional hyperplane to these points is required to estimate the solution x.

The LS approach (for the case $n = 2$, see Fig. 2.2(a)) then looks for a best approximation b' to b, satisfying (2.17) and (2.16), such that the space $R_r([A;b'])$ generated by the rows of the LS approximation $[A;b']$ is a lower-dimensional subspace (two-dimensional in the case of Fig. 2.2(a)). Only

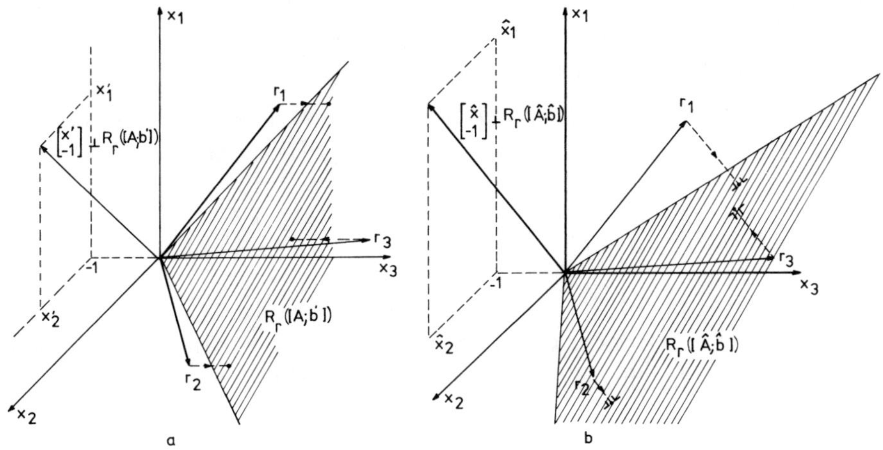

FIG. 2.2. *The difference between* (a) *the* LS *solution* x' *and* (b) *the* TLS *solution* $\widehat{x}$ *in the* $n+1=3$*-dimensional row space* $\mathcal{R}^{n+1}$. $r_1, r_2,$ *and* r_3 *are the rows of* $[A;b]$. $R_r([A;b'])$ *is the two-dimensional hyperplane best fitting according to* LS. $R_r([\widehat{A};\widehat{b}])$ *is the two-dimensional hyperplane best fitting according to* TLS.

the *last* components of $r_1, \cdots, r_m$ can vary. The solution $[x'^T; -1]^T$ is then found at the intersection of the orthogonal complement of $R_r([A;b'])$ with the hyperplane $x_{n+1} = -1$. The coefficients of the solution determine the equation of the hyperplane $R_r([A;b'])$. If the observed row vectors have additive random disturbances in all coordinate directions, the ordinary LS method, which assumes random errors along *one* coordinate axis only, is not appropriate.

The TLS approach (Fig. 2.2(b)) looks for a lower-dimensional row space $R_r([\widehat{A};\widehat{b}])$, closest to the row vectors $r_1, \cdots, r_m$ of $[A;b]$, such that (2.19) and (2.20) are satisfied. Similarly, the solution $[\widehat{x}^T; -1]^T$ is found at the intersection of the orthogonal complement of $R_r([\widehat{A};\widehat{b}])$ with the hyperplane $x_{n+1} = -1$. In contrast to LS, the data changes are *not restricted* to be along *one* coordinate axis x_{n+1}. Also, the equation of the hyperplane $R_r([\widehat{A};\widehat{b}])$ defined by

$$R_r([\widehat{A};\widehat{b}]) = \{\begin{bmatrix} \widetilde{a} \\ \widetilde{b} \end{bmatrix} \mid \widetilde{a} \in \mathcal{R}^n, \widetilde{b} \in \mathcal{R}, \widetilde{b} = \widehat{x}^T \widetilde{a}\}$$

is given by the coefficients of the TLS solution $\widehat{x}$.

Observe that all TLS correction vectors Δr_i, applied to the rows of $[A;b]$

and given by the rows of $[\Delta\widehat{A};\Delta\widehat{b}]$, are *parallel* with the solution vector $[\widehat{x}^T;-1]^T$. Indeed, (2.25) and (2.26) prove that all these vectors are proportional to the $(n+1)$th right singular vector v_{n+1} of $[A;b]$ and are orthogonal to $R_r([\widehat{A};\widehat{b}])$. Note also that the sum of squared horizontal distances is minimized in the LS problem while in the TLS problem the perpendicular distances are critical. Indeed, since TLS minimizes $\| [\Delta\widehat{A};\Delta\widehat{b}] \|_F = \sqrt{\sum_{i=1}^m \| \Delta r_i \|_2^2} = \sigma_{n+1}$, $\| \Delta r_i \|_2$ must be the *perpendicular* distance from the row vector r_i to the nearest n-dimensional hyperplane $R_r([\widehat{A};\widehat{b}])$. Using the relation

$$\sigma_{n+1}^2 = \frac{\| [A;b][\widehat{x}^T;-1]^T \|_2^2}{\| [\widehat{x}^T;-1]^T \|_2^2} = \frac{\sum_{i=1}^m |a_i^T\widehat{x} - b_i|^2}{\widehat{x}^T\widehat{x}+1}$$

with a_i^T the ith row of A [68], these distances $\| \Delta r_i \|_2$ are given by $|a_i^T\widehat{x} - b_i|/\sqrt{\widehat{x}^T\widehat{x}+1}$. Hence, the TLS solution $\widehat{x}$ thus minimizes the squared Frobenius norm of the "normalized" residual matrix, defined as

$$\psi(x) = \sum_{i=1}^m \frac{|a_i x - b_i|^2}{x^T x + 1}.$$

Remember that the LS solution x' minimizes the squared Frobenius norm of the "ordinary" residual matrix, i.e., $\phi(x) = \sum_{i=1}^m |a_i x - b_i|^2$. For the simplest case $n = 1$, i.e., the fitting of a straight line, the LS and TLS measures of goodness-of-fit have been described in §1.1 and are shown in Fig. 1.1.

2.4.3. An extreme example. It is easily proved that the TLS correction (2.19) is always *smaller* in norm than the LS correction (2.16).

THEOREM 2.8. *Let* $Ax = b'$ *and* $\widehat{A}x = \widehat{b}$ *be, respectively, the* LS *and* TLS *approximation of the set* $Ax \approx b$, *then*

$$\|b - b'\|_2 \geq \|[A;b] - [\widehat{A};\widehat{b}]\|_F.$$

Proof. The TLS problem is converted into the LS problem if $A = \widehat{A}$ is fixed. This restricted minimization cannot produce a lower minimum. □

This difference between LS and TLS corrections is nicely illustrated in the following extreme example.

Example 2.3. For any set of equations $Ax \approx b$ of the form

$$\begin{bmatrix} N-1 & -1 & \cdots & -1 \\ -1 & N-1 & \cdots & -1 \\ \cdot & & & \\ \cdot & & & \\ \cdot & & & \\ -1 & -1 & \cdots & N-1 \\ -1 & -1 & \cdots & -1 \\ -1 & -1 & \cdots & -1 \end{bmatrix}_{N,N-2} \begin{bmatrix} x_1 \\ \cdot \\ \cdot \\ \cdot \\ x_{N-3} \\ x_{N-2} \end{bmatrix} \approx \begin{bmatrix} -1 \\ -1 \\ \cdot \\ \cdot \\ \cdot \\ -1 \\ N-1 \\ -1 \end{bmatrix},$$

the LS solution x' and TLS solution $\widehat{x}$ can be computed analytically [156] and are given by

$$x' = -0.5[1, \cdots, 1]^T \qquad \text{and} \qquad \widehat{x} = -[1, \cdots, 1]^T.$$

The corresponding corrections are

$$\begin{aligned} \Delta b' &= b - b' = 0.5[0, \cdots, 0, N, -N]^T, \\ [\Delta \widehat{A}; \Delta \widehat{b}] &= [A; b] - [\widehat{A}; \widehat{b}] = \frac{1}{N-1}[-1, \cdots, -1, N-1]^T [1, \cdots, 1]. \end{aligned}$$

Hence, the ratio of the length of the deviations

$$\frac{\| \Delta b' \|_2^2}{\| [\Delta \widehat{A}; \Delta \widehat{b}] \|_F^2} = \frac{N}{2}$$

becomes large for large dimension N.

$$\begin{aligned} \text{As} \quad \lim_{N \to \infty} \frac{\| \Delta b' \|_2}{\| b \|_2} &= \lim_{N \to \infty} \sqrt{\frac{N}{2(N-1)}} = \frac{1}{\sqrt{2}} \\ \text{and} \quad \lim_{N \to \infty} \frac{\| \Delta \widehat{b} \|_2}{\| b \|_2} &= \lim_{N \to \infty} \frac{\| \Delta \widehat{A} \|_F}{\| A \|_F} = \lim_{N \to \infty} \frac{1}{N-1} = 0 \end{aligned}$$

we observe that, if only the vector b is allowed to vary (the LS approach), b must be modified over approximately 70 percent for large N. If the matrix A is also allowed to vary (TLS approach), a small change of $1/(N-1)$ is sufficient. The reader can appreciate the difference between LS and TLS for this example with $N = 4$ from the geometric representation in Fig. 2.3.

2.5. Other TLS problem formulations

The basic principle of TLS is that the noisy data $C = [A; b]$, while not satisfying a linear relation, are modified with minimal effort, as measured by the Frobenius norm, in a "nearby" matrix $\widehat{C} = [\widehat{A}; \widehat{b}]$ which is rank-deficient so that its columns are linearly related. Here, *all* data are modified, in contrast to LS, which modifies only *one* column of $[A; b]$. This principle can, in fact, be expressed in different ways. The TLS problem formulation (2.19)–(2.21) typically applies to problems in which we are primarily interested in a linear relation between the columns of A on one hand and b at the other hand. For example, in system identification we want to find a linear relationship between the inputs and the output, not between one of the inputs and the other inputs plus the output.

However, other settings are possible (and may be more appropriate) depending on the linear relation in which we are interested. In particular, we may be interested in estimating a linear relationship between the columns

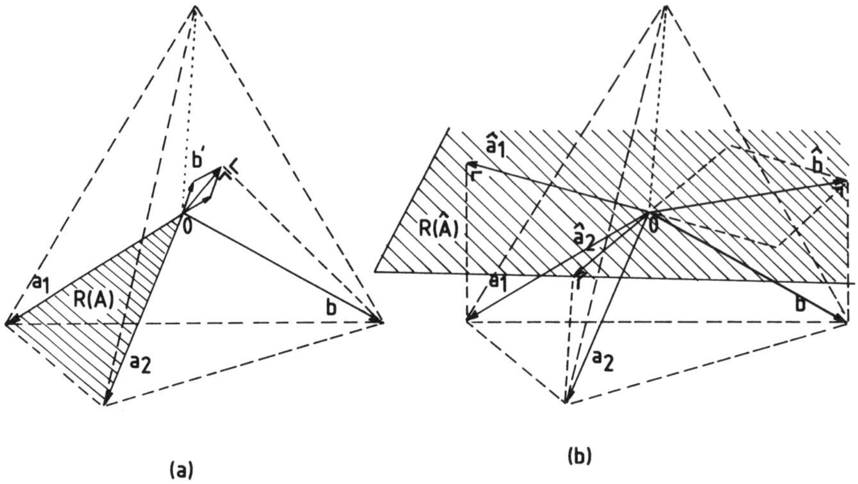

FIG. 2.3. *Geometric illustration of Example* 2.3 *for* $N = 4$.(a) *The* LS *solution is obtained by solving* $Ax = b'$ *with* b' *the orthogonal projection of* b *onto* $R(A)$, *generated by the columns* a_1 *and* a_2. (b) *The* TLS *solution is obtained by approximating the columns* a_1 *and* a_2 *of* A *by* $\widehat{a}_1$ *and* $\widehat{a}_2$ *and* b *by* $\widehat{b}$ *until* $\widehat{b}$ *is in the space* $R(\widehat{A})$, *generated by the columns* $\widehat{a}_1$ *and* $\widehat{a}_2$, *and solving* $\widehat{A}x = \widehat{b}$.

of $C = [A; b]$ no matter which column is used as right-hand side. In other words, there is no preference for writing a particular column of C as a linear combination of the others. In these applications the following alternative TLS problem formulation may be appropriate.

DEFINITION 2.4 (Alternative TLS problem). Consider the set of equations $C_{m\times n} y \approx 0$. The TLS problem seeks to

$$\text{(2.30)} \qquad \underset{\widehat{C} \in \mathcal{R}^{m\times n},\ y \in \mathcal{R}^n}{\text{minimize}} \quad \|C - \widehat{C}\|_F$$

$$\text{(2.31)} \qquad \text{subject to} \quad \widehat{C}y = 0,$$

$$\text{(2.32)} \qquad y^T y = 1.$$

The first constraint ensures the rank-deficiency of $\widehat{C}$, if $y \neq 0$ which is ensured by the second constraint. Using the Eckart–Young–Mirsky Theorem 2.3 or a Lagrangian approach [61], [39], it is straightforward to show that the optimal solution is given by

$$\begin{aligned}\widehat{C} &= C - u_n \sigma_n v_n^T,\\ y &= v_n,\end{aligned}$$

where σ_n is the smallest singular value of C, and u_n and v_n are the corresponding left and right singular vectors. Geometrically, this means that all columns of the matrix C are modified with a vector that is proportional to the left singular vector u_n.

The quadratic constraint $y^T y = 1$ may be replaced by a linear constraint on one of the components, for instance, the nth one: $y_n = -1$. This corresponds to the requirement that the nth modified column $\widehat{c}_n$ must lie in the column space generated by the remaining $n-1$ modified columns $\widehat{c}_i$. The solution to the TLS problem $A_{m\times(n-1)}x \approx b$ defined by (2.19)–(2.21) with $C_{m\times n} = [A; b]$ is then essentially the same. Indeed, generically, the TLS solution $[\widehat{x}^T; -1]^T$ is also the right singular vector v_n, normalized in such a way that its nth component is -1.

Observe that in these applications the linear relation $\widehat{C}v_n = 0$ described by the lowest singular vector v_n, without a particular normalization, is sufficient and contains all necessary information.

Here is another important remark. If there is *structure* in the matrix C, it may be desirable to require the *same* structure for the approximation $\widehat{C}$. For instance, in many signal processing and system identification problems, the matrix C has (block) Hankel or (block) Toeplitz structure. If this is the case, the set of possible matrices $\mathcal{R}^{m\times n}$ should be restricted to the set $S = \{\widehat{C} | \widehat{C} \in \mathcal{R}^{m\times n}$, and $\widehat{C}$ has the same structure as $C\}$ in the TLS problem formulated above. However, the solution to this structured problem is no longer given by the SVD of the matrix C. In general, the Karush–Kuhn–Tucker conditions are difficult, if not impossible, to solve for such structured problems, see, e.g., [2]–[5].

There is, however, another interpretation possible, as shown below. Therefore, we consider the following optimization problem, called the *orthogonal l_2 approximation* problem [149], [198], which is another possibility of obtaining an approximate linear relation between the columns of a matrix C.

DEFINITION 2.5 (Orthogonal l_2 approximation problem). Consider the data matrix $C \in \mathcal{R}^{m\times n}$. The orthogonal l_2 approximation problem seeks to

$$\underset{y \in \mathcal{R}^n}{\text{minimize}} \quad \| Cy \|_2 \tag{2.33}$$

$$\text{subject to} \quad y^T y = 1. \tag{2.34}$$

The vector e defined as $e = Cy$ can be considered as a vector of residuals or errors on each equation. Therefore, the orthogonal l_2 approximation problem

(2.33) and (2.34) can also be interpreted as an *equation error* identification problem.

A constraint in y is needed to avoid the trivial solution $y = 0$. It is straightforward to find that the solution to (2.33) and (2.34) is given by the eigenvector of $C^T C$, corresponding to the smallest eigenvalue. This is precisely the nth right singular vector of C, corresponding to the smallest singular value, and hence coincides with the TLS solution $[\widehat{x}^T; -1]^T$, up to a scaling factor.

The difference between the TLS problem, defined by (2.19)–(2.21) or (2.30)–(2.32), and the present minimization procedure lies in the data modification idea. In the first approach, we use as an intermediate step a matrix $\widehat{C} = [\widehat{A}; \widehat{b}]$ that represents the modified data. If, however, the matrix C is structured and we require $\widehat{C}$ to have the same structure, the optimal solution using the idea of TLS (which can be found using a Lagrangian approach) is no longer given by the SVD of the matrix C, while in the orthogonal l_2 approximation approach (2.33) and (2.34), the answer is still given by the SVD of C (independent of the structure). The TLS problem as formulated in (2.19) and (2.20) or (2.30)–(2.32) remains, in general, unsolved for structured matrices.

In this book, we focus on the first TLS problem formulation given by (2.19)–(2.21). Note, however, that although different TLS problem formulations (as given above) may be used, they generically all compute the same solution. Indeed, if the TLS problem $A_{m \times n} x \approx b$ with (1.22) the SVD of $C = [A; b]$, is *generic* (which implies that $v_{n+1,n+1} \neq 0$ if $\sigma_n > \sigma_{n+1}$), then the solution to all TLS problems formulated in this chapter is essentially the same. Differences only arise when $v_{n+1,n+1} = 0$. In these so-called nongeneric problems, the TLS problem defined by (2.19)–(2.21) fails to have a solution while the problems defined in this section do not.

2.6. Concluding remarks

Solving sets of linear equations is a basic issue in many applications. If these sets $Ax \approx b$ are overdetermined, then generally only an approximate solution can be computed, for instance by using a LS or TLS approach.

The basic principle of TLS is that the noisy data $[A; b]$, while not satisfying a linear relation, are modified with minimal effort, as measured by the Frobenius norm, in a "nearby" matrix $[\widehat{A}; \widehat{b}]$ that is rank-deficient so that the set $\widehat{A}x = \widehat{b}$ is compatible. This matrix $[\widehat{A}; \widehat{b}]$ is a rank one modification of the data matrix $[A; b]$. The solution to the TLS problem can be determined from the SVD of the matrix $[A; b]$ (§2.3). A simple algorithm outlines the computations of the solution of the basic TLS problem. By "basic" we mean that only one right-hand side vector b is considered (one-dimensional) and that the TLS problem is solvable (generic) and has a unique solution. Extensions of this basic TLS problem are investigated in Chapter 3.

The LS approach for determining a linear relation between the columns of

A and b consists in modifying only the right-hand side vector b with minimal effort, as measured by the Euclidean norm, in a "nearby" vector b' such that $Ax = b'$ is compatible (§2.2). LS also modifies the data by a rank one matrix and it is shown how to compute the LS solution by means of the SVD of the matrix A (although computationally simpler but numerically less reliable methods can be used).

Furthermore, the basic principle and solution of LS and TLS problems are compared geometrically (§2.4). The TLS problem is tantamount to finding a "closest" subspace to the rows of the data matrix $[A; b]$. By depicting the LS and TLS measures of goodness-of-fit, it is shown that in the LS problem minimization of the sum of squared distances along one coordinate axis is performed while the TLS problem seeks to minimize the sum of squared perpendicular distances to the best subspace. In particular, it is proved that the data modifications applied by TLS are always less than those applied by LS to make the set $Ax \approx b$ compatible.

Note that the TLS problem, as formulated in §2.3 and used throughout this book, aims to estimate a linear relationship between the columns of A and b, but other formulations may be used depending on the particular linear relationship among the data in which we are interested (§2.5). Nevertheless, although formulations differ, the solution of the basic TLS problem is essentially the same.

Chapter 3

Extensions of the Basic Total Least Squares Problem

3.1. Introduction

In the previous chapter, the basic total least squares (TLS) problem $Ax \approx b$ was presented and fully analyzed. "Basic" means that only one right-hand side vector is considered (one-dimensional) and that a solution of the TLS problem exists (generic) and is unique.

In this chapter, various extensions of this basic TLS problem are considered and investigated. First, in §3.2, the TLS solution is given for problems $AX \approx B$ in which more than one right-hand side vector is associated with the same data matrix A (*multidimensional* TLS problem). Second, problems for which the TLS solution is *no longer unique* are treated in §3.3. This happens if the smallest singular values of the associated matrix $[A; B]$ coincide. To preserve uniqueness, TLS seeks for the minimum norm solution. Third, in §3.4, TLS problems that fail to have a solution altogether are investigated. The properties of these so-called nongeneric problems are described and an extension of the generic TLS problem, called *nongeneric* TLS, is proposed to make these problems solvable. It is proved that the solution to these nongeneric TLS problems is still optimal with respect to the TLS criteria for any number of observation vectors in B if additional constraints are imposed on the TLS solution space. Fourth, in §3.5, the TLS problem is extended to *mixed* LS-TLS problems where it is assumed that *some* of the columns in the data matrix A are *error free.*

Finally in §3.6, the TLS computations are summarized in one *algorithm* that takes into account all the extensions described above. First, the TLS algorithm is given for erroneous data. This is called the *classical* TLS algorithm because it is based on computing the classical SVD of $[A; B]$, as done in the algorithm of Golub and Van Loan. A fully documented program of this algorithm is provided in [176], [178]. Second, a *mixed* LS-TLS algorithm, which takes account of error-free columns in the data matrix A, is presented. This algorithm can handle the ordinary LS problem (A is assumed to be entirely error free) and TLS problem (A is erroneous), as well as a variety of mixed

LS-TLS problems. However, these algorithms need to compute the whole SVD of a matrix. A computationally *more efficient* algorithm, called *partial* TLS (PTLS), is developed in Chapter 4. Moreover, if a priori information about the solution of the problem to be solved is available, the TLS computations can be speeded up in an *iterative* way (see Chapter 5).

A nice *overview* of all TLS problems, discussed in this chapter, is given in the following diagram. For each case, it describes the main properties of the TLS solution. The SVD of $[A; B]$ is given by (1.22) and

$$V_\gamma = \begin{bmatrix} v_{n+1,p+1} & \cdots & v_{n+1,n+d} \\ \cdot & & \cdot \\ \cdot & & \cdot \\ \cdot & & \cdot \\ v_{n+d,p+1} & \cdots & v_{n+d,n+d} \end{bmatrix}$$

with $p \leq n$ and $v_{j,i}$ the jth component of v_i. Hence, for $p = n$, $V_\gamma = V_{22}$.

TLS problem — $d = 1$: one-dimensional

$A_{m \times n} X \approx B, \ m > n$ — $d > 1$: multidimensional

↙ ↘

$\sigma_n > \sigma_{n+1}$ $\qquad$ $\sigma_p > \sigma_{p+1} = \cdots = \sigma_{n+1} \ (p < n)$

↙ ↘ ↙ ↘

V_{22} nonsingular $\qquad$ V_{22} singular $\qquad$ V_γ rank-deficient $\qquad$ V_γ has full rank d

↙ ↘ ↙ ↘

V_{22} nonsingular	V_{22} singular / V_γ rank-deficient	V_γ has full rank d
TLS solution is *generic* and *unique*.	TLS solution is *nongeneric*	TLS solution is *generic* and *not* unique.
	If not unique, minimum norm TLS solution is computed.	*Minimum norm* TLS solution is computed.
(Theorems 2.6 and 3.1)	Theorems 3.12 and 3.15 describe the conditions. Theorems 3.11 and 3.14 describe the properties.	(Theorems 3.7 and 3.9)
(see §§2.3 and 3.2)	(see §3.4)	(see §3.3)

3.2. The multidimensional TLS problem with unique solution

The TLS solution of a set of equations $AX \approx B$ where B is an $m \times d$ observation matrix follows from a straightforward generalization of the basic TLS problem, given in §2.3. If we allow *more than one* observation vector, the TLS problem definition can be generalized as follows.

DEFINITION 3.1 (Multidimensional TLS problem). Given an overdetermined set of m linear equations $AX \approx B, B \in \mathcal{R}^{m \times d}$, in $n \times d$ unknowns X.

The multidimensional total least squares (TLS) problem seeks to

$$\text{(3.1)} \qquad \underset{[\hat{A};\hat{B}] \in \mathcal{R}^{m\times(n+d)}}{\text{minimize}} \quad \| [A;B] - [\hat{A};\hat{B}] \|_F$$

$$\text{(3.2)} \qquad \text{subject to} \quad R(\hat{B}) \subseteq R(\hat{A}).$$

Once a minimizing $[\hat{A};\hat{B}]$ is found, then any X satisfying

$$\text{(3.3)} \qquad \hat{A}X = \hat{B}$$

is called a TLS *solution* and $[\Delta\hat{A};\Delta\hat{B}] = [A;B] - [\hat{A};\hat{B}]$ the corresponding TLS correction.

The TLS solution is denoted by $\hat{X}$.

To solve this problem, we bring $AX \approx B$ into the following form:

$$\text{(3.4)} \qquad [A;B][X^T;-I_d]^T \approx 0.$$

To obtain a solution, the rank of $[A;B]$ must be reduced to n. Using the Eckart–Young–Mirsky Theorem 2.3, the best rank n approximation $[\hat{A};\hat{B}]$ of $[A;B]$ that minimizes (3.1) is obtained by making the smallest singular values σ_{n+i}, $1 \le i \le d$, of $[A;B]$ zero. If the SVD of $[A;B]$ is given by (1.22), then the *lower* rank approximation of (3.4) is

$$\text{(3.5)} \qquad [\hat{A};\hat{B}][X^T;-I_d]^T = 0 \quad \text{with } [\hat{A};\hat{B}] = U_1\Sigma_1[V_{11}^T;V_{21}^T],$$

the TLS approximation of $[A;B]$. The corresponding TLS correction matrix is given by

$$[\Delta\hat{A};\Delta\hat{B}] = U_2\Sigma_2[V_{12}^T;V_{22}^T]$$

and its Frobenius norm

$$\text{(3.6)} \quad \| [\Delta\hat{A};\Delta\hat{B}] \|_F = \| \Sigma_2 \|_F = \sqrt{\sum_{i=1}^{t} \sigma_{n+i}^2}, \quad \text{where} \quad t = \min\{m-n, d\}$$

represents the *minimal* TLS *correction.* The solution of (3.5) is then given by the d right singular vectors $\begin{bmatrix} V_{12} \\ V_{22} \end{bmatrix}$, which belong to $N([\hat{A};\hat{B}])$. Indeed,

(3.7)

$$\begin{aligned} R\left(\begin{bmatrix} \hat{X} \\ -I \end{bmatrix}\right) &= R\left(\begin{bmatrix} V_{12} \\ V_{22} \end{bmatrix}\right) \\ &\Longleftrightarrow \exists \text{ nonsingular } S \in \mathcal{R}^{d\times d} : \begin{bmatrix} \hat{X} \\ -I \end{bmatrix} = \begin{bmatrix} V_{12} \\ V_{22} \end{bmatrix} S. \end{aligned}$$

If V_{22} is *nonsingular*, the TLS solution follows immediately:

$$\text{(3.8)} \qquad \hat{X} = -V_{12}V_{22}^{-1}.$$

The TLS solution can also be computed directly from (3.3). Indeed, with (3.5), (3.3) becomes

$$U_1\Sigma_1 V_{11}^T \widehat{X} = U_1\Sigma_1 V_{21}^T. \tag{3.9}$$

Hence if V_{11} is nonsingular, the TLS solution is obtained from:

$$\widehat{X} = (V_{11}^T)^{-1}V_{21}^T = (V_{21}V_{11}^{-1})^T. \tag{3.10}$$

From [69, Thm. 2.6.1] and the orthonormality of V, we know that $\| V_{11}^{-1} \|_2 = \| V_{22}^{-1} \|_2$. This implies that the minimal singular values of V_{11} and V_{22} are equal. Hence the singularity of V_{22} implies that of V_{11}. If $n < d$, this procedure would be more efficient than the TLS computation given by (3.8). In practice, the inverses in (3.8) and (3.10) are never computed explicitly but the solution $\widehat{X}$ is obtained by solving a set of equations (see §3.6.1).

The conditions V_{22} (or equivalently V_{11}) nonsingular guarantee that a solution to the multidimensional TLS problem (3.1) and (3.2) exists. This solution is, moreover, unique if in addition $\sigma_n > \sigma_{n+1}$ holds, as proved in the next theorem.

THEOREM 3.1 (Solution of the multidimensional TLS problem $AX \approx B$). *Suppose* (1.22) *is the* SVD *of* $[A; B]$. *If* $\sigma_n > \sigma_{n+1}$ *and* V_{22} *nonsingular, then*

$$[\widehat{A}; \widehat{B}] = U\mathrm{diag}(\sigma_1, \cdots, \sigma_n, 0, \cdots, 0)V^T = U_1\Sigma_1[V_{11}^T; V_{21}^T], \tag{3.11}$$

with corresponding TLS *correction matrix*

$$[\Delta\widehat{A}; \Delta\widehat{B}] = [A; B] - [\widehat{A}; \widehat{B}] = U_2\Sigma_2[V_{12}^T; V_{22}^T], \tag{3.12}$$

solves the TLS *problem* (3.1) *and* (3.2) *and*

$$\widehat{X} = -V_{12}V_{22}^{-1} \tag{3.13}$$

exists and is the unique solution to $\widehat{A}X = \widehat{B}$.

Proof. The TLS condition $R(\widehat{B}) \subseteq R(\widehat{A})$ implies that there exists an $X \in \mathcal{R}^{n\times d}$ such that

$$\widehat{A}X = \widehat{B}, \quad \text{i.e.,} \quad [\widehat{A}; \widehat{B}][X^T; -I]^T = 0. \tag{3.14}$$

Thus, $[\widehat{A}; \widehat{B}]$ has at most rank n. Using the Eckart–Young–Mirsky Theorem 2.3, it follows that

$$\| [A; B] - [\widehat{A}; \widehat{B}] \|_F^2 = \| [\Delta\widehat{A}; \Delta\widehat{B}] \|_F^2 \geq \sum_{i=1}^{t} \sigma_{n+i}^2 \tag{3.15}$$

and that equality results if $[\Delta\widehat{A}; \Delta\widehat{B}]$ is given by (3.12). The assumption $\sigma_n > \sigma_{n+1}$ ensures that $[\Delta\widehat{A}; \Delta\widehat{B}]$ given by (3.12) is the unique minimizer. Using the properties of the SVD, the expression (3.11) of $[\widehat{A}; \widehat{B}] = [A; B] - [\Delta\widehat{A}; \Delta\widehat{B}]$ then follows immediately. Moreover, the null space of $[\widehat{A}; \widehat{B}] = U_1\Sigma_1[V_{11}^T; V_{21}^T]$ is the range of $[V_{12}^T; V_{22}^T]^T$. Thus, from (3.14), we have $\begin{bmatrix} X \\ -I_d \end{bmatrix} = \begin{bmatrix} V_{12} \\ V_{22} \end{bmatrix} S$ for some $d \times d$ matrix S. From the equations $X = V_{12}S$ and $-I = V_{22}S$ we have $S = -V_{22}^{-1}$. Since V_{22} is nonsingular, its inverse exists and hence $\widehat{X} = V_{12}S = -V_{12}V_{22}^{-1}$ also exists and solves the TLS problem. □

The rank of the TLS correction matrix (3.12) is given by

$$\text{rank}([\Delta\widehat{A}; \Delta\widehat{B}]) = d - \text{corank}([A; B]).$$

Usually, $m \geq n + d$ and $\sigma_{n+d} > 0$. In these cases $[\Delta\widehat{A}; \Delta\widehat{B}]$ is a rank d matrix.

A closed-form expression of the multidimensional TLS solution follows from a straightforward generalization of Theorem 2.7.

COROLLARY 3.1 (Closed-form expression of the multidimensional TLS solution). *Let* (1.21) (*respectively,* (1.22)) *be the* SVD *of* A (*respectively,* $[A; B]$). *If* $\sigma'_n > \sigma_{n+1} = \cdots = \sigma_{n+d}$, *then*

$$\widehat{X} = (A^T A - \sigma_{n+1}^2 I)^{-1} A^T B. \tag{3.16}$$

Proof. Since this corollary is a special case of Theorem 3.10 (set $p = n$), we refer to the proof of that theorem (see §3.3.2). □

This closed form is only an analytical expression of the multidimensional TLS solution in the following problems. If $m \geq n + d$, then the d smallest singular values of $[A; B]$ must coincide but if $n < m < n + d$, the restriction $\sigma_{n+1} = \cdots = \sigma_{n+d}$ means that the smallest $m - n$ singular values $\sigma_{n+1} = \cdots = \sigma_m$ of $[A; B]$ must be zero (since $\sigma_{m+1} = \cdots = \sigma_{n+d} = 0$). It is important to note that the more general estimator $(A^T A - (1/d)\sum_{i=1}^{d} \sigma_{n+i}^2 I_n)^{-1} A^T B$ no longer coincides with the TLS solution of $AX \approx B$ if the restriction $\sigma_{n+1} = \cdots = \sigma_{n+d}$ is omitted.

Sufficient conditions for the existence and uniqueness of a solution to the multidimensional TLS problem (3.1)–(3.3) are proven in the next theorem.

THEOREM 3.2 (Existence and uniqueness condition for the TLS solution). *Let the* SVD *of* $[A; B]$ *be given by* (1.22) *and the* SVD *of* A *by* (1.21), *then*

$$\sigma'_n > \sigma_{n+1} \Longrightarrow \sigma_n > \sigma_{n+1} \text{ and } V_{22} \text{ is nonsingular.} \tag{3.17}$$

Proof. The proof that $\sigma' > \sigma_{n+1} \Longrightarrow V_{22}$ is nonsingular is a special case of Lemma 3.1 given in §3.3.2 (set $p = 2$). Alternative proofs are given in [69, Thm. 12.3.1] and [177, Thm. 1.1]. Since $\sigma_n \geq \sigma'_n$, according to the interlacing Theorem 2.4 for singular values, the statement $\sigma'_n > \sigma_{n+1} \Longrightarrow \sigma_n > \sigma_{n+1}$ follows immediately. □

If the conditions of Theorems 3.1 or 3.2 are not satisfied, a solution to the TLS problem (3.1)–(3.3) may not exist or may not be unique. In particular, we have the following: If σ_{n+1} coincides with σ_n or, more generally,

If $\sigma_{p+1} = \cdots = \sigma_{n+1}$ with $p < n$, the TLS problem (3.1) and (3.2) may still have a solution (i.e., the problem is generic) but this solution is no longer unique (see §3.3). If V_{22} is *singular*, then the multidimensional TLS problem (3.1) and (3.2) fails to have a solution. These problems are called nongeneric. By adding additional constraints, these problems are again solvable [185] (see §3.4).

Other useful conditions that ensure the existence of a TLS solution have been proved in [175] for the one-dimensional TLS problem ($d = 1$). These results even extend to TLS problems with more than one TLS solution, as the following theorem shows.

THEOREM 3.3 (Existence of a TLS solution). *Assume that* $A_0 x = b_0$ *and denote* $\tilde{\sigma}_r(P) = \sqrt{\sum_{i \geq r} \sigma_i^2(P)}$, *where* $\sigma_i(P)$ *are the singular values of the matrix* P *in nonincreasing order of magnitude. If* $\| [A; b] - [A_0; b_0] \|_F \leq \varepsilon < \frac{1}{2}\tilde{\sigma}_r(A_0)$, *then there exists a rank* r *approximation* $[\hat{A}; \hat{b}]$ *nearest to* $[A; b]$ *such that* $\hat{b} \in R(\hat{A})$, *i.e., the* TLS *problem* (2.19)–(2.21) *has a solution. Moreover,* $\| [\Delta\hat{A}; \Delta\hat{b}] \|_F \leq \varepsilon$.

A more practical theorem, in particular with conditions (3.19), that uses only quantities that are known in principle, is given below.

THEOREM 3.4 (Existence of a TLS solution). *Let* $[A; b]$ *have rank* $\geq r$, *let* $\tilde{A}$ *have rank* r, *and let* $\tilde{b}$ *be the orthogonal projection of* b *onto* $R(\tilde{A})$. *Define* $\tilde{\sigma}_r(P)$ *as in Theorem* 3.3. *If one of the conditions*

$$\| [A - \tilde{A}; b - \tilde{b}] \|_F^2 < \tilde{\sigma}_r^2(A) \tag{3.18}$$

or

$$\tilde{\sigma}_r^2(\tilde{A}) - 2\tilde{\sigma}_r(\tilde{A}) \| A - \tilde{A} \|_F - \| b - \tilde{b} \|_2^2 > 0 \tag{3.19}$$

is satisfied, then the TLS *problem* (2.19)–(2.21) *has a solution.*

Theorems 3.3 and 3.4 remain true if all Frobenius norms are replaced by 2-norms and $\tilde{\sigma}_r$ is replaced by σ_r.

If $\tilde{A}$ equals the TLS approximation $\hat{A}, r = n$ and the 2-norm is used, then (3.18) reduces to condition (3.17), i.e., $\| [\Delta\hat{A}; \Delta\hat{b}] \|_2 = \sigma_{n+1} < \sigma'_n$.

If $\tilde{A} = A$ (LS case), (3.18) and (3.19) reduce to the condition $\| b - b' \|_2 < \tilde{\sigma}_r$, which is more severe than condition (3.17) when $r = n$.

Contrary to this algebraic approach, Gleser [59] used a statistical approach to prove under which conditions V_{22} is nonsingular. In his proofs we can find the following interesting property (the superscript m on a sample quantity indicates that the quantity is calculated from the first m rows of $[A; B]$).

THEOREM 3.5. *Consider any matrix* $[A; B]$ *of* m *observations of the exact but unobservable relation* $A_0 X = B_0$ *such that* $[A; B] = [A_0; B_0] + [\Delta A; \Delta B]$. *Let* (1.22) *be the* SVD *of* $[A; B]$ *and assume that the rows of the error matrix* $[\Delta A; \Delta B]$ *are independently and identically distributed with zero mean and common covariance matrix* $\sigma_\nu^2 I_{n+d}$ (σ_ν *unknown scalar*). *Assume further that*

$\lim_{m\to\infty}(1/m)A_0^T A_0$ *exists and is positive definite then, with probability one*:

$$(3.20)\quad \exists m_0,\ \ 0 < m_0 < \infty : V_{11}^{(m)} \ \textit{and}\ V_{22}^{(m)} \ \ \textit{are nonsingular} \quad \forall m \geq m_0.$$

Proof. See [59, Lemma 3.3] for the proof. □

Equation (3.20) implies that the TLS solution $\widehat{X}^{(m)} = -V_{12}^{(m)} V_{22}^{(m)-1}$ exists for all $m \geq m_0$. Hence, the property guarantees that, if the assumptions hold, the TLS problem (3.1)–(3.3) has a solution when the set of equations $AX \approx B$ is sufficiently overdetermined; in other words, a *nongeneric* TLS *problem can be made generic* by adding *more* equations. This is the case if the data $[A; B]$ are observations of an exact but unobservable relation $A_0 X = B_0$ with A_0 of full row rank, and the observation errors $[\Delta A; \Delta B]$ are statistically independent and equally sized (same variance).

These statistical results agree with our algebraic approach: nongeneric TLS problems occur whenever A_0 is (nearly) rank-deficient or when the set $A_0 X \approx B_0$ is highly incompatible. Although "exact" nongeneric TLS problems seldom occur, close-to-nongeneric TLS problems are not uncommon. Moreover, from a numerical point of view, it is very interesting to investigate their properties and generalize the TLS computations to solve these problems in the TLS sense. Indeed, as will be shown in Theorems 3.11 and 3.14 of §3.4, the TLS problem becomes close to nongeneric when σ_n' approaches σ_{n+1}. In those cases, the generic TLS solution can still be computed, but it is unstable and even becomes very sensitive to data errors when $\sigma_n' - \sigma_{n+1}$ is very close to zero [68]. Identifying the problem as nongeneric and computing the nongeneric TLS solution, stabilizes the TLS solution and makes it rather insensitive to data errors as shown in §7.2.3.

Exact expressions for the 2-*norm* and *Frobenius norm* of the multidimensional TLS solution (3.13) are proved in the following theorem using the C-S decomposition [69, Thm. 2.6.1].

THEOREM 3.6 (2-norm and Frobenius norm of the TLS solution). *Let* (1.22) *be the* SVD *of* $[A; B]$. *If* $\sigma_n > \sigma_{n+1}$ *and if* V_{22} *is nonsingular with singular values* $\gamma_i, i = 1, \cdots, d$, *then the solution* $\widehat{X}$ *of the* TLS *problem* (3.1)–(3.3) *satisfies the following*:

$$(3.21)\qquad \| \widehat{X} \|_2^2 = (1 - \gamma_d^2)\gamma_d^{-2},$$

$$(3.22)\qquad \| \widehat{X} \|_F^2 = \sum_{i=1}^{d} (1 - \gamma_i^2)\gamma_i^{-2} = \| V_{22}^{-1} \|_F^2 - d.$$

Proof. Since

$$V = \begin{bmatrix} V_{11} & V_{12} \\ V_{21} & V_{22} \end{bmatrix} \begin{matrix} n \\ d \end{matrix}$$
$$\qquad\quad n \qquad\ d$$

is orthogonal, we can use the C-S *decomposition* [69, Thm. 2.6.1], i.e., there exist orthogonal matrices $W_1, T_1 \in \mathcal{R}^{n\times n}$ and orthogonal matrices $W_2, T_2 \in \mathcal{R}^{d\times d}$ such that (assume $n \geq d$):

$$(3.23)\quad \begin{bmatrix} W_1 & 0 \\ 0 & W_2 \end{bmatrix}^T \begin{bmatrix} V_{11} & V_{12} \\ V_{21} & V_{22} \end{bmatrix} \begin{bmatrix} T_1 & 0 \\ 0 & T_2 \end{bmatrix} = \begin{bmatrix} I & 0 & 0 \\ 0 & C & S \\ 0 & -S & C \end{bmatrix} \begin{matrix} n-d \\ d \\ d, \end{matrix}$$
$$\qquad\qquad\qquad\qquad\qquad\qquad\qquad\qquad n-d \quad d \quad d$$

where $C = \operatorname{diag}(c_1, \cdots, c_d)$ and $S = \operatorname{diag}(s_1, \cdots, s_d)$ and $c_i = \cos\theta_i, s_i = \sin\theta_i, c_i^2 + s_i^2 = 1$ for all $i = 1, \cdots, d$ and $0 \leq \theta_1 \leq \cdots \leq \theta_d \leq \pi/2$. The assumption $n \geq d$ is made for notational convenience only.

With $\widehat{X}$ given by (3.13) and since the 2-norm is invariant with respect to orthogonal transformations [69, §2.3], we obtain, using the C-S decomposition (3.23) of V (partitioned as in (1.22)),

$$(3.24)\quad \| \widehat{X} \|_2^2 = \| V_{12}V_{22}^{-1} \|_2^2 = \| W_1^T \; V_{12} \; T_2 \; T_2^T \; V_{22}^{-1} \; W_2 \|_2^2 = \| SC^{-1} \|_2^2 \,.$$

Since $c_1 \geq \cdots \geq c_d$ and $s_1 \leq \cdots \leq s_d$, (3.24) yields

$$(3.25)\qquad \| SC^{-1} \|_2^2 = s_d^2/c_d^2 = (1-c_d^2)/c_d^2.$$

Since V_{22} and C have the same singular values, (3.21) follows directly from (3.25). The Frobenius norm is also invariant with respect to orthogonal transformations. Hence, we derive analogously as in (3.24):

$$(3.26)\qquad \| \widehat{X} \|_F^2 = \| SC^{-1} \|_F^2 = \sum_{i=1}^{d} (1-c_i^2)/c_i^2 = \sum_{i=1}^{d} (1/c_i^2) - d.$$

Since C and V_{22} have the same singular values and $c_i^{-1}, i = 1, \cdots, d$ are the singular values of V_{22}^{-1}, (3.22) follows directly from (3.26). □

Finally, using expressions (3.13) and (3.10) of the TLS solution, the following relationships between the TLS related matrices V_{11}, V_{22}, V_{12}, and the TLS solution $\hat{X}$ are proved.

COROLLARY 3.2. *Let* (1.22) *be the* SVD *of* $[A; B]$. *If* $\sigma_n > \sigma_{n+1}$ *and* V_{22} *nonsingular, then*

$$(3.27)\qquad V_{11}V_{11}^T = (I + \widehat{X}\widehat{X}^T)^{-1},$$
$$(3.28)\qquad V_{22}V_{22}^T = (I + \widehat{X}^T\widehat{X})^{-1},$$
$$(3.29)\qquad V_{12}V_{12}^T = \widehat{X}(I + \widehat{X}^T\widehat{X})^{-1}\widehat{X}^T.$$

Proof. From (3.10) and the orthonormality of V, we obtain

$$I + \widehat{X}\widehat{X}^T = [I; \widehat{X}] \begin{bmatrix} I \\ \widehat{X}^T \end{bmatrix} = (V_{11}^{-1})^T [V_{11}^T; V_{21}^T] \begin{bmatrix} V_{11} \\ V_{21} \end{bmatrix} V_{11}^{-1} = (V_{11}V_{11}^T)^{-1},$$

which implies (3.27). From (3.13) and the orthonormality of V, we also obtain

$$I + \widehat{X}^T\widehat{X} = [\widehat{X}^T; I]\begin{bmatrix} \widehat{X} \\ I \end{bmatrix} = (V_{22}^{-1})^T[V_{12}^T; V_{22}^T]\begin{bmatrix} V_{12} \\ V_{22} \end{bmatrix}V_{22}^{-1} = (V_{22}V_{22}^T)^{-1},$$

which proves (3.28). From (3.13) and (3.28), (3.29) follows immediately. □

Equations (3.27) and (3.28) are quite useful in the derivation of the asymptotic properties of the TLS problem (see Chapter 8). Using (3.27) the relationship between the data matrix A and its TLS approximation $\widehat{A}$ is easily deduced.

COROLLARY 3.3 (TLS approximation of A.). *Let* (1.22) *be the* SVD *of* $[A; B]$. *If* $\sigma_n > \sigma_{n+1}$ *and* V_{22} *is nonsingular, then*

$$\widehat{A} = (A + B\widehat{X}^T)(I + \widehat{X}\widehat{X}^T)^{-1}. \tag{3.30}$$

Proof. From (3.11), we obtain $\widehat{A} = U_1\Sigma_1 V_{11}^T$, which also equals

$$\widehat{A} = [A; B][V_{11}^T; V_{21}^T]^T V_{11}^T$$

and thus

$$\widehat{A} = (A + BV_{21}V_{11}^{-1})V_{11}V_{11}^T. \tag{3.31}$$

Substituting (3.10) and (3.27) into (3.31) gives (3.30). □

Equation (3.30) suggests that the TLS approximation $\widehat{A}$ is not a consistent estimate of the true but unknown data matrix A_0, observed by A. This is proved in §8.4.

3.3. Nonuniqueness of the TLS solution

3.3.1. The one-dimensional case.

As mentioned before, the TLS solution of $Ax \approx b$ with (1.22) the SVD of $[A; b]$ is no longer unique if $\sigma_n = \sigma_{n+1}$. Indeed, if $\sigma_p > \sigma_{p+1} = \cdots = \sigma_{n+1}$ with $p < n$ then any $[\widehat{A}; \widehat{b}] = [A; b] - [\Delta\widehat{A}; \Delta\widehat{b}]$ with

$$[\Delta\widehat{A}; \Delta\widehat{b}] = [A; b]\begin{bmatrix} w \\ \beta \end{bmatrix}[w^T; \beta]/\|[w^T; \beta]^T\|_2^2 \text{ and } \begin{bmatrix} w \\ \beta \end{bmatrix} \in R([v_{p+1}, \cdots, v_{n+1}])$$

solves the TLS problem (2.19) and (2.20) provided $\beta \neq 0$. Indeed, assume $\begin{bmatrix} w \\ \beta \end{bmatrix} = \sum_{i=p+1}^{n+1} \lambda_i v_i$, then

$$\|[\Delta\widehat{A}; \Delta\widehat{b}]\|_F^2 = \sum_{i=p+1}^{n+1} \sigma_i^2\lambda_i^2/\|[w^T; \beta]^T\|_2^2 = \sigma_{n+1}^2$$

is minimal. The corresponding TLS solution $\widehat{x} = -w/\beta$ satisfies

$$[\widehat{A}; \widehat{b}]\begin{bmatrix} \widehat{x} \\ -1 \end{bmatrix} = -\frac{1}{\beta}\{[A; b] - [\Delta\widehat{A}; \Delta\widehat{b}]\}\begin{bmatrix} w \\ \beta \end{bmatrix}$$

$$= -\frac{1}{\beta}[A;b]\begin{bmatrix} w \\ \beta \end{bmatrix}\left\{1 - \frac{[w^T;\beta]\begin{bmatrix} w \\ \beta \end{bmatrix}}{\|[w^T;\beta]\|_2^2}\right\} = 0.$$

Hence, there exists an $\widehat{x}$ such that $\widehat{A}\widehat{x} = \widehat{b}$, i.e., $\widehat{b} \in R(\widehat{A})$.

For reasons of stability and minimal sensitivity, it is desirable to single out the TLS solution with minimal 2-norm or Frobenius norm. To this end, consider the expressions of $\|\widehat{x}\|_2$ and $\|\widehat{x}\|_F$ proven in Theorem 3.6. This suggests maximizing $v_{n+1,n+1}$ in Theorem 2.6 and can be accomplished as follows.

THEOREM 3.7 (Minimum norm TLS solution of $Ax \approx b$). *Consider the* TLS *problem* (2.19)–(2.21). *Let* (1.22) *be the* SVD *of* $[A;b]$ *and assume that* $\sigma_p > \sigma_{p+1} = \cdots = \sigma_{n+1}$ *with* $p \leq n$. *Let* Q *be an orthonormal matrix such that*

$$[v_{p+1},\cdots,v_{n+1}]Q = \begin{bmatrix} Y & z \\ 0\cdots 0 & \gamma \end{bmatrix}\begin{matrix} n \\ 1 \end{matrix}. \tag{3.32}$$
$$\qquad\qquad\quad n-p \quad\; 1$$

If $\gamma \neq 0$, *then the* TLS *solution*

$$\widehat{x} = -z/\gamma \tag{3.33}$$

has minimal 2*-norm* $\| \widehat{x} \|_2 = \sqrt{1-\gamma^2}/\gamma$.

The TLS *approximation and corresponding* TLS *correction matrix are given by*

$$[\widehat{A};\widehat{b}] = [A - \Delta\widehat{A}; b - \Delta\widehat{b}] \text{ with } [\Delta\widehat{A};\Delta\widehat{b}] = [A;b]\begin{bmatrix} z \\ \gamma \end{bmatrix}[z^T;\gamma]. \tag{3.34}$$

Proof. Since $\sigma_p > \sigma_{p+1} = \cdots = \sigma_{n+1}$, $p \leq n$, any vector of the form $-(w/\beta)$ where $\left[\begin{smallmatrix} w \\ \beta \end{smallmatrix}\right]$ is a unit vector of $R([v_{p+1},\cdots,v_{n+1}])$, $\beta \neq 0$, is a TLS solution $\widehat{x}$ of the problem (2.19)–(2.21), as shown above. The norm of $x = -(w/\beta)$, given by $\|x\|_2^2 = (1-\beta^2)/\beta^2$, is minimized if β is maximal. This is accomplished as follows. If Q is an orthonormal matrix such that (3.32) is obtained, e.g., by using Householder transformations, then $\left[\begin{smallmatrix} z \\ \gamma \end{smallmatrix}\right]$ has the largest $(n+1)$th component of all unit vectors in $R([v_{p+1},\cdots,v_{n+1}])$. Since $\left[\begin{smallmatrix} \widehat{x} \\ -1 \end{smallmatrix}\right] /\!/ \left[\begin{smallmatrix} z \\ \gamma \end{smallmatrix}\right]$ and $\gamma \neq 0$, the TLS solution (3.33) has minimal norm $\| \widehat{x} \|_2 = \| \widehat{x} \|_F$. Moreover,

$$\begin{bmatrix} I_p & 0 & 0 \\ 0 & Q & 0 \\ 0 & 0 & I_{m-n-1} \end{bmatrix}^T U^T[A;b]V\begin{bmatrix} I_p & 0 \\ 0 & Q \end{bmatrix} = \Sigma$$

and so $[\Delta\widehat{A};\Delta\widehat{b}]$ is given by (3.34) and has minimal $\|[\Delta\widehat{A};\Delta\widehat{b}]\|_F = \sigma_{n+1}$. $\square$

Closed-form expressions of the minimum norm TLS solution $\widehat{x}$ are given below.

THEOREM 3.8 (Closed-form expressions of the minimum norm TLS solution). *Consider the* TLS *problem* (2.19)–(2.21). *Let* (1.22) *be the* SVD *of* $[A;b]$ *and assume that* $\sigma_p > \sigma_{p+1} = \cdots = \sigma_{n+1}$ *with* $p \leq n$. *If not all* $v_{n+1,i} = 0,\ i = p+1,\cdots,n+1$, *then the minimum norm* TLS *solution* $\widehat{x}$ *is given by*

$$\widehat{x} = -\Big(\sum_{i=p+1}^{n+1} v_{n+1,i}[v_{1,i},\cdots,v_{n,i}]^T\Big)/\sum_{i=p+1}^{n+1} v_{n+1,i}^2 \tag{3.35}$$

$$= -\Big(\sum_{i=1}^{p} v_{n+1,i}[v_{1,i},\cdots,v_{n,i}]^T\Big)/\Big(1-\sum_{i=1}^{p} v_{n+1,i}^2\Big). \tag{3.36}$$

Proof. Let

$$[\widehat{x}^T;-1]^T = \sum_{i=p+1}^{n+1} g_i v_i = \bar{V}_p g,$$

where g is a $p \times 1$ vector and $\bar{V}_p = [v_{p+1},\cdots,v_{n+1}]$ and let $e_{n+1} = [0,\cdots,0,1]^T$ be the $(n+1)\times 1$ vector comprised of all zeros except for a one in the $(n+1)$th element position. Then, the problem of finding the minimum norm solution in the TLS problem (2.19)–(2.21) can be cast as the following constrained optimization problem [217]:

$$\min_g \|\bar{V}_p g\|_2^2 = g^T g \quad \text{subject to} \quad e_{n+1}^T \bar{V}_p g = -1.$$

Using the Lagrange multipliers, it is easily derived that the unique solution is $g = \lambda \bar{V}_p^T e_{n+1}$ where $-(1/\lambda) = e_{n+1}^T \bar{V}_p \bar{V}_p^T e_{n+1}$. Hence, the minimum norm TLS solution is then given by

$$\begin{bmatrix} \widehat{x} \\ -1 \end{bmatrix} = \frac{-1}{e_{n+1}^T \bar{V}_p \bar{V}_p^T e_{n+1}} \bar{V}_p \bar{V}_p^T e_{n+1}. \tag{3.37}$$

Working out (3.37), we obtain (3.35).

Also note that if $p < n-p$, it is more efficient to work with the first right singular vectors $V_p = [v_1,\cdots,v_p]$. Since $\bar{V}_p\bar{V}_p^T = I - V_p V_p^T$, the closed-form expression (3.36) for the minimum norm TLS solution is immediately derived from (3.37). □

If $\text{rank}([A;b]) = p < n+1$ then $\sigma_p > \sigma_{p+1} = \cdots = \sigma_{n+1} = 0$ or $n-p+1$ singular values coincide at 0. In this case, $[A;b] = [\widehat{A};\widehat{b}]$ and the exact minimum norm TLS solution can be found in the same way as described above.

In practice, singular values of $[A;b]$ will *rarely* coincide. Due to *errors* in the data, coinciding singular values of the corresponding unobservable exact relation $A_0 x_0 = b_0$ become *distinct*. In this case it is realistic to define an error bound ε such that all singular values σ_i, satisfying $|\sigma_i - \sigma_{n+1}| < \varepsilon$, are considered as coinciding with σ_{n+1} and the TLS solution is obtained as above (see also §3.6.1).

3.3.2. The multidimensional case. In the *multidimensional* TLS problem $AX \approx B$, i.e., $d > 1$, the minimum norm TLS solution can be computed *analogously* provided the smallest singular values σ_{n+i}, $i \leq d$, of $[A; B]$ coincide. To minimize $\|\widehat{X}\|_2$ and $\|\widehat{X}\|_F$, given in Theorem 3.6, V_{22} in Theorem 3.1 must be chosen so that its minimal singular value is maximized. This can be accomplished as follows.

THEOREM 3.9 (Minimum norm TLS solution of $AX \approx B$). *Consider the* TLS *problem* (3.1)–(3.3). *Let* (1.22) *be the* SVD *of* $[A; B]$ *and assume that* $\sigma_p > \sigma_{p+1} = \cdots = \sigma_{n+d}$ *with* $p \leq n$. *Let* Q *be an orthonormal matrix such that*

$$[v_{p+1}, \cdots, v_{n+d}]Q = \underset{\quad n-p \qquad d}{\begin{bmatrix} Y & Z \\ 0 & \Gamma \end{bmatrix}} \begin{matrix} n \\ d \end{matrix} .$$

If Γ *is nonsingular, then the generic* TLS *solution*

$$\widehat{X} = -Z\Gamma^{-1} \tag{3.38}$$

has minimal 2*-norm* $\| \widehat{X} \|_2$ *and minimal Frobenius norm* $\| \widehat{X} \|_F$, *given by*

$$\| \widehat{X} \|_2^2 = (1 - \sigma_d^2(\Gamma))/\sigma_d^2(\Gamma) \quad \text{and} \quad \| \widehat{X} \|_F^2 = \| \Gamma^{-1} \|_F^2 - d, \tag{3.39}$$

where $\sigma_d(\Gamma)$ *is the minimal singular value of* Γ.

The TLS *approximation and corresponding* TLS *correction matrix are given by*

$$[\widehat{A}; \widehat{B}] = [A - \Delta\widehat{A}; B - \Delta\widehat{B}] \quad \text{with} \quad [\Delta\widehat{A}; \Delta\widehat{B}] = [A; B] \begin{bmatrix} Z \\ \Gamma \end{bmatrix} [Z^T; \Gamma^T]. \tag{3.40}$$

Proof. Since $\sigma_p > \sigma_{p+1} = \cdots = \sigma_{n+d}$, any $[\widehat{A}; \widehat{B}] = [A; B] - [\Delta\widehat{A}; \Delta\widehat{B}]$ with $[\Delta\widehat{A}; \Delta\widehat{B}]$ given by (3.40) and

$$\underset{d}{\begin{bmatrix} Z \\ \Gamma \end{bmatrix}} \begin{matrix} n \\ d \end{matrix}$$

an orthonormal basis $\in R([v_{p+1}, \cdots, v_{n+d}])$ solves the TLS problem (3.1)–(3.3) provided Γ is nonsingular. Indeed,

$$\exists Q = [\underset{n-p}{Q_1} \; ; \; \underset{d}{Q_2}] \in \mathcal{R}^{(n+d-p)\times(n+d-p)}$$

orthonormal such that

$$\begin{bmatrix} Z \\ \Gamma \end{bmatrix} = [v_{p+1}, \cdots, v_{n+d}]Q_2 = \begin{bmatrix} V_1 \\ V_2 \end{bmatrix} \begin{matrix} n \\ d \end{matrix} Q_2.$$

Then, (3.40) can be written as

$$\begin{aligned} [\Delta\hat{A}; \Delta\hat{B}] &= [A; B] \begin{bmatrix} V_1 \\ V_2 \end{bmatrix} Q_2 Q_2^T [V_1^T; V_2^T] \\ &= U_2 \mathrm{diag}(\sigma_{p+1}, \cdots, \sigma_{n+t}) Q_2 [(V_1 Q_2)^T; (V_2 Q_2)^T] \end{aligned} \tag{3.41}$$

with $U_2 = [u_{p+1}, \cdots, u_m]$ and $t = \min\{m - n, d\}$.

Assume first $m \geq n + d$. Since $\sigma_{n+1} = \ldots = \sigma_{n+d}$, (3.41) equals

$$[\Delta\hat{A}; \Delta\hat{B}] = \sigma_{n+1}[u_{p+1}, \ldots, u_{n+d}]Q_2[(V_1Q_2)^T; (V_2Q_2)^T].$$

It is easy to check that $\|[\Delta\hat{A}; \Delta\hat{B}]\|_F = \sigma_{n+1}\sqrt{d}$.

If $m < n + d$, the condition $\sigma_{n+1} = \ldots = \sigma_{n+d}$ requires that all $\sigma_i = 0, i = p+1, \cdots, m$; hence $\|[\Delta\hat{A}; \Delta\hat{B}]\|_F = 0$.

Thus, (3.1) is satisfied in either case. Furthermore, there exists $\widehat{X}$ such that $\widehat{A}\widehat{X} = \widehat{B}$, i.e., $R(\widehat{B}) \subseteq R(\widehat{A})$. Indeed, $\widehat{X} = -Z\Gamma^{-1}$ satisfies

$$[\hat{A}; \hat{B}] \begin{bmatrix} \widehat{X} \\ -I \end{bmatrix} = [A; B] \begin{bmatrix} Z \\ \Gamma \end{bmatrix} \left\{ I - [Z^T; \Gamma^T] \begin{bmatrix} Z \\ \Gamma \end{bmatrix} \right\} (-\Gamma^{-1}) = 0.$$

Since the solution $\left[\begin{smallmatrix} \widehat{X} \\ -I \end{smallmatrix}\right]$ is deduced from an orthonormal basis $\left[\begin{smallmatrix} Z \\ \Gamma \end{smallmatrix}\right]^n_d$, it is immediately proved, following the proof of Theorem 3.6, that $\| \widehat{X} \|_2$ and $\| \widehat{X} \|_F$ of (3.38) are given by (3.39). To minimize these expressions we have to select d basis vectors $\left[\begin{smallmatrix} Z \\ \Gamma \end{smallmatrix}\right]^n_d$ within $R([v_{p+1}, \cdots, v_{n+d}])$ such that $\| \Gamma^{-1} \|_F$ is minimized and the minimal singular value $\sigma_d(\Gamma)$ of Γ is maximized. This is done by computing an orthonormal matrix Q (e.g., by using Householder transformations) such that

$$[v_{p+1}, \cdots, v_{n+d}]Q = \begin{bmatrix} V_1 Q \\ V_2 Q \end{bmatrix} \begin{matrix} n \\ d \end{matrix} = \begin{bmatrix} Y & Z \\ 0 & \Gamma \end{bmatrix} \begin{matrix} n \\ d \end{matrix} . \tag{3.42}$$
$$\qquad\qquad\qquad\qquad n - p \quad d$$

V_2 and V_2Q have the same singular values. Denote by $\widetilde{\gamma}_i, i = 1, \cdots, d$ the singular values of a submatrix $\widetilde{\Gamma}$ of V_2 or V_2Q (obtained by deleting $n - p$ columns) and by $\sigma_i, i = 1, \cdots, d$ the singular values of V_2 or V_2Q in nonincreasing order of magnitude. Then, the interlacing Theorem 2.4 for singular values yields

$$\widetilde{\gamma}_i \leq \sigma_i \text{ or, equivalently, } \widetilde{\gamma}_i^{-1} \geq \sigma_i^{-1} \quad \forall i = 1, \cdots, d.$$

Hence, $\| \widetilde{\Gamma}^{-1} \|_F^2 - \sum_{i=1}^{d} \widetilde{\gamma}_i^{-2}$ and $\widetilde{\gamma}_d^{-2}$ are minimized if $\widetilde{\gamma}_i = \sigma_i$, for all i. Since the d by d submatrix Γ, defined above, has the same singular values as V_2, it follows directly that the TLS solution $\widehat{X} = -Z\Gamma^{-1}$ in (3.38) has minimal norm $\| \widehat{X} \|_2$ and $\| \widehat{X} \|_F$. □

Observe that the conditions $\sigma_{p+1} = \cdots = \sigma_{n+d}$ require that all σ_i, $i > p$, to be zero if $m < n + d$. The rank of the TLS correction matrix (3.40) equals $d -$ corank$([A; B])$. Usually, $m \geq n + d$ and $\sigma_{n+d} > 0$. In these cases $[\Delta\widehat{A}; \Delta\widehat{B}]$ has rank d.

If the assumptions of Theorem 3.9 hold, we can prove, analogously to Corollary 3.2, the following relationship between Γ and the minimum norm TLS solution $\widehat{X}$:

$$\Gamma\Gamma^T = (I + \widehat{X}^T\widehat{X})^{-1}. \tag{3.43}$$

Neat closed-form expressions for the multidimensional minimum norm TLS solution can also be proved [218], [207], as shown below.

THEOREM 3.10 (Closed-form expressions of the minimum norm TLS solution). *Consider the* TLS *problem* (3.1)–(3.3) *and assume that* $\sigma_p > \sigma_{p+1} = \cdots = \sigma_{n+d}$ *with* $p \leq n$. *Let* (1.22) *be the* SVD *of* $[A; B]$ *but partition* V *differently as follows*:

$$V = \begin{bmatrix} V_{11} & V_{12} \\ V_{21} & V_{22} \end{bmatrix} \begin{matrix} n \\ d \end{matrix} . \qquad \begin{matrix} p & n-p+d \end{matrix}$$

If V_{22} *is of full row rank, the multidimensional minimum norm* TLS *solution* $\widehat{X}$ *is given by*

$$\begin{aligned} \widehat{X} &= -V_{12}V_{22}^{\dagger} = -V_{12}V_{22}^T(V_{22}V_{22}^T)^{-1} \\ &= V_{11}V_{21}^T(I - V_{21}V_{21}^T)^{-1} = (V_{11}^T)^{\dagger}V_{21}^T && (3.44) \\ &= (A^TA - \sigma_{n+1}^2 I)^{\dagger}A^TB. && (3.45) \end{aligned}$$

Proof. Since V_{22} is of full (row) rank, V_{11} is also of full (column) rank. This is a direct consequence of the application of the extended CS decomposition theorem given in [117, p. 402]. We first prove equality of all formulas in (3.44). Let

$$X_1 = (V_{11}^T)^{\dagger}V_{21}^T. \tag{3.46}$$

Since V_{11} and V_{22} are of full rank and V is orthonormal, we have

$$V_{11}^TV_{12} + V_{21}^TV_{22} = 0, \quad V_{11}V_{21}^T + V_{12}V_{22}^T = 0 \tag{3.47}$$

and

$$(V_{11}^T)^{\dagger} = V_{11}(V_{11}^TV_{11})^{-1}, \quad V_{22}^{\dagger} = V_{22}^T(V_{22}V_{22}^T)^{-1}. \tag{3.48}$$

From these,

$$V_{21}^T = V_{21}^T(V_{22}V_{22}^{\dagger}) = (V_{21}^TV_{22})V_{22}^{\dagger} = -V_{11}^TV_{12}V_{22}^{\dagger}. \tag{3.49}$$

Substituting (3.49) into (3.46), we get

$$X_1 = -(V_{11}^T)^\dagger V_{11}^T V_{12} V_{22}^\dagger$$

but, from (3.47) and (3.48),

$$V_{12}V_{22}^\dagger = V_{12}V_{22}^T(V_{22}V_{22}^T)^{-1} = -V_{11}V_{21}^T(V_{22}V_{22}^T)^{-1}.$$

So, X_1 can be written as

$$\begin{aligned} X_1 &= (V_{11}^T)^\dagger V_{11}^T V_{11} V_{21}^T (V_{22}V_{22}^T)^{-1} = V_{11}V_{21}^T(V_{22}V_{22}^T)^{-1} \\ &= -V_{12}V_{22}^T(V_{22}V_{22}^T)^{-1} = -V_{12}V_{22}^\dagger. \end{aligned} \tag{3.50}$$

This is the first formula in (3.44). Substituting (3.48) in (3.50) gives

$$X_1 = -V_{12}V_{22}^T(V_{22}V_{22}^T)^{-1}. \tag{3.51}$$

This is the second formula in (3.44). Since V is orthonormal, $V_{22}V_{22}^T + V_{21}V_{21}^T = I$ and hence, (3.51) yields

$$X_1 = -V_{12}V_{22}^T(I - V_{21}V_{21}^T)^{-1}. \tag{3.52}$$

From (3.47), we have $-V_{12}V_{22}^T = V_{11}V_{21}^T$. Substituting this in (3.52) proves the third formula of (3.44). Now, to prove (3.45), note that

$$A = U\Sigma[V_{11}; V_{12}]^T \text{ and } B = U\Sigma[V_{21}; V_{22}]^T.$$

So, by setting $D = \text{diag}(\sigma_1^2 - \sigma_{n+1}^2, \cdots, \sigma_p^2 - \sigma_{n+1}^2)$, we obtain

$$\begin{aligned} (A^TA - \sigma_{n+1}^2 I)^\dagger &= ([V_{11}; V_{12}]\,\text{diag}(D,0)\,[V_{11}; V_{12}]^T)^\dagger = (V_{11}DV_{11}^T)^\dagger \\ &= (V_{11}^T)^\dagger D^{-1} V_{11}^\dagger \end{aligned} \tag{3.53}$$

and

$$\begin{aligned} A^TB &= [V_{11}; V_{12}]\,\text{diag}(\sigma_1^2, \cdots, \sigma_p^2, \sigma_{n+1}^2, \cdots, \sigma_{n+1}^2)\,[V_{21}; V_{22}]^T \\ &= V_{11}\,\text{diag}(\sigma_1^2, \cdots, \sigma_p^2)V_{21}^T + \sigma_{n+1}^2 V_{12}V_{22}^T \\ &= V_{11}\,\text{diag}(\sigma_1^2, \cdots, \sigma_p^2)V_{21}^T - \sigma_{n+1}^2 V_{11}V_{21}^T \\ &= V_{11}DV_{21}^T. \end{aligned} \tag{3.54}$$

Substituting (3.54) and (3.53) in (3.45) and using $V_{11}^\dagger = (V_{11}^T V_{11})^{-1} V_{11}^T$, we obtain

$$(A^TA - \sigma_{n+1}^2 I)^\dagger A^T B = (V_{11}^T)^\dagger V_{21}^T = X_1.$$

Now to prove that X_1 is the minimum norm solution of the TLS problem (3.1)–(3.3), note that all TLS solutions X satisfy

$$\begin{bmatrix} V_{11} \\ V_{21} \end{bmatrix}^T \begin{bmatrix} X \\ -I \end{bmatrix} = V_{11}^T X - V_{21}^T = 0.$$

Among these, $X_1 = (V_{11}^T)^\dagger V_{21}^T = -V_{12}V_{22}^\dagger$ has minimal length and exists since V_{22} is of full rank. This completes the proof of the theorem. □

These forms are of theoretical interest but, for computational purposes, we should use the numerically more reliable procedure, as presented in Theorem 3.9.

Theorem 3.9 proves that a solution to the TLS problem (3.1)–(3.3) exists provided that

$$\sigma_p > \sigma_{p+1} = \cdots = \sigma_{n+d},\ p \le n, \ \text{ and } V_{22} \text{ has full rank.}$$

These conditions are equivalent to the requirement $\sigma_p' > \sigma_{p+1} = \cdots = \sigma_{n+d}$, as proven by Wei [207]. We first prove the following lemma, due to Wei [207].

LEMMA 3.1 (Existence condition for a solution to TLS problems). *Let* (1.21) *be the* SVD *of* $A \in \mathcal{R}^{m\times n}$. *Consider* $B \in \mathcal{R}^{m\times d}$ *and let* (1.22) *be the* SVD *of* $[A; B]$ *but partition* V *into the following form:*

$$V = \begin{bmatrix} V_{11} & V_{12} \\ V_{21} & V_{22} \end{bmatrix} \begin{matrix} n \\ d \end{matrix} \quad \text{for some } p \le n. \tag{3.55}$$
$$\qquad p \qquad n-p+d$$

Then

$$\sigma_p' > \sigma_{p+1} \Rightarrow V_{11} \text{ and } V_{22} \text{ are of full rank.}$$

Proof. If $\sigma_p' > \sigma_{p+1}$ but $\operatorname{rank}(V_{11}) = p_1 < p$, then from the SVD (1.22) of $[A; B]$ and the partitioning (1.22) of V, we have

$$A = U_1\Sigma_1 V_{11}^T + U_2\Sigma_2 V_{12}^T, \quad B = U_1\Sigma_1 V_{21}^T + U_2\Sigma_2 V_{22}^T, \tag{3.56}$$

where U_1 and U_2 are the first p and the last $m-p$ columns of U, respectively, $\Sigma_1 = \operatorname{diag}(\sigma_1, \cdots, \sigma_p)$, $\Sigma_2 = \operatorname{diag}(\sigma_{p+1}, \cdots, \sigma_{n+t})$ with $t = \min\{m-n, d\}$. Applying the SVD property [69, Cor. 8.3.2] (σ_j denotes the jth singular value):

$$\begin{aligned} |\sigma_j(G) - \sigma_j(H)| \le \|G-H\|_2 &= \sigma_1(G-H) \text{ for } j = 1, \cdots, n \\ & \qquad G, H \in \mathcal{R}^{m\times n} \text{ and } \sigma_j \ge \sigma_{j+1} \end{aligned}$$

to $G = A$ and $H = U_1\Sigma_1 V_{11}^T$ in (3.56), we obtain

$$|\sigma_j' - \sigma_j(U_1\Sigma_1 V_{11}^T)| \le \|U_2\Sigma_2 V_{12}^T\|_2 = \sigma_{p+1}, \quad j = 1, \cdots, n.$$

Now $p_1 < p$ by assumption, so $\sigma_j(U_1\Sigma_1 V_{11}^T) = 0$ for $j = p_1+1, \cdots, p, \cdots, n$. In particular, $\sigma'_p \leq \sigma_{p+1}$, which is in contradiction with our assumptions. So, $\text{rank}(V_{11}) = p$. Since V_{11} is of full rank, V_{22} also has full rank d. This is a direct consequence of the application of the extended CS decomposition theorem given in [117, p. 402]. □

Using this lemma, the following equivalence is obtained.

COROLLARY 3.4 (Existence conditions for a solution to TLS problems). *Let* (1.21) *be the* SVD *of* A. *Let* (1.22) *be the* SVD *of* $[A; B]$ *but partition* V *as in* (3.55). *Then*

$$\sigma_p > \sigma_{p+1} = \cdots = \sigma_{n+d} \text{ and } V_{11},\ V_{22} \text{ of full rank}$$
$$\Updownarrow$$
$$\sigma'_p > \sigma_{p+1} = \cdots = \sigma_{n+d}.$$

Proof. The $\Uparrow$ is proved by Lemma 3.1 and by noting that $\sigma_p \geq \sigma'_p$ according to the interlacing Theorem 2.4 for singular values.

Now let us prove the $\Downarrow$. Consider $A^TA - \sigma_{n+1}^2 I$. On one hand, from (3.53), we have $A^TA - \sigma_{n+1}^2 I = V_{11}DV_{11}^T$ with $D = \text{diag}(\sigma_1^2 - \sigma_{n+1}^2, \cdots, \sigma_p^2 - \sigma_{n+1}^2)$ being nonsingular. Then, from (3.48),

$$p = \text{rank}(D) \geq \text{rank}(V_{11}DV_{11}^T) \geq \text{rank}(V_{11}^\dagger V_{11}DV_{11}^T(V_{11}^T)^\dagger) = \text{rank}(D),$$

i.e., $A^TA - \sigma_{n+1}^2 I$ has rank p. On the other hand, by the interlacing Theorem 2.4 for singular values and the assumptions $\sigma_p > \sigma_{p+1} = \cdots = \sigma_{n+d}$, we have $\sigma'_{p+1} = \cdots = \sigma'_n = \sigma_{n+1}$. Then, using (1.21),

$$\begin{aligned} A^TA - \sigma_{n+1}^2 I &= V'\text{diag}(\sigma_1'^2, \cdots, \sigma_n'^2)V'^T - \sigma_{n+1}^2 V'V'^T \\ &= V'\text{diag}(\sigma_1'^2 - \sigma_{n+1}^2, \cdots, \sigma_p'^2 - \sigma_{n+1}^2, 0, \cdots, 0)V'^T. \end{aligned}$$

Since $A^TA - \sigma_{n+1}^2 I$ must be of rank p, the condition $\sigma'_p > \sigma_{n+1}$ follows. □

Now assume that $\sigma_{n+1} \neq \sigma_{n+d}$, then we can, of course, still compute a minimum norm solution $\widehat{X}$, as deduced in Theorem 3.9. According to Lemma 3.1, this solution still exists provided $\sigma'_p > \sigma_{p+1}$. However, this solution is no longer the minimum norm TLS solution of the TLS problem (3.1)–(3.3) with the Frobenius norm as measure but it is still the minimum norm TLS solution of the TLS problem with the 2-*norm* as measure. Indeed, in these problems, the corresponding correction matrix (3.40) has minimal 2-norm, $\| [\Delta\widehat{A}; \Delta\widehat{B}] \|_2$ given by σ_{n+1}, but it no longer has minimal Frobenius norm $\| [\Delta\widehat{A}; \Delta\widehat{B}] \|_F$ (i.e., (3.1) is not satisfied). Only if $\sigma_{n+1} = \cdots \sigma_{n+d}$, any orthonormal d-dimensional basis $\left[\begin{smallmatrix} Z \\ \Gamma \end{smallmatrix}\right] \in R([v_{p+1}, \cdots, v_{n+d}])$ is a right singular subspace of $[A; B]$, associated with the same singular value $\sigma_{p+1} = \cdots = \sigma_n = \sigma_{n+1} = \cdots = \sigma_{n+d}$ such that the Frobenius norm $\sqrt{d}\,\sigma_{n+1}$ of (3.40) satisfies (3.1).

But if *not all* $\sigma_{n+i}, i = 1, \cdots, d$ *coincide*, then the minimum norm solution computed by (3.38) may have an associated correction matrix (3.40) with Frobenius norm *larger* than the minimal value $\sqrt{\sum_{i=1}^d \sigma_{n+i}^2}$. In fact, this solution is the minimum norm TLS solution of a "*nearby*" system $\widetilde{A}X \approx \widetilde{B}$ where $[\widetilde{A};\widetilde{B}] = \sum_{i=1}^{p} \sigma_i u_i v_i^T + \sigma_{n+1} \sum_{i=p+1}^{n+t} u_i v_i^T$, i.e., the $t = \min\{m-n, d\}$ smallest singular values are made equal to σ_{n+1} (if $m < n+d$, then $\sigma_{n+1} = \sigma_{n+d}$ requires $\sigma_{n+1} = 0$). Note that this minimum norm solution $\widehat{X}$, as deduced in Theorem 3.9, still satisfies (3.44) and (3.43) if $\sigma_{n+1} \neq \sigma_{n+d}$ but expression (3.45) is no longer valid.

When the conditions $\sigma_{n+1} = \cdots = \sigma_{n+d}$ are *not* satisfied, *two options* are possible:

- The minimization of the norm of $\widehat{X}$ is more relevant than that of $[\Delta\widehat{A};\Delta\widehat{B}]$. Then, the TLS procedure described in Theorem 3.9 is appropriate.
- The minimization of $\| [\Delta\widehat{A};\Delta\widehat{B}] \|_F$ is more important.

The *first* option is *more realistic* if the set of equations $AX \approx B$ observe an exact but unknown rank-deficient set $A_0X_0 = B_0$. This implies that the smallest zero singular value of $[A_0;B_0]$ has multiplicity $n-p+d$, $p \leq n$. The $n-p+d$ smallest singular values of $[A;B]$ are in fact "observations" of the exact minimal singular value of $[A_0;B_0]$ with multiplicity $n-p+d$ but, due to errors, they no longer coincide. In this case the $n-p+d$ smallest singular values of $[A;B]$ can indeed be considered as equal and a strict minimization of $\| [\Delta\widehat{A};\Delta\widehat{B}] \|_F$ makes no sense.

The *second* option only makes sense if the singular values $\sigma_{n+1}, \cdots, \sigma_{n+d}$ differ *significantly*. However, this situation implies that the equations $AX \approx B$ are highly incompatible. If the user should be confronted with such a situation, he might simply conclude that the data are not appropriate for linear modeling and reject the problem as *irrelevant* from a linear modeling point of view. Nevertheless, the situation may occur. Let (1.22) be the SVD of $[A;B]$ and assume $\sigma_p > \sigma_{p+1} = \cdots = \sigma_{n+1} = \cdots = \sigma_{n+e} > \sigma_{n+e+1} \geq \cdots \geq \sigma_{n+d}$. To minimize the corrections $\| [\Delta\widehat{A};\Delta\widehat{B}] \|_F$, the $n \times d$ TLS solution $\widehat{X}$ must be obtained from the $d-e$ right singular vectors $v_{n+e+1}, \cdots, v_{n+d}$, associated with the smallest singular values of $[A;B]$, and an e-dimensional subset of the $(n-p+e)$-dimensional right singular subspace associated with σ_{n+1} such that $\| \widehat{X} \|_2$ and $\| \widehat{X} \|_F$, given in (3.39), are minimized. Procedures for selecting this subset have not yet been fully analyzed.

3.4. The nongeneric TLS problem and solution

3.4.1. The one-dimensional case. If the existence conditions given in Theorems 2.6, 3.1, 3.2–3.5, and 3.9 are not satisfied, the TLS problem (2.19) and (2.20) may fail to have a solution. For instance, if $\sigma_n > \sigma_{n+1}$ this occurs when $v_{n+1,n+1} = 0$. More generally, assume that $\sigma_p > \sigma_{p+1} = \cdots = \sigma_{n+1}$, $p \leq n$, then we call the TLS problem *nongeneric* if all $v_{n+1,j} = 0$ $j = p+1, \cdots, n+1$. Theorem 3.11 describes the properties of these problems.

THEOREM 3.11 (Properties of the nongeneric one-dimensional TLS problem). *Let* (1.21) (*respectively,* (1.22)) *be the* SVD *of* A (*respectively,* $[A;b]$). *Let* b' *be the orthogonal projection of* b *onto* $R(A)$ *and* $[\widehat{A};\widehat{b}]$ *the rank* n *approximation of* $[A;b]$, *as given by* (2.24) *or* (3.34). *If* $V'(\sigma_j)$ (*respectively,* $U'(\sigma_j)$) *is the right* (*respectively, left*) *singular subspace of* A *associated with* σ_j, *then the following relations can be proven:*

(3.57)

$$\begin{array}{llll}
\text{(a) } v_{n+1,j}=0 & \Longleftrightarrow & v_j=\begin{bmatrix} v' \\ 0 \end{bmatrix} & \textit{with } v'\in V'(\sigma_j),\\
\text{(b) } v_{n+1,j}=0 & \Longrightarrow & \sigma_j=\sigma'_k & \textit{with } k=j-1 \textit{ or } k=j \textit{ and } 1\le k\le n,\\
\text{(c) } v_{n+1,j}=0 & \Longrightarrow & b\perp u' & \textit{with } u'\in U'(\sigma_j),\\
\text{(d) } v_{n+1,j}=0 & \Longleftrightarrow & u_j=u' & \textit{with } u'\in U'(\sigma_j),\\
\text{(e) } v_{n+1,j}=0 & \Longrightarrow & b'\perp u' & \textit{with } u'\in U'(\sigma_j),\\
\text{(f) } v_{n+1,j}=0 & \Longrightarrow & \widehat{b}\perp u' & \textit{with } u'\in U'(\sigma_j).
\end{array}$$

If σ_j *is an isolated singular value, the converse of relations* (c) *and* (e) *also holds.*

Proof. For the proof see [177, Thm. 1-3]. □

In particular, assume that $\sigma_n > \sigma_{n+1}$ and $v_{n+1,n+1} = 0$. Then, Theorem 3.11 proves that in this situation we have

$$(3.58)\qquad \left\{\begin{array}{l} \sigma_{n+1}=\sigma'_n,\ u_{n+1}=\pm u'_n,\ v_{n+1}=\begin{bmatrix} \pm v'_n \\ 0 \end{bmatrix} \\ \text{and } b,\ b' \text{ as well as } \widehat{b} \text{ are orthogonal to } u'_n \end{array}\right. .$$

It is easy to see that the generic TLS approximation $[\widehat{A};\widehat{b}] = \sum_{i=1}^{n} \sigma_i u_i v_i^T$ and corresponding TLS correction matrix $[\Delta\widehat{A};\Delta\widehat{b}] = \sigma_{n+1}u_{n+1}v_{n+1}^T$ (see Theorem 2.6) minimizes $\|[\Delta\widehat{A};\Delta\widehat{b}]\|_F$ but does *not* satisfy the constraint $\widehat{b}\in R(\widehat{A})$ and therefore, this $[\widehat{A};\widehat{b}]$ does not solve the TLS problem (2.19) and (2.20). Indeed, using (1.21), (1.22), and (3.58), we obtain

$$\begin{aligned}
[\widehat{A};\widehat{b}] &= [A;b]-[\Delta\widehat{A};\Delta\widehat{b}] = [A;b]-\sigma_{n+1}u_{n+1}v_{n+1}^T \\
&= [\sum_{i=1}^{n} \sigma'_i u'_i v_i'^T; b] - \sigma'_n u'_n [v_n'^T; 0] \\
&= [\sum_{i=1}^{n-1} \sigma'_i u'_i v_i'^T; b].
\end{aligned}$$

Observe that this approximation makes $\widehat{A}$ rank-deficient! Indeed, A is reduced to a matrix $\widehat{A}$ of rank $n-1$ while b is not changed at all. Since $\text{rank}(\widehat{A}) = n-1 < \text{rank}([\widehat{A};\widehat{b}]) = n$, it follows that $\widehat{b}\notin R(\widehat{A})$. Moreover, (3.58) also yields

(3.59)

$$\begin{aligned}
&[\widehat{A};\widehat{b}]v_{n+1} = 0 \;\Rightarrow\; \widehat{A}v'_n = 0 \\
&[A;b]v_{n+1} = \sigma_{n+1}u_{n+1} \Rightarrow Av'_n + 0\cdot b = Av'_n = \sigma'_n u'_n \approx 0, \quad \text{if} \sigma'_n = 0.
\end{aligned}$$

This means that the solution v'_n of the set $\widehat{A}x = 0$ describes an approximate linear relation among the columns of A instead of estimating the desired linear relation between A and b.

In these situations, described by Theorem 3.11, the generic TLS problem (2.19) and (2.20) as such has no meaningful solution. Indeed, to satisfy $\widehat{b} \in R(\widehat{A})$, we must make one direction $v \in R_r([A;b])$ orthogonal to $R_r([\widehat{A};\widehat{b}])$, i.e., $[\widehat{A};\widehat{b}]v = 0$, and such that $\widehat{A}x = \widehat{b}$ is compatible. If $v = \varepsilon v_n + \sqrt{1-\varepsilon^2}\, v_{n+1}$, then for every $\varepsilon > 0$, the correction matrix:

$$\begin{aligned} [\Delta\widehat{A};\Delta\widehat{b}] &= [A;b](\varepsilon v_n + \sqrt{1-\varepsilon^2}\, v_{n+1})(\varepsilon v_n + \sqrt{1-\varepsilon^2}\, v_{n+1})^T \\ &= (\varepsilon\sigma_n u_n + \sqrt{1-\varepsilon^2}\,\sigma_{n+1}u_{n+1})(\varepsilon v_n + \sqrt{1-\varepsilon^2}\, v_{n+1})^T \end{aligned} \tag{3.60}$$

satisfies $b - \Delta\widehat{b} \in R(A - \Delta\widehat{A})$. This implies that the data A, b are projected in the lower-dimensional subspace orthogonal to $\varepsilon\sigma_n u_n + \sqrt{1-\varepsilon^2}\,\sigma_{n+1}u_{n+1}$. The TLS corrections are given by

$$\|[\Delta\widehat{A};\Delta\widehat{b}]\|_F = \sqrt{\varepsilon^2\sigma_n^2 + (1-\varepsilon^2)\sigma_{n+1}^2}. \tag{3.61}$$

However, to satisfy (2.19) simultaneously, ε must be as small as possible. This "smallest" ε cannot be determined and therefore, there is *no "smallest"* $\|[\Delta\widehat{A};\Delta\widehat{b}]\|_F$ for which $\widehat{b} \in R(\widehat{A})$. Moreover, this approach makes the problem *very ill-conditioned.* Indeed, using (3.60), the solution $\widehat{x}$ of $(A - \Delta\widehat{A})x = b - \Delta\widehat{b}$ is given by

$$[\widehat{x}^T;-1]^T = -(\varepsilon v_n + \sqrt{1-\varepsilon^2}\, v_{n+1})/(\varepsilon v_{n+1,n} + \sqrt{1-\varepsilon^2}\, v_{n+1,n+1}) \tag{3.62}$$

$$= -\frac{v_n}{v_{n+1,n}} - \frac{\sqrt{1-\varepsilon^2}\, v_{n+1}}{\varepsilon v_{n+1,n}} \qquad \text{if } v_{n+1,n+1} = 0. \tag{3.63}$$

If ε is small, then the coefficients of $\widehat{x}$ are inflated and have *large* variances due to the influence of the second term in (3.63). This makes the generic TLS solution very sensitive to errors.

Therefore, we propose here a *nongeneric* solution that still satisfies the TLS criteria (2.19)–(2.20) under the *additional constraint* that $[\Delta\widehat{A};\Delta\widehat{b}]\ v_{n+1} = 0$ or equivalently $\begin{bmatrix} \widehat{x} \\ -1 \end{bmatrix} \perp v_{n+1}$.

DEFINITION 3.2 (Nongeneric one-dimensional total least squares problem). Given an overdetermined set of m linear equations $Ax \approx b$ in n unknowns x. Let (1.22) be the SVD of $[A;b]$. The nongeneric one-dimensional total least squares (TLS) problem seeks to

$$\underset{[\widehat{A};\widehat{b}] \in \mathcal{R}^{m\times(n+1)}}{\text{minimize}} \quad \|\,[A;b] - [\widehat{A};\widehat{b}]\,\|_F \tag{3.64}$$

$$\text{subject to} \quad \widehat{b} \in R(\widehat{A}) \tag{3.65}$$

$$\text{and} \quad [\Delta\widehat{A};\Delta\widehat{b}]v_j = 0 \quad \forall j : \max_{v_{n+1,p}\neq 0}\{p\} < j \leq n+1. \tag{3.66}$$

Once a minimizing $[\widehat{A};\widehat{b}]$ is found, then any x satisfying

$$\widehat{A}x = \widehat{b} \tag{3.67}$$

is called a *nongeneric* TLS *solution*, and $[\Delta\widehat{A};\Delta\widehat{b}] = [A;b] - [\widehat{A};\widehat{b}]$ is the corresponding nongeneric TLS correction.

As will be shown later, the nongeneric TLS solution also equals the *minimum norm* TLS solution in the *lower*-dimensional subspace $[\widehat{\widehat{A}};\widehat{\widehat{b}}]$, obtained by making one more σ_i in (2.24) zero, i.e., $\sigma_n = \sigma_{n+1} = 0$. Moreover, LS searches its solution in a subspace of the same dimensionality, i.e., there does not exist an LS solution in a subspace of higher dimensionality than that of the nongeneric TLS approximation. Let us first motivate our nongeneric approach with an example.

Example 3.1. Consider the set of three equations in two unknowns (see Fig. 3.1):

$$\begin{bmatrix} \sqrt{6} & \sqrt{6} \\ \sqrt{2}/4 & -\sqrt{2}/4 \\ \sqrt{2}/2 & \sqrt{2}/2 \end{bmatrix} \begin{bmatrix} x_1 \\ x_2 \end{bmatrix} \approx \begin{bmatrix} 2 \\ 0 \\ -\sqrt{3} \end{bmatrix}. \tag{3.68}$$

Taking the SVD of $[A;b]$, we obtain

$$U\Sigma V^T = \begin{bmatrix} 1 & 0 & 0 \\ 0 & 0 & 1 \\ 0 & 1 & 0 \end{bmatrix} \begin{bmatrix} 4 & 0 & 0 \\ 0 & 2 & 0 \\ 0 & 0 & .5 \end{bmatrix} \begin{bmatrix} \sqrt{6}/4 & \sqrt{6}/4 & 1/2 \\ \sqrt{2}/4 & \sqrt{2}/4 & -\sqrt{3}/2 \\ \sqrt{2}/2 & -\sqrt{2}/2 & 0 \end{bmatrix}. \tag{3.69}$$

Note that $v_{n+1,n+1} = v_{3,3} = 0$. Since $b = \sigma_1 v_{3,1} u_1 + \sigma_2 v_{3,2} u_2 + 0 \cdot u_3$ this implies that $b \perp u_3$. By taking the rank 2 approximation (2.24), we obtain the following approximating vectors:

$$\widehat{a}_1 = \widehat{a}_2 = \sqrt{6}\, u_1 + 0.5\sqrt{2} u_2 \quad \text{and} \quad \widehat{b} = b. \tag{3.70}$$

Since $\widehat{b} \notin R(\widehat{A})$, (2.20) is not satisfied. Hence, $[\widehat{A};\widehat{b}]$, as defined in (3.70), is not a valid TLS approximation. To satisfy (2.20), A and b must be projected orthogonally into a plane that is not orthogonal to u_3. According to (3.60), this plane is chosen orthogonal to $\varepsilon\sigma_2 u_2 + \sqrt{1-\varepsilon^2}\,\sigma_3 u_3$. However, to satisfy (2.19) simultaneously, ε must be as small as possible; i.e., this plane must be *infinitesimally close* to the plane $\perp u_3$. This "*smallest*" ε cannot be determined and, moreover, makes the solution $\widehat{x}$, given in (3.63), unstable.

To stabilize the coefficients of $\widehat{x}$ and *reduce the sensitivity* of the TLS problem considerably, we impose the *additional constraint* that $\begin{bmatrix} \widehat{x} \\ -1 \end{bmatrix} \perp v_3$ or, equivalently, that the TLS approximation $\widehat{b} \perp u_3$. The nongeneric TLS solution satisfying the TLS criteria under this constraint is obtained by making σ_2 in (3.69) zero such that the nongeneric TLS solution $\begin{bmatrix} \widehat{x} \\ -1 \end{bmatrix}$ is now parallel with v_2 and is given by

$$\widehat{x} = [1/\sqrt{6}, 1/\sqrt{6}]^T. \tag{3.71}$$

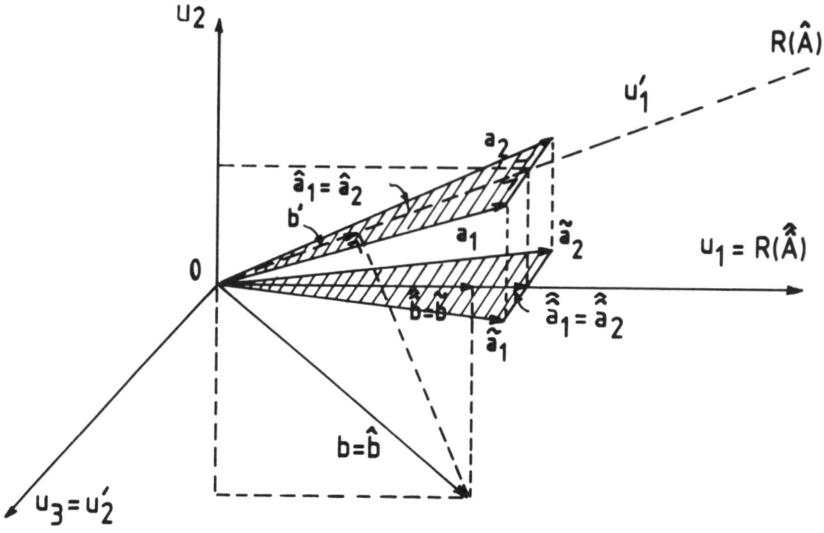

FIG. 3.1. *Geometric representation of Example* 3.1 $[A_{3\times 2}; b] = \sum_{i=1}^{3} \sigma_i u_i v_i^T$ *and* $v_{3,3} = 0$, *i.e.*, $b \perp u_3$. b' *is the* LS *approximation of* b. $[\widehat{A}; \widehat{b}] = [\widehat{a}_1, \widehat{a}_2, \widehat{b}]$ *(respectively,* $[\widehat{\widehat{A}}; \widehat{\widehat{b}}] = [\widehat{\widehat{a}}_1, \widehat{\widehat{a}}_2, \widehat{\widehat{b}}]$*) are rank* 2 *(respectively, rank* 1*) approximations of* $[A; b]$ *with* $\widehat{b} \notin R(\widehat{A})$ *(respectively,* $\widehat{\widehat{b}} \in R(\widehat{\widehat{A}})$*) and* $[\tilde{A}; \tilde{b}] = [\tilde{a}_1, \tilde{a}_2, \tilde{b}]$ *is the nongeneric* TLS *approximation of* $[A; b]$ *with minimal correction* $\|[A - \tilde{A}; b - \tilde{b}]\|_F$ *such that* $\tilde{b} \in R(\tilde{A})$ *and* $[A - \tilde{A}; b - \tilde{b}]v_3 = 0$ *or* $u_3^T[A - \tilde{A}; b - \tilde{b}] = 0$.

Observe that $[\widehat{x}^T; -1]^T$ is orthogonal to v_3. The corresponding nongeneric TLS approximation is given by

$$[\tilde{A}; \tilde{b}] = U\widehat{\Sigma}V^T \text{ with } \widehat{\Sigma} = \operatorname{diag}(\sigma_1, 0, \sigma_3), \tag{3.72}$$

and the corresponding correction matrix of rank 1 is given by

$$[\Delta\tilde{A}; \Delta\tilde{b}] = \sigma_2 u_2 v_2^T = [A; b]v_2 v_2^T. \tag{3.73}$$

This implies that the data A, b are projected in a plane orthogonal to u_2. Note that the nongeneric TLS solution (3.71) also equals the minimum norm solution in the lower-dimensional subspace $R([\widehat{\widehat{A}}; \widehat{\widehat{b}}])$ obtained by replacing σ_2 and σ_3 in (3.69) by zero. Then we obtain the TLS approximation $[\widehat{\widehat{A}}; \widehat{\widehat{b}}]$ such that $\widehat{\widehat{b}} \in R(\widehat{\widehat{A}})$. $\widehat{\widehat{a}}_1 = \widehat{\widehat{a}}_2 = \sqrt{6}\, u_1$ and $\widehat{\widehat{b}} = 2u_1$ and $\widehat{\widehat{b}}$ remains orthogonal to u_3.

Now let us compute the LS solution of (3.68). Using the SVD of A,

$$(3.74)\ A = U'\Sigma'V'^T = \begin{bmatrix} 2\sqrt{3/13} & 0 \\ 0 & 1 \\ 1/\sqrt{13} & 0 \end{bmatrix} \begin{bmatrix} \sqrt{13} & 0 \\ 0 & 0.5 \end{bmatrix} \begin{bmatrix} \sqrt{2}/2 & \sqrt{2}/2 \\ \sqrt{2}/2 & -\sqrt{2}/2 \end{bmatrix},$$

we obtain the LS solution x' as

$$(3.75) \qquad x' = V'\Sigma'^{-1}U'^T b = \frac{3}{13}\sqrt{\frac{3}{2}} \begin{bmatrix} 1 \\ 1 \end{bmatrix}.$$

Observe that $\sigma_2' = \sigma_3 = .5$ and $u_2' = u_3$. Hence $b \perp u_2'$ and the LS approximation b' remains orthogonal to u_2'. All those facts are generalized and proven in Theorem 3.11.

The introduction of the *additional constraints* $\begin{bmatrix} \hat{x} \\ -1 \end{bmatrix} \perp v_j$, for all j with $v_{n+1,j} = 0$ in the nongeneric TLS approach can be motivated as follows. The singular vectors v_j with negligible or zero last component and associated with a small singular value, are called *nonpredictive multicollinearities* in linear regression since they reveal multicollinearities (i.e., approximate linear dependencies between the columns of A) in A that are of no (or negligible) value in predicting the response b. In fact, if $v_{n+1,n+1} = 0$, then using Theorem 3.11 yields (3.59). Since $\sigma_n' = \sigma_{n+1}$ (according to Theorem 3.11) is the smallest singular value, this means that A is nearly rank-deficient, or else the set of equations is highly incompatible. Also, since $b \perp u_n' = u_{n+1}$, according to Theorem 3.11, there is *no correlation* between A and b in the direction of $u_n' = u_{n+1}$. Hence, it does not make sense here to induce a correlation between $\hat{A}$ and $\hat{b}$ in the direction u_{n+1}. This would induce a wrong interpretation about the exact relation between A and b. If $b \perp u_{n+1}$ then also keep the TLS approximation $\hat{b} \perp u_{n+1}$ as done in our nongeneric TLS approach. Therefore, the strategy of nongeneric TLS is to eliminate those directions in A that are *not* at all correlated with the observation vector b. Indeed, solving a set of equations is computing the relation between A and b. If there is no correlation between b and a left singular vector u_j' of A, i.e., $b \perp u_j'$, it is better to eliminate that direction rather than forcing a solution in that direction.

Latent root regression [203] follows a similar approach. It minimizes the residual sum of squares $\|Ax - b\|_2^2$ subject to the *same* constraints, i.e., $\begin{bmatrix} x \\ -1 \end{bmatrix} \perp$ all nonpredictive v_j. Hence, if there are no nonpredictive v_j, then the latent root regression estimate equals the LS solution. This is a way of *stabilizing* the solution in linear regression when the data in A are nearly multicollinear ($\sigma_n' \approx 0$). It is well known that then the ordinary LS solution tends to be inflated and predicted values may be unreasonable.

The nongeneric TLS solution is another way of handling this multicollinearity problem and stabilizing the TLS solution. Nongeneric TLS minimizes the applied corrections $\|[A - \widehat{A}, b - \widehat{b}]\|_F$ subject to $\widehat{A}x = \widehat{b}$ and the *same* constraints as used in latent root regression. If there are *no* nonpredictive v_j, then the nongeneric TLS solution equals the generic TLS solution. This is proven in our next theorem (which is a correction of [177, Thm. 1.4]).

THEOREM 3.12 (Nongeneric one-dimensional TLS solution). *Let* (1.22) *be the* SVD *of* $[A; b]$ *and assume* $v_{n+1,j} = 0$ *for* $j = p+1, \cdots, n+1$, $p \leq n$. *If* $\sigma_{p-1} > \sigma_p$ *and* $v_{n+1,p} \neq 0$, *then*

$$[\widehat{A}; \widehat{b}] = U\widehat{\Sigma}V^T \text{ with } \widehat{\Sigma} = \operatorname{diag}(\sigma_1, \cdots, \sigma_{p-1}, 0, \sigma_{p+1}, \cdots, \sigma_{n+1}) \tag{3.76}$$

with corresponding nongeneric TLS *correction matrix*

$$[\Delta\widehat{A}; \Delta\widehat{b}] = [A; b] - [\widehat{A}; \widehat{b}] = \sigma_p u_p v_p^T \tag{3.77}$$

solves the nongeneric TLS *problem* (3.64)–(3.66) *and*

$$\widehat{x} = -\frac{1}{v_{n+1,p}}[v_{1,p}, \cdots, v_{n,p}]^T \tag{3.78}$$

exists and is the unique solution to $\widehat{A}x = \widehat{b}$.

Proof. The nongeneric TLS approach searches in the $(n+1)$-dimensional row space of $[A; b]$ for an approximation $[\widehat{A}; \widehat{b}]$ of rank n with minimal deviation $[\Delta\widehat{A}; \Delta\widehat{b}]$ such that (3.65) is satisfied and $[\Delta\widehat{A}; \Delta\widehat{b}]v_j = 0 \quad j = p+1, \cdots, n+1$. Since $[A; b] = \sum_{i=1}^{p} \sigma_i u_i v_i^T + \sum_{i=p+1}^{n+1} \sigma_i u_i v_i^T$ and the rank n TLS approximation $[\widehat{A}; \widehat{b}] = [A; b] - [\Delta\widehat{A}; \Delta\widehat{b}]$, these last constraints imply that

$$[\widehat{A}; \widehat{b}] = C + \sum_{i=p+1}^{n+1} \sigma_i u_i v_i^T, \tag{3.79}$$

where C is a rank $(p-1)$ approximation of $[A; b]_p = \sum_{i=1}^{p} \sigma_i u_i v_i^T$. Hence, the nongeneric TLS approach must look in the p-dimensional space $R([A; b]_p)$ for an approximation C of rank $p-1$ with minimal approximation effort $\|[\Delta\widehat{A}; \Delta\widehat{b}]\|_F$ such that (3.65) is satisfied. Using the Eckart–Young–Mirsky Theorem 2.3 and the properties of the SVD, this minimizing $[\Delta\widehat{A}; \Delta\widehat{b}]$ is given by (3.77). The condition $\sigma_{p-1} > \sigma_p$ ensures that (3.77) is the unique minimizer. The nongeneric TLS approximation $[\widehat{A}; \widehat{b}]$ of rank n is then given by (3.76). Since v_p belongs to the null space of $[\widehat{A}; \widehat{b}]$, i.e., $[\widehat{A}; \widehat{b}]v_p = 0$, and since $v_{n+1,p} \neq 0$, (3.65) is satisfied, i.e., $\widehat{b} \in R(\widehat{A})$. The solution $\widehat{x}$ of the nongeneric TLS problem is then obtained by scaling v_p until its last component is -1, which proves (3.78). The uniqueness of $\widehat{x}$ follows from the condition that $\sigma_{p-1} > \sigma_p$. □

The following comments are in order:

- Since (3.66) $\Longleftrightarrow \begin{bmatrix} \widehat{x} \\ -1 \end{bmatrix} \perp v_j$, $j = p+1, \cdots, n+1$ (provided $v_{n+1,p} \neq 0$), we could as well use the latter constraints instead of (3.66).

- There is no restriction in imposing the condition $\sigma_{p-1} > \sigma_p$. Indeed, if $\sigma_{p-1} = \sigma_p$, then the nongeneric TLS solution still exists but is possibly not unique. This is the case when σ_p has multiplicity r such that $\sigma_{p-r+1} = \cdots = \sigma_p$ and several $v_{n+1,j} \neq 0$ for $j = p-r+1, \cdots, p$. If desirable, we can single out the *minimum norm* solution $[\widehat{x}^T; -1]^T$ in the right singular subspace $R([v_{p-r+1}, \cdots, v_{n+1}])$, as outlined in Theorem 3.7.

- The nongeneric TLS solution (3.78) also equals the minimum norm TLS solution computed in the restricted row space $R_r([\widehat{\widehat{A}}; \widehat{\widehat{b}}]) = R_r([A; b]) \setminus R([v_p, \cdots, v_{n+1}])$. This result follows directly from (3.32) using the assumptions $v_{n+1,j} = 0$ for $j = p+1, \cdots, n+1$.

- If $v_{n+1,n+1} = 0, v_{n+1,n} \neq 0$ and $\sigma_{n-1} > \sigma_n$, the nongeneric TLS solution (3.78) in Theorem 3.12 is given by

  (3.80) $$[\widehat{x}^T; -1]^T = -v_n / v_{n+1,n}$$

  with corresponding correction matrix $[\Delta\widehat{A}; \Delta\widehat{b}] = \sigma_n u_n v_n^T$ having minimal Frobenius norm σ_n such that $b - \Delta\widehat{b} \in R(A - \Delta\widehat{A})$ and $[\Delta\widehat{A}; \Delta\widehat{b}]v_{n+1} = 0$. Observe that in this case a smaller $\|[\Delta\widehat{A}; \Delta\widehat{b}]\|_F$ can always be found if we omit the constraint $[\Delta\widehat{A}; \Delta\widehat{b}]v_{n+1} = 0$. Indeed, every $[\Delta\widehat{A}; \Delta\widehat{b}]$ of the form (3.60) with $0 < \varepsilon < 1$ has smaller Frobenius norm than $[\Delta\widehat{A}; \Delta\widehat{b}] = \sigma_n u_n v_n^T$ of our nongeneric TLS problem. However, as said before, these solutions are not meaningful since they induce false correlations between A and b and make the TLS solution very sensitive to errors.

Let us also *compare* LS *and* TLS in the nongeneric case. LS looks for a solution x' such that $\sum_{i=1}^n \sigma_i' \ u_i' \ v_i'^T \ x' = b'$. If $v_j = \begin{bmatrix} v_j'' \\ v_{n+1,j} \end{bmatrix}$ with $v_{n+1,j} = 0$ for $j = p+1, \cdots, n+1$, then, using Theorem 3.11, this implies that $\sigma_{j-1}' = \sigma_j$ and b as well as b' are orthogonal onto $u_j = u'$ with $u' \in U'(\sigma_j)$ for $j = p+1, \cdots, n+1$. Or, $[A; b]_p = [A_{p-1}; b]$ with $[A; b]_p$ (respectively, A_{p-1}) the rank p (respectively, rank $(p-1)$) approximation of $[A; b]$ (respectively, A), as derived from the Eckart–Young–Mirsky Theorem 2.3. Also, $x' \perp v_j'', j = p+1, \cdots, n+1$. Hence, LS must also look for a solution x' in the $(p-1)$-dimensional row space $R_r(A_{p-1})$ of A_{p-1}, orthogonal to $R([v_{p+1}'', \cdots, v_{n+1}'']) = R([v_p', \cdots, v_n'])$. Thus b, as well as $\widehat{b}$ and b', are orthogonal onto the same $n - p + 1$ singular vectors $u_j, j = p+1, \cdots, n+1$. TLS and LS search for an optimal solution in a subspace of same dimension $p - 1$ with p the largest index such that $v_{n+1,p} \neq 0$, i.e., $\dim(R[A_{p-1}; b']) = \dim(R[A_{p-1}; b]_{p-1}) = p - 1$. Hence, it follows that

(3.81)
$$\| b - b' \|_2 = \| [A_{p-1}; b] - [A_{p-1}; b'] \|_F \geq \min_{\text{rank } (C) = p-1} \| [A_{p-1}; b] - C \|_F$$

$$= \sigma_p = \| [\Delta\widehat{A}; \Delta\widehat{b}] \|_F$$

with $[\Delta\widehat{A}; \Delta\widehat{b}]$ given by (3.77). The LS correction remains *larger* than the TLS correction. Observe that the *generic* TLS solution does *not* exist if the set of equations $Ax \approx b$ is *highly incompatible* so that $\| b - b' \|_2$ is large and exceeds σ'_p of A (as illustrated in Fig. 3.1 for $p = n$), or if A is *nearly rank-deficient* (i.e., $\sigma'_p \approx \cdots \approx \sigma'_n \approx 0$).

In the extreme case where b is orthogonal to $R(A)$, nongeneric TLS and LS give the *zero* solution. Indeed, as $b \perp R(A), b' = 0$. Hence, the LS solution x' is zero. The zero nongeneric TLS solution follows from direct application of Theorem 3.12. Indeed, the set of left singular vectors of $[A; b]$ is precisely $\{u'_1, u'_2, \cdots, b/\|b\|_2, u'_i, \cdots, u'_n\}$. This is the set of left singular vectors of A where b is introduced at the ith location corresponding to its size $\sigma'_{i-1} \geq \|b\|_2 \geq \sigma'_i$. Correspondingly, the ith column v_i of V is $[0, \cdots, 0, 1]$ and the last row of V is $[0, \cdots, 0, 1, \cdots, 0]$ with its ith element equal to 1. Since $v_{n+1,j} = 0$ for $j = i+1, \cdots, n+1$ and $v_{n+1,i} \neq 0$, the TLS solution $[\widehat{x}^T; -1]^T$ is given by $-v_i/v_{n+1,i}$. Hence, $\widehat{x}$ is zero. In this case, it makes no sense to force a solution other than zero that expresses the absence of correlation between A and b.

In numerical practice $v_{n+1,j}$ will seldom be zero. Indeed, due to errors in the observations $[A; b]$, a zero-valued $v_{n+1,j}$ in the SVD of the unobservable true data matrix $[A_0; b_0]$ will differ from zero in the SVD of $[A; b]$. Hence, it is advisable to define an error bound ε such that all $|v_{n+1,j}| < \varepsilon$ are considered to be zero (see §3.6.1).

Using Theorem 3.11, the following useful characterization of the nongeneric TLS solution (3.78) can be proven.

THEOREM 3.13 (Closed-form expressions of the nongeneric TLS solution). *Let* (1.21) (*respectively,* (1.22)) *be the* SVD *of* A (*respectively,* $[A; b]$) *and assume* $v_{n+1,j} = 0$ *for* $j = p+1, \cdots, n+1$, $p \leq n$. *If* $\sigma_{p-1} > \sigma_p$ *and* $v_{n+1,p} \neq 0$, *the nongeneric* TLS *solution equals*

$$\widehat{x} = (A^T A - \sigma_p^2 I_n)^{-1} A^T b \tag{3.82}$$

$$= \sum_{i=1}^{p-1} (\sigma_j'^2 - \sigma_p^2)^{-1} \sigma'_j v'_j u_j'^T b. \tag{3.83}$$

Proof. Since $\begin{bmatrix} \widehat{x} \\ -1 \end{bmatrix} // v_p$, it is an eigenvector of $[A; b]^T [A; b]$ corresponding to the eigenvalue σ_p^2. Hence, it satisfies the eigenvector equations:

$$[A; b]^T [A; b] \begin{bmatrix} \widehat{x} \\ -1 \end{bmatrix} = \begin{bmatrix} A^T A & A^T b \\ b^T A & b^T b \end{bmatrix} \begin{bmatrix} \widehat{x} \\ -1 \end{bmatrix} = \sigma_p^2 \begin{bmatrix} \widehat{x} \\ -1 \end{bmatrix} \tag{3.84}$$

or

$$(A^T A - \sigma_p^2 I_n)\widehat{x} = A^T b, \tag{3.85}$$

$$b^T A \widehat{x} + \sigma_p^2 = b^T b. \tag{3.86}$$

We first prove that $A^TA - \sigma_p^2 I$ is nonsingular. Therefore, $\sigma'_{p-1} > \sigma_p > \sigma'_p$ must hold. Since the solution is nongeneric, it follows that $\sigma_p > \sigma_{p+1}$. Indeed, if to the contrary $\sigma_p \leq \sigma_{p+1}$, we necessarily have $\sigma_p = \sigma_{p+1}$ since $\sigma_i \geq \sigma_{i+1}$, for all i. In this case we would compute the minimum norm TLS solution $\left[\begin{smallmatrix} \widehat{x} \\ -1 \end{smallmatrix}\right]$ in $R([v_p, \cdots, v_{n+1}])$. Since $v_{n+1,p} \neq 0$, we can compute a generic TLS solution, a contradiction since $\widehat{x}$ is nongeneric. Thus $\sigma_p > \sigma_{p+1}$. Since we also assumed $\sigma_{p-1} > \sigma_p, \sigma_p$ is not a repeated singular value and hence v_p is also unique, up to a scaling factor.

Now, the interlacing Theorem 2.4 for singular values implies that

$$\sigma_1 \geq \sigma'_1 \geq \cdots \geq \sigma'_{p-1} \geq \sigma_p \geq \sigma'_p \geq \sigma_{p+1} \geq \cdots \geq \sigma'_n \geq \sigma_{n+1}. \tag{3.87}$$

Since $v_{n+1,j} = 0$ for $j = p+1, \cdots, n+1$, it follows from Theorem 3.11(b) that $\sigma_j = \sigma'_{j-1}$. Hence, we have $\sigma_{p+1} = \sigma'_p$. Since $\sigma_p > \sigma_{p+1}$ is also assumed, we must have $\sigma_p > \sigma'_p$.

We next prove that $\sigma'_{p-1} > \sigma_p$ by contradiction. Assume that $\sigma'_{p-1} \leq \sigma_p$, then, from (3.87), we necessarily have that $\sigma'_{p-1} = \sigma_p$. Since σ'_{p-1} is a singular value of A with corresponding singular vector v'_{p-1}, we obtain from the corresponding eigenvector equations:

$$A^T A v'_{p-1} = \sigma'^2_{p-1} v'_{p-1} = \sigma_p^2 v'_{p-1}$$

or

$$[A;b]^T [A;b] \begin{bmatrix} v'_{p-1} \\ 0 \end{bmatrix} = \sigma_p^2 \begin{bmatrix} v'_{p-1} \\ 0 \end{bmatrix}. \tag{3.88}$$

Equation (3.88) implies that $\left[\begin{smallmatrix} v'_{p-1} \\ 0 \end{smallmatrix}\right]$ is also a right singular vector of $[A;b]$ corresponding to σ_p, a contradiction since the right singular vector v_p, corresponding to σ_p, is unique up to a scaling factor and has a nonzero last component: $v_{n+1,p} \neq 0$. Thus $\sigma'_{p-1} > \sigma_p$.

Since $\sigma'_{p-1} > \sigma_p > \sigma'_p$ is now proven, $(A^TA - \sigma_p^2 I_n)$ is nonsingular. Hence, (3.82) then follows immediately from (3.85).

Now take the SVD of (3.82):

$$\widehat{x} = \sum_{j=1}^{n} (\sigma'^2_j - \sigma_p^2)^{-1} \sigma'_j v'_j u'^T_j b. \tag{3.89}$$

Since $v_{n+1,j} = 0$ for $j = p+1, \cdots, n+1$, we have from Theorem 3.11(c) that $b \perp u'_{j-1}$, i.e., $u'^T_i b = 0$ for $i = p, \cdots, n$. Substituting this into (3.89) proves (3.83). □

Comparing expressions (2.27) and (3.82) of the generic and nongeneric TLS solutions, respectively, we observe that both solutions coincide if $p = n+1$, i.e., when the additional constraints (3.66) vanish.

3.4.2. The multidimensional case. When the set of equations $AX \approx B$ has *more than one* observation vector, the generic TLS solution $\widehat{X}$ does not exist if V_{22} is *singular*. This happens if $\sigma_{n+d} \leq \sigma'_n \leq \sigma_{n+1}$, as shown below.

LEMMA 3.2 (Existence condition for nongeneric TLS problems). *Let* (1.22) *be the* SVD *of* $[A; B]$ *and the* SVD *of* A *be given by* (1.21). *Then*

$$V_{22} \text{ singular} \Rightarrow \sigma_{n+d} \leq \sigma'_n \leq \sigma_{n+1}. \tag{3.90}$$

Proof. Using Theorem 3.2, we obtain by contraposition:

$$V_{22} \text{ singular} \Rightarrow \sigma'_n \leq \sigma_{n+1}. \tag{3.91}$$

Using the interlacing Theorem 2.4 for singular values, σ'_n is bounded by

$$\sigma_{n+d} \leq \sigma'_n \leq \sigma_n \tag{3.92}$$

From (3.91) and (3.92), we obtain (3.90). □

Condition (3.90) implies that the sum of squared projections σ'_n of all columns a_i of A onto its lowest singular vector is smaller than the sum of squared projections of all columns of $[A; B]$ onto its $(n+1)$th singular vector, associated with σ_{n+1}. This is the case if A is (*nearly*) *rank-deficient*, i.e., $\sigma'_n \approx 0$, or when the set of equations $AX \approx B$ (or at least one subset $Ax_i \approx b_i$) is highly incompatible. Before generalizing Theorem 3.11 for $d > 1$, it is convenient to describe the singularity of V_{22} in terms of its *null space* $N(V_{22})$.

LEMMA 3.3. *Let* (1.22) *be the* SVD *of* $[A; B]$. *If* $N(V_{22})$ *is the null space of* V_{22}, *then*

(3.93)

$$V_{22} \text{ singular} \iff \exists \text{ nonzero } x \in N(V_{22}) : \begin{bmatrix} V_{12} \\ V_{22} \end{bmatrix} x = \begin{bmatrix} y \\ 0 \end{bmatrix} \begin{matrix} n \\ d \end{matrix} \text{ with } \| x \|_2 = \| y \|_2$$

Proof. The proof follows straightforwardly from the assumption that V_{22} is singular, the definition of the null space of V_{22}, and the orthonormality of V. □

The following theorem generalizes the properties given in Theorem 3.11 to multidimensional nongeneric TLS problems. Since conditions concerning the multiplicity of σ_{n+1} are imposed, this theorem does not describe the properties of *all* nongeneric TLS problems. Let us first assume that $\sigma_n > \sigma_{n+1}$ and V_{22} is singular and thus, the generic multidimensional TLS solution (3.13) does *not exist*.

THEOREM 3.14 (Properties of the nongeneric multidimensional TLS problem). *Let* (1.21) (*respectively,* (1.22)) *be the* SVD *of* A (*respectively,* $[A; B]$). *Let* $\sigma_n > \sigma_{n+1}$ *and* V_{22} *be singular. Let* B' *be the orthogonal projection of* B *onto* $R(A)$ *and* $[\widehat{A}; \widehat{B}]$ *the rank* n *approximation of* $[A; B]$, *as given by* (3.11). *Call* $V'(\sigma_j)$ (*respectively,* $U'(\sigma_j)$) *the right* (*respectively, left*) *singular* *subspace of A associated with σ_j, and $V(\sigma_j)$ (respectively, $U(\sigma_j)$) the right (respectively, left) singular subspace of $[A; B]$, associated with σ_j. Consider any unit vector $x \in N(V_{22})$ and let s (respectively, e) be the index of the first (respectively, last) nonzero component of x. Then the following relations can be proven.*

If $\sigma_{n+s} = \cdots = \sigma_j = \cdots = \sigma_{n+e}$ then

$$
(3.94) \quad
\begin{array}{lrcl}
\text{(a)} & \exists v \in V(\sigma_j), \exists v' \in V'(\sigma_j) & : & \begin{bmatrix} V_{12} \\ V_{22} \end{bmatrix} x = v = \begin{bmatrix} v' \\ 0 \end{bmatrix}, \\
\text{(b)} & \exists i & : & \sigma_j = \sigma'_i, \\
\text{(c)} & \exists u' \in U'(\sigma_j) & : & B \perp u', \\
\text{(d)} & \exists u \in U(\sigma_j), \exists u' \in U'(\sigma_j) & : & u = u', \\
\text{(e)} & \exists u' \in U'(\sigma_j) & : & B' \perp u', \\
\text{(f)} & \exists u' \in U'(\sigma_j) & : & \widehat{B} \perp u'.
\end{array}
$$

Proof. For the proof see [177, Thm. 1-8]. □

The following comments are in order:

- If all components of $x \in N(V_{22})$ differ from zero, Theorem 3.14 describes the properties of nongeneric TLS problems $AX \approx B$ characterized by $\sigma_{n+1} = \cdots = \sigma_{n+d}$. In this case, every linear combination of the columns of $\left[\begin{smallmatrix} V_{12} \\ V_{22} \end{smallmatrix}\right]$ is a right singular vector of $[A; B]$, associated with σ_{n+1}.

- For the one-dimensional problem ($d = 1$) with $\sigma_n > \sigma_{n+1}$ and $v_{n+1,n+1} = 0$, the conditions vanish and Theorem 3.14 reduces to a special case of Theorem 3.11.

- If $\text{corank}(V_{22}) = \kappa \geq 1$, Theorem 3.14 can be applied to each vector of $N(V_{22})$. Or more conveniently, Theorem 3.14 can be formulated for κ-dimensional subspaces (see [177, p. 51]). Also, Theorem 3.14 can be straightforwardly extended to describe the properties of the case that the minimum norm generic TLS solution (3.38) does not exist, i.e., $\sigma_p > \sigma_{p+1} = \cdots = \sigma_{n+1}$, $p \leq n$ and Γ in (3.42) singular (see [177, p. 53]).

- The conditions $\sigma_{n+s} = \cdots = \sigma_{n+e}$ may appear somewhat *restrictive*. However, solving the set of equations $AX \approx B$ with TLS only makes sense when the data A and B are observations of an exact but unobservable relation $A_0 X = B_0$. Due to errors, the singular values $\sigma_{n+1}, \cdots, \sigma_{n+d}$ of $[A; B]$ will differ from the exact singular values $\sigma^0_{n+1} = \cdots = \sigma^0_{n+d} = 0$ of $[A_0; B_0]$. Since these differences do not convey any useful information, it is advisable to consider $\sigma_{n+1}, \cdots, \sigma_{n+d}$ of $[A; B]$ as *coinciding* and thus assume that the conditions of Theorem 3.14 are indeed satisfied.

Assuming that the conditions of Theorem 3.14 are satisfied, let us consider the *properties* of the nongeneric TLS problem for the case that $\text{corank}(V_{22}) = 1$. Assume that $\sigma'_{n-1} > \sigma'_n$ and $\sigma_{n-1} > \sigma_n$, then Theorem 3.14 yields:

$$
(3.95) \quad \begin{cases} \exists v \in V(\sigma_j), \exists u \in U(\sigma_j) : \begin{bmatrix} V_{12} \\ V_{22} \end{bmatrix} x = v = \begin{bmatrix} v'_n \\ 0 \end{bmatrix}, \sigma_j = \sigma'_n, u = u'_n, \\ B \text{ (and } \widehat{B}) \perp u = u'_n \in U(\sigma_j) \quad \text{and} \quad n+1 \le j \le n+d. \end{cases}
$$

Since $v = \left[\begin{smallmatrix} v'_n \\ 0 \end{smallmatrix}\right] \in R([v_{n+1}, \cdots, v_{n+d}])$, V_{22} is singular such that the generic TLS solution (3.13) does not exist. This means that there is no generic TLS solution $[\widehat{X}^T; -I_d]^T$ along v.

Analogously to the one-dimensional problem, we could construct a multidimensional analogue $[\Delta\widehat{A}; \Delta\widehat{B}]$ of (3.60) such that $R(\widehat{B}) \subseteq R(\widehat{A})$. Since $\|[\Delta\widehat{A}; \Delta\widehat{B}]\|_F$ depends on ε, no "smallest" $\|[\Delta\widehat{A}; \Delta\widehat{B}]\|_F$ can be computed and hence the TLS problem (3.1)–(3.2) has no solution. To solve these problems and compute a solution with reduced sensitivity, the nongeneric one-dimensional problem (3.64)–(3.66) is generalized as follows.

DEFINITION 3.3 (Nongeneric multidimensional total least squares problem). Given an overdetermined set of m linear equations $AX \approx B$, $B \in \mathcal{R}^{m \times d}$, in $n \times d$ unknowns X. Let (1.22) be the SVD of $[A; B]$. The nongeneric multidimensional total least squares (TLS) problem seeks to

$$
(3.96) \qquad \underset{[\widehat{A}; \widehat{B}] \in \mathcal{R}^{m \times (n+d)}}{\text{minimize}} \quad \| [A; B] - [\widehat{A}; \widehat{B}] \|_F
$$

$$
(3.97) \text{ subject to} \qquad R(\widehat{B}) \subseteq R(\widehat{A})
$$

$$
(3.98) \qquad \text{and} \qquad [\Delta\widehat{A}; \Delta\widehat{B}] \begin{bmatrix} w \\ 0 \end{bmatrix} \begin{matrix} n \\ d \end{matrix} = 0 \quad \forall \begin{bmatrix} w \\ 0 \end{bmatrix} \in R\left(\begin{bmatrix} V_1 \\ V_2 \end{bmatrix} \right),
$$

where $\left[\begin{smallmatrix} V_1 \\ V_2 \end{smallmatrix}\right] \begin{smallmatrix} n \\ d \end{smallmatrix} = [v_{q+1}, \cdots, v_{n+d}]$ and q maximal such that V_2 is of full rank. Once a minimizing $[\widehat{A}; \widehat{B}]$ is found, then any X satisfying

$$
\widehat{A}X = \widehat{B}
$$

is called a *nongeneric* TLS *solution* and $[\Delta\widehat{A}; \Delta\widehat{B}] = [A; B] - [\widehat{A}; \widehat{B}]$ is the corresponding nongeneric TLS correction.

It is worthwhile to discuss here in more detail the significance of the additional constraints (3.98). As mentioned in §3.4.1, the vectors $\left[\begin{smallmatrix} w \\ 0 \end{smallmatrix}\right]$ identify nonpredictive multicollinearities, i.e., approximate linear dependencies between the columns of A that do not reveal anything about the relation between A and B and that are of no (or little) value in predicting B. This is because

(3.99)

$$([A;B]\begin{bmatrix} w \\ 0 \end{bmatrix})^T([A;B]\begin{bmatrix} w \\ 0 \end{bmatrix}) = \alpha^2 \iff w^TA^TAw = \alpha^2 \quad \sigma_{n+d} \le \alpha \le \sigma_{q+1}.$$

If σ_{q+1} is small then $\alpha \approx 0$ and (3.99) implies

$$(3.100) \qquad Aw \approx 0 \quad \forall \begin{bmatrix} w \\ 0 \end{bmatrix} \begin{matrix} n \\ d \end{matrix} \in R([v_{q+1}, \cdots, v_{n+d}]),$$

i.e., there exist $n - q$ independent approximate linear dependencies between the columns of A or, A is close to a rank q matrix.

The closer the linear relations in (3.100) are to zero, the stronger are the multicollinearities and the more damaging are their effects on the generic TLS solution. This has prompted research into nongeneric TLS, a biased errors-in-variables estimator that attempts to introduce a small bias into the generic TLS solution while greatly reducing the variance appearing in the generic TLS. Despite their relative paucity of exact distributional theory, biased estimators can provide more realistic parameter estimates. They are, in general, more accurate when the data are multicollinear.

To stabilize the coefficients of the generic TLS solution, nongeneric TLS imposes the constraints (3.98). Since $R_r([\Delta\widehat{A}; \Delta\widehat{B}]) = R\left(\begin{bmatrix} \widehat{X} \\ -I_d \end{bmatrix}\right)$, these imply that the nongeneric TLS solution $\widehat{X}$ must satisfy

$$(3.101) \qquad w^T\widehat{X} = 0 \quad \forall \begin{bmatrix} w \\ 0 \end{bmatrix} \in R([v_{q+1}, \cdots, v_{n+d}]),$$

i.e., all nonpredictive multicollinearities $\left[\begin{smallmatrix} w \\ 0 \end{smallmatrix}\right]$ are eliminated from the solution or estimator.

The same is true for the LS solution. Small singular values σ_i' of A and their corresponding right singular vectors identify multicollinearities whose effects on the LS solution are more damaging as σ_i' goes to zero. To stabilize the LS solution, these small singular values are set to zero, i.e., A is reduced to a rank r matrix and the minimum norm solution $X' = \sum_{i=1}^{r} \sigma_i'^{-1} v_i' u_i'^T B$, called truncated SVD solution [28] in numerical analysis and principal component estimator in linear regression [47], is computed in the reduced r-dimensional subspace of A. In a sense, we could thus say that the *nongeneric* TLS solution stabilizes the generic TLS solution in *errors-in-variables* models just as the truncated SVD solution (or *principal component estimator*) stabilizes the LS solution in *regression* models when *multicollinearities* are present in the data A, B (see also Chapter 9). A more thorough statistical analysis is needed to reveal the merits of the nongeneric TLS estimator in multicollinearity problems.

Equation (3.99) shows that the nongeneric TLS estimator will be more accurate as σ_{q+1} goes to zero. Large σ_{q+1} imply highly incompatible TLS problems and hence the accuracy of the nongeneric TLS estimator, as well as the relevance of the problem to be solved, may be questioned.

If the conditions of Theorem 3.14 are satisfied, then the vectors w in the additional constraints (3.98), identifying nonpredictive multicollinearities, are given by right singular vectors v' of A. However, if V_{22} (or Γ) is singular and the conditions $\sigma_{n+s} = \cdots = \sigma_{n+e}$ are *not* satisfied, the nice properties described in Theorem 3.14 are no longer valid and no precise comparison with the LS solution can be made. Nevertheless, the nongeneric TLS solution still exists as proven below. In these cases, w can be any vector within the row space of A. These vectors still identify nonpredictive multicollinearities if the sum of squared projections of the rows of A onto w, i.e., $\alpha^2 = w^T A^T A w$, is small. This occurs when σ_{q+1} is not much larger than σ_{n+d} that implies that w is most likely oriented along the lowest right singular vectors $v'_n, \cdots$ of A. However, large values of α will occur if at least one subset $Ax_i \approx b_i$ in (1.17) is highly incompatible. It affects the accuracy of the other solutions when solving the d-dimensional TLS problem. In this case, it is expected to obtain more accurate TLS solutions $\widehat{x}_i, i = 1, \cdots, d$, by solving d separate one-dimensional TLS problems $Ax_i \approx b_i, i = 1, \cdots, d$.

We now show how to solve the nongeneric multidimensional TLS problem (3.96)–(3.98) and compute the nongeneric TLS solution $\widehat{X}$. This theorem generalizes Theorem 3.12 and is a correction of [177, Thm. 1-7]).

THEOREM 3.15 (Nongeneric multidimensional TLS solution). *Let* (1.22) *be the* SVD *of* $[A; B]$. *Assume* $\sigma_p > \sigma_{p+1} = \cdots = \sigma_{n+1}$ *and* $p \leq n$. *Let* Q_1 *be an orthonormal matrix such that*

$$[v_{p+1}, \cdots, v_{n+d}]Q_1 = \begin{bmatrix} Y_1 & Z_1 \\ 0 & \Gamma_1 \end{bmatrix} \begin{matrix} n \\ d \end{matrix} \qquad (3.102)$$
$$\qquad\qquad n-p \quad\; d$$

with Γ_1 *singular and* $\operatorname{corank}(\Gamma_1) = \kappa$.

Let Q *be an orthonormal matrix such that for the extended set of vectors* v_i $(q < p)$

$$[v_{q+1}, \cdots, v_{n+d}]Q = \begin{bmatrix} Y & Z \\ 0 & \Gamma \end{bmatrix} \begin{matrix} n \\ d \end{matrix} \qquad (3.103)$$
$$\qquad\qquad n-q \quad\; d$$

with minimal $n - q$ *such that* Γ *is nonsingular* $(p - q \geq \kappa)$.

Then, if $\sigma_q > \sigma_{q+1}$,

$$[\widehat{A}; \widehat{B}] = [A; B] - [\Delta\widehat{A}; \Delta\widehat{B}] \;\; \textit{with} \;\; [\Delta\widehat{A}; \Delta\widehat{B}] = [A; B]\begin{bmatrix} Z \\ \Gamma \end{bmatrix}[Z^T; \Gamma^T] \qquad (3.104)$$

solves the nongeneric multidimensional TLS *problem* (3.96)–(3.98) *and*

$$\widehat{X} = -Z\Gamma^{-1} \tag{3.105}$$

exists and is the unique solution to $(A - \Delta\widehat{A})\widehat{X} = B - \Delta\widehat{B}$.

Proof. As $\sigma_p > \sigma_{p+1} = \cdots = \sigma_{n+1}$, the solution space, i.e., the orthogonal row complement of the approximation $[A; B]_p = \sum_{i=1}^{p} \sigma_i u_i v_i^T$, is $R([v_{p+1}, \cdots, v_{n+d}])$ within which the d-dimensional generic minimum norm TLS solution $[\widehat{X}^T; -I_d]^T$ must be singled out. To satisfy (3.2), this range must contain a d-dimensional subspace $R(\left[\begin{smallmatrix} Z \\ \Gamma \end{smallmatrix}\right])$ such that Γ is nonsingular. However, the assumption Γ_1 singular means that no d-dimensional subspace $R(\left[\begin{smallmatrix} Z_1 \\ \Gamma_1 \end{smallmatrix}\right])$ of $R([v_{p+1}, \cdots, v_{n+d}])$ exists with nonsingular Γ_1. Since Γ_1^{-1} does not exist, no solution $\widehat{X} = -Z_1\Gamma_1^{-1}$ can be deduced. Hence (3.2) cannot be satisfied and thus, a solution to the generic TLS problem (3.1)-(3.2) does not exist. Hence, we must extend the solution space until it contains a d-dimensional subspace $R(\left[\begin{smallmatrix} Z \\ \Gamma \end{smallmatrix}\right])$ with Γ nonsingular.

Since corank$(\Gamma_1) = \kappa$, the space generated by

$$\left\{ \begin{bmatrix} w \\ 0 \end{bmatrix} \begin{matrix} n \\ d \end{matrix} \;\middle|\; \begin{bmatrix} w \\ 0 \end{bmatrix} \in R\left(\begin{bmatrix} Z_1 \\ \Gamma_1 \end{bmatrix} \right) \right\}$$

has dimension κ. Hence, we must add at least κ additional vectors v_i to $[v_{p+1}, \cdots, v_{n+d}]$ in (3.102) to obtain a submatrix $\left[\begin{smallmatrix} Z_1 \\ \Gamma_1 \end{smallmatrix}\right]$ with nonsingular Γ_1. Let Q be an orthonormal matrix such that

$$[v_{q+1}, \cdots, v_{n+d}]Q = \begin{bmatrix} Y & Z \\ 0 & \Gamma \end{bmatrix} \begin{matrix} n \\ d \end{matrix}$$
$$\qquad\qquad\qquad\qquad n-q \quad d$$

with minimal $n - q$ such that Γ is nonsingular ($p - q \geq \kappa$). Then the nongeneric TLS approximation $[\widehat{A}; \widehat{B}] = [A; B] - [\Delta\widehat{A}; \Delta\widehat{B}]$, with TLS correction $[\Delta\widehat{A}; \Delta\widehat{B}]$ as given by (3.104), solves the nongeneric TLS problem (3.96)–(3.98).

Indeed, let us first prove that $[\Delta\widehat{A}; \Delta\widehat{B}]$ in (3.104) satisfies the additional constraints (3.98). Consider $[v_{q+1}, \cdots, v_{n+d}] = \left[\begin{smallmatrix} V_1 \\ V_2 \end{smallmatrix}\right] \begin{smallmatrix} n \\ d \end{smallmatrix}$. Since Q is orthonormal, V_2 and $V_2 Q = [0\ ; \Gamma]$ have the same singular values and hence Γ nonsingular $\iff$ V_2 nonsingular. Therefore, the index q defined by (3.98) is identical to q defined in (3.103). Since $[v_{q+1}, \cdots, v_{n+d}]$ is an orthonormal basis and Q is orthonormal, the columns of $\left[\begin{smallmatrix} Y & Z \\ 0 & \Gamma \end{smallmatrix}\right]$ in (3.103) are also an orthonormal basis of $R([v_{q+1}, \cdots, v_{n+d}])$. Now, $R_r([\Delta\widehat{A}; \Delta\widehat{B}]) = R(\left[\begin{smallmatrix} Z \\ \Gamma \end{smallmatrix}\right])$. Γ nonsingular implies that no $y \neq 0$ exists such that $\left[\begin{smallmatrix} Z \\ \Gamma \end{smallmatrix}\right] y = \left[\begin{smallmatrix} w \\ 0 \end{smallmatrix}\right]$, according to Lemma 3.3, and thus, all $\left[\begin{smallmatrix} w \\ 0 \end{smallmatrix}\right] \begin{smallmatrix} n \\ d \end{smallmatrix} \in R([v_{q+1}, \cdots, v_{n+d}])$ must be within the subspace $R\left(\left[\begin{smallmatrix} Y \\ 0 \end{smallmatrix}\right]\right)$. Since

$R\left(\begin{bmatrix} Z \\ \Gamma \end{bmatrix}\right) \perp R\left(\begin{bmatrix} Y \\ 0 \end{bmatrix}\right)$, it follows that (3.104) satisfies (3.98). Furthermore, since $\operatorname{corank}(\Gamma_1) = \kappa$, the null space of Γ_1 has κ basis vectors. Therefore, the dimension $n-q$ of $[Y^T;0]^T$ in (3.103) must be at least $n-p+\kappa$, i.e., $p-q \geq \kappa$.

This $[\Delta\widehat{A};\Delta\widehat{B}]$ given in (3.104) also minimizes (3.96) subject to the additional constraints (3.98). Indeed, since $[A;B] = \sum_{i=1}^{q} \sigma_i u_i v_i^T + [A;B][Y^T;0]^T[Y^T;0] + [A;B][Z^T;\Gamma^T]^T[Z^T;\Gamma^T]$ and the rank n nongeneric TLS approximation $[\widehat{A};\widehat{B}] = [A;B] - [\Delta\widehat{A};\Delta\widehat{B}]$, the additional constraints $[\Delta\widehat{A};\Delta\widehat{B}][Y^T;0]^T = 0$ imply that

$$[\widehat{A};\widehat{B}] = C + [A;B]\begin{bmatrix} Y \\ 0 \end{bmatrix}\left[Y^T;0\right],$$

where C is the rank q approximation of $\sum_{i=1}^{q} \sigma_i u_i v_i^T + [A;B][Z^T;\Gamma^T]^T[Z^T;\Gamma^T]$. This implies that the nongeneric TLS approach must look in the $(q+d)$-dimensional space $R(\sum_{i=1}^{q} \sigma_i u_i v_i^T + [A;B][Z^T;\Gamma^T]^T[Z^T;\Gamma^T])$ for an approximation C of rank q with minimal approximation effort $\|[\Delta\widehat{A};\Delta\widehat{B}]\|_F$ such that (3.97) is satisfied. Using the Eckart–Young–Mirsky Theorem 2.3 and the properties of the SVD, this minimizing $[\Delta\widehat{A};\Delta\widehat{B}]$ is given by (3.104). Since $\sigma_q > \sigma_{q+1}$, $[\Delta\widehat{A};\Delta\widehat{B}]$ is the unique minimizer of (3.96) satisfying the additional constraints $[\Delta\widehat{A};\Delta\widehat{B}][Y^T;0]^T = 0$. Since the null space of the nongeneric TLS approximation $[\widehat{A};\widehat{B}]$ of rank n equals $R(\begin{bmatrix} Z \\ \Gamma \end{bmatrix})$, i.e., $[\widehat{A};\widehat{B}]\begin{bmatrix} Z \\ \Gamma \end{bmatrix} = 0$, and since Γ is nonsingular, a solution $\widehat{X} = -Z\Gamma^{-1}$ can be deduced such that $R(\widehat{B}) \subseteq R(\widehat{A})$, i.e., (3.97) is satisfied. The uniqueness of $\widehat{X}$ follows from the assumption $\sigma_q > \sigma_{q+1}$. □

The following comments are in order:

- Since
$$(3.98) \iff R\left(\begin{bmatrix} \widehat{X} \\ -I \end{bmatrix}\right) \perp \begin{bmatrix} w \\ 0 \end{bmatrix}\begin{matrix} n \\ d \end{matrix}, \forall \begin{bmatrix} w \\ 0 \end{bmatrix} \in R([v_{q+1},\cdots,v_{n+d}]),$$
we could as well use the latter constraints instead of (3.98).

- Theorem 3.15 proves that a TLS solution, called nongeneric and satisfying the TLS criteria (3.96) and (3.97) and the additional constraints (3.98), still exists and how it is computed: if $\operatorname{corank}(V_{22}) = \kappa$ or $\operatorname{corank}(\Gamma_1) \geq 1$, we must add the minimal number (at least κ) of right singular vectors associated with the smallest singular values of $[A;B]$, until a nonsingular Γ in (3.103) is obtained.

- The nongeneric TLS solution (3.105) of the multidimensional TLS problem also equals the *minimum norm* TLS solution computed in the restricted row space $R_r([\widehat{A};\widehat{B}])$ given by $R_r([A;B]) \setminus R([v_{q+1},\cdots,v_{n+d}])$. This result follows directly from Theorem 3.9 using the assumptions of Theorem 3.15.

- As in the one-dimensional case (see Theorem 3.12), there is no restriction in imposing the condition $\sigma_q > \sigma_{q+1}$. Indeed, if $\sigma_q = \sigma_{q+1}$ then the nongeneric TLS solution still exists but is possibly not unique. In order to preserve uniqueness, the *minimum norm* solution $[\widehat{X}^T; -I_d]^T$ can be singled out in $R([v_{q-r+2}, \cdots, v_{n+d}])$ (as outlined in Theorem 3.9), where r denotes the multiplicity of σ_{q+1} such that $\sigma_{q-r+2} = \cdots = \sigma_{q+1}$. As for the multidimensional generic problem, this solution still satisfies (3.97) and (3.98), but (3.96) is only satisfied for the 2-norm unless the smallest $n - q + d$ singular values coincide, i.e., $\sigma_{q+1} = \cdots = \sigma_{n+1} = \cdots = \sigma_{n+d}$.

Observe that in generic problems the additional constraints (3.98) can be neglected. Consider the SVD (1.22) of $[A; B]$. If $\sigma_n > \sigma_{n+1}$, the TLS problem (3.1)–(3.3) is solvable provided V_{22} is nonsingular. Hence, $q = n$ in (3.98). Since V_{22} is nonsingular, there does not exist $\left[\begin{smallmatrix} w \\ 0 \end{smallmatrix}\right] \in R\left(\left[\begin{smallmatrix} V_{12} \\ V_{22} \end{smallmatrix}\right]\right)$ according to Lemma 3.3 and hence, the constraints (3.98) vanish. In the case of nonuniqueness (assume that $\sigma_p > \sigma_{p+1} = \cdots = \sigma_{n+1}$ with $p \leq n$), the minimum norm generic TLS solution (3.38) is singled out with Γ nonsingular. Since all $\left[\begin{smallmatrix} w \\ 0 \end{smallmatrix}\right] \in R([v_{p+1}, \cdots, v_{n+d}])$ belong to $R(\left[\begin{smallmatrix} Y \\ 0 \end{smallmatrix}\right])$ and $R(\left[\begin{smallmatrix} Y \\ 0 \end{smallmatrix}\right]) \perp R(\left[\begin{smallmatrix} Z \\ \Gamma \end{smallmatrix}\right]) = R(\left[\begin{smallmatrix} \widehat{X} \\ -I \end{smallmatrix}\right])$ because of the orthonormality of $[v_{p+1}, \cdots, v_{n+d}]Q$, the constraints (3.98) are automatically satisfied and need not be imposed.

Finally, we compare the nongeneric multidimensional TLS problem with the LS problem in the following situation. Assume that the conditions of Theorem 3.14 are satisfied and that a nonsingular Γ can be obtained in (3.103) of Theorem 3.15 for $q = n - 1$ and $\sigma_{n-1} > \sigma_n$. In this case, corank$(V_{22}) = 1$. Also if $\sigma'_{n-1} > \sigma'_n$, then (3.95) describes the properties of this problem. Since $v = \left[\begin{smallmatrix} v'_n \\ 0 \end{smallmatrix}\right] \in R([v_{n+1}, \cdots, v_{n+d}])$, V_{22} is singular and the generic TLS solution (3.13) does not exist. We then look for a nongeneric $[\widehat{X}^T; -I_d]^T$ which is orthogonal to v. The nongeneric TLS correction matrix (3.104) satisfying $[\Delta\widehat{A}; \Delta\widehat{B}]v = 0$ has minimal norm under this condition and is given by

$$\| [\Delta\widehat{A}; \Delta\widehat{B}] \|_F = \sqrt{-\sigma_j^2 + \sum_{k=n}^{n+d} \sigma_k^2}. \tag{3.106}$$

LS looks for a solution X' such that $\sum_{i=1}^{n} \sigma'_i u'_i v_i'^T X' = B'$ but in the considered problem, B, as well as B', are orthogonal onto $u'_n, \sigma'_n = \sigma_j, n+1 \leq j \leq n+d$, according to Theorem 3.14. This means that LS and TLS look for a lower rank approximation of $[A_{n-1}; B]$ with A_{n-1} the approximation of rank $n-1$ of A obtained by making σ'_n zero. Indeed, since $B \perp u'_n, X'$ is orthogonal onto v'_n and LS also looks for a solution X' in the $(n-1)$-dimensional subspace, $R([v'_1, \cdots, v'_{n-1}])$ is orthogonal to v'_n. This solution is given by $X' = \sum_{i=1}^{n-1} \sigma_i'^{-1} v'_i u_i'^T B$. Thus B, as well as $\widehat{B}$ and B', are orthogonal onto the same singular vector u'_n. TLS, as well as LS, searches for an optimal solution in a subspace of the same dimension $n-1$, approximating $[A_{n-1}; B]$. Hence

$$\begin{aligned} \| B - B' \|_2 &= \| [A_{n-1}; B] - [A_{n-1}; B'] \|_2 \\ &\geq \min_{\text{rank}(C)=n-1} \| [A_{n-1}; B] - C \|_2 = \sigma_n = \| [\Delta\widehat{A}; \Delta\widehat{B}] \|_2 \end{aligned}$$

and

$$\begin{aligned} \| B - B' \|_F &= \| [A_{n-1}; B] - [A_{n-1}; B'] \|_F \\ &\geq \min_{\text{rank}(C)=n-1} \| [A_{n-1}; B] - C \|_F = \| [\Delta\widehat{A}; \Delta\widehat{B}] \|_F, \end{aligned}$$

where $[\Delta\widehat{A}; \Delta\widehat{B}]$ is given by (3.104), and its Frobenius norm by (3.106). As in the one-dimensional case, we observe that the generic TLS solution does not exist if $AX \approx B$ is highly incompatible (so that $\| B - B' \|_2 > \sigma'_n$), or if A is (nearly) rank-deficient (i.e., $\sigma'_n \approx 0$).

3.5. The mixed LS-TLS problem and solution

In certain linear modeling problems $AX \approx B$, *some* of the columns of A may be *known exactly*. For instance, every intercept model

$$\alpha + a_1 x_1 + \cdots + a_n x_n = b$$

gives rise to an overdetermined set of equations

$$[1_m; A] \begin{bmatrix} \alpha \\ x \end{bmatrix} = b$$

with $1_m = [1, \cdots, 1]^T$ in which the first column of the left-hand side matrix is known exactly [59], [62]. Other examples occur in system identification, for instance, when a system can be modeled as a transfer function model. If the inputs to the system can be measured without error and enough input-output measurements are available, then estimating the parameters of these models gives rise to TLS problems with exactly known columns [191], [168].

To maximize the accuracy of the estimated parameters X, it is natural to require that the corresponding columns of A be unperturbed since they are known exactly. Therefore, the classical TLS problem formulation, as given in §3.2, is generalized as follows.

DEFINITION 3.4 (Multidimensional mixed LS-TLS problem). Given a set of m linear equations in $n \times d$ unknowns X:

$$AX \approx B, \qquad A \in \mathcal{R}^{m\times n}, B \in \mathcal{R}^{m\times d} \text{ and } X \in \mathcal{R}^{n\times d}, \tag{3.107}$$

Partition

$$\begin{aligned} A &= [A_1; A_2], & A_1 \in \mathcal{R}^{m\times n_1} \text{ and } A_2 \in \mathcal{R}^{m\times n_2}, \\ X &= [X_1^T; X_2^T]^T, & X_1 \in \mathcal{R}^{n_1\times d} \text{ and } X_2 \in \mathcal{R}^{n_2\times d}, \end{aligned} \tag{3.108}$$

and assume that the columns of A_1 are known exactly and $n = n_1 + n_2$.

Then, the mixed LS-TLS problem seeks to

$$\text{(3.109)} \qquad \underset{[\hat{A}_2;\hat{B}] \in \mathcal{R}^{m\times(n_2+d)}}{\text{minimize}} \qquad \| [A_2; B] - [\hat{A}_2; \hat{B}] \|_F$$

$$\text{(3.110)} \qquad \text{subject to} \quad R(\hat{B}) \subseteq R(\hat{A}) = R([A_1; \hat{A}_2]).$$

Once a minimizing $[\hat{A}_2; \hat{B}]$ is found, then any $X = [X_1^T; X_2^T]^T$ satisfying

$$\text{(3.111)} \qquad \hat{A}X = A_1X_1 + \hat{A}_2X_2 = \hat{B}$$

is called a *mixed* LS-TLS *solution* and $[\Delta\hat{A}_2; \Delta\hat{B}] = [A_2; B] - [\hat{A}_2; \hat{B}]$ is the corresponding mixed LS-TLS correction.

Whenever the solution is *not unique*, the *minimum norm* solution, denoted by $\hat{X} = [\hat{X}_1^T; \hat{X}_2^T]^T$, can be singled out.

More general TLS formulations, which allow for error-free rows and correlations between the errors in each row or column of $[A_2; B]$ are given in §10.3. It is worth noting that when all columns of A are known exactly, the mixed LS-TLS solution reduces to the ordinary LS estimate. By varying n_1 from zero to n, this formulation can handle the ordinary LS problem, any mixed TLS-LS problem, and the ordinary TLS problem.

We now show how to solve the above problem when the solution is unique or $d = 1$. The technique, due to Golub, involves computing a QR *factorization* of the "known" columns and then solving a TLS problem of *reduced* dimension.

Let a matrix $A = [A_1; A_2]$ be given whose first p columns A_1 have no error and have full column rank. Then, the algorithm is as follows. Perform p Householder transformations Q on the matrix $[A; B]$ so that

$$\text{(3.112)} \qquad Q^T[A_1; A_2; B] = \underset{\quad p \qquad n-p \qquad d}{\begin{bmatrix} R_{11} & R_{12} & R_{1b} \\ 0 & R_{22} & R_{2b} \end{bmatrix}} \begin{matrix} p \\ m-p \end{matrix},$$

where R_{11} is a $p \times p$ upper triangular matrix. Then, compute the TLS solution $\hat{X}_2$ of $R_{22}X \approx R_{2b}$. $\hat{X}_2$ yields the last $n-p$ components of each solution vector $\hat{x}_i$. To find the first p rows $\hat{X}_1$ of the solution matrix $\hat{X}$, solve

$$\text{(3.113)} \qquad R_{11}\hat{X}_1 = R_{1b} - R_{12}\hat{X}_2.$$

This is simply the LS solution obtained by projecting the reduced observation vector $B - A_2\hat{X}_2$ into the space $R(A_1)$ generated by the "known" columns of A.

Using a *generalization* of the Eckart–Young–Mirsky-Theorem 2.3, Golub, Hoffman, and Stewart [62] proved that the above procedure, which consists in computing the mixed LS-TLS estimate $[\widehat{X}_1^T; \widehat{X}_2^T]^T$, finds the best rank r approximation ($r \leq n$) $[A_1; \widehat{A}_2; \widehat{B}]$ to $[A; B]$ that leaves A_1 fixed such that

(3.114)
$$\| [A_1; A_2; B] - [A_1; \widehat{A}_2; \widehat{B}] \|_F = \min_{\text{rank } [A_1; \widetilde{A}_2; \widetilde{B}] \leq r} \| [A_1; A_2; B] - [A_1; \widetilde{A}_2; \widetilde{B}] \|_F .$$

To guarantee that the computed solution always has minimal norm, this procedure must be slightly modified, as outlined in §3.6.3.

Conditions for existence and uniqueness of the mixed LS-TLS solution are given below.

COROLLARY 3.5 (Existence and uniqueness of the mixed LS-TLS solution). *Consider the mixed* LS-TLS *problem* (3.107)–(3.110) *and assume that* A_1 *has full rank* n_1. *Denote by* σ' (*respectively,* σ) *the smallest* (*respectively,* $(n_2+1)th$) *singular value of* R_{22} (*respectively,* $[R_{22}; R_{2b}]$) *as defined by* (3.112), *then*

$$\sigma' > \sigma \quad \Rightarrow \quad \textit{solution } \widehat{X} = \begin{bmatrix} \widehat{X}_1 \\ \widehat{X}_2 \end{bmatrix} \textit{ of the mixed LS-TLS problem exists and is unique.}$$

Proof. Corollary 3.5 is a special case of [188, Thm. 3]. □

Closed-form expressions of the mixed LS-TLS solution can also be derived, as shown below.

COROLLARY 3.6 (Closed-form expression of the mixed LS-TLS solution). *Consider the notation and assumptions of Corollary* 3.5 *and assume further that the smallest singular values* $\sigma = \sigma_{n_2+1} = \cdots = \sigma_{n_2+d}$ *of* $[R_{22}; R_{2b}]$ *coincide. If* $\sigma' > \sigma$, *then the mixed* LS-TLS *solution is given by*

$$\widehat{X} = \left(A^T A - \sigma^2 \begin{bmatrix} 0 & 0 \\ 0 & I_{n_2} \end{bmatrix} \right)^{-1} A^T B.$$

Proof. Corollary 3.6 is a special case of [188, Thm. 4]. □

Another closed-form expression was derived by Zoltowski [218].

COROLLARY 3.7 (CLOSED-FORM EXPRESSION OF THE MIXED LS-TLS SOLUTION). *Consider the notation and assumptions of Corollary* 3.5. $A_1^\dagger = (A_1^T A_1)^{-1} A_1^T$ *is the pseudo-inverse of* A_1 *and let* $\widetilde{U}\widetilde{\Sigma}\widetilde{V}^T$ *be the* SVD *of* $[R_{22}; R_{2b}]$, *where* $\widetilde{V} = \begin{bmatrix} \widetilde{V}_{11} & \widetilde{V}_{12} \\ \widetilde{V}_{21} & \widetilde{V}_{22} \end{bmatrix} \begin{matrix} n_2 \\ d \end{matrix}$ (column blocks n_2, d). *If* $\sigma' > \sigma$, *the mixed* LS-TLS *solution is given by*

$$(3.115) \qquad \widehat{X} = \begin{bmatrix} \widehat{X}_1 \\ \widehat{X}_2 \end{bmatrix} = \begin{bmatrix} A_1^\dagger A_2 \\ -I_{n_2} \end{bmatrix} \widetilde{V}_{12} \widetilde{V}_{22}^{-1} + \begin{bmatrix} A_1^\dagger B \\ 0_{n_2 \times n_2} \end{bmatrix}.$$

Proof. See [218]. The proof follows the lines of the derivation given above. $\widehat{X}_2 = -\widetilde{V}_{12}\widetilde{V}_{22}^{-1}$ is the TLS solution of the reduced set $R_{22}X_2 \approx R_{2b}$. Since A_1 is of full rank, R_{11}^{-1} exists and, hence, (3.113) yields

$$\widehat{X}_1 = R_{11}^{-1}R_{1b} - R_{11}^{-1}R_{12}\widehat{X}_2 = A_1^{\dagger}B - A_1^{\dagger}A_2\widehat{X}_2$$

and proves (3.115). □

3.6. Outline of the algorithms

3.6.1. Classical TLS algorithm. In this section, we want to summarize the practical TLS computations into one algorithm that solves any generic and nongeneric TLS problem of dimension d. In the case of nonuniqueness, the minimum norm TLS solution is computed. A fully documented Fortran 77 program of this classical TLS algorithm is given in [178], and [176]. The code can be obtained through netlib [46].

ALGORITHM 3.1 (Classical TLS computation of the solution $\widehat{X}$ of $AX \approx B$).

Given: An $m \times n$ data matrix A and an $m \times d$ observation matrix B.

Step 1. SVD of $[A; B]$

1(a). If $m > \frac{5}{3}(n+d)$, transform $[A; B]$ into upper triangular form R by Householder transformations.

1(b). Compute the Singular Value Decomposition (1.22) of $[A; B]$ (or R).

Step 2. Rank determination

If not user-determined, compute the numerical rank $r(\leq n)$ of $[A; B]$ as follows:

(3.116) $$\sigma_1 \geq \cdots \geq \sigma_r > R_\nu \geq \sigma_{r+1} \geq \cdots \geq \sigma_{n+d}$$

with R_ν an appropriate rank determinator.

Step 3. TLS solution $\widehat{X}$

Compute with Householder transformations the orthogonal matrix Q such that

$$[v_{r+1}, \cdots, v_{n+d}]Q = \begin{bmatrix} Y & Z \\ 0 & \Gamma \end{bmatrix} \begin{matrix} n \\ d \end{matrix} \quad \text{and } \Gamma\ d \times d \text{ upper triangular}$$
$$\qquad\qquad\qquad\qquad n-r \quad d$$

(3.117)

If Γ singular then begin {TLS *solution is nongeneric*}
lower the rank r with the multiplicity of σ_r
go back to the beginning of Step 3

end

(3.118) else solve by forward elimination : $\widehat{X}\Gamma = -Z$

END

The following comments are in order:

- If $m < n$, the set of equations $AX \approx B$ is always *underdetermined.* This implies that the solution is generically never unique. In this case, Algorithm 3.1 can still be applied to compute the minimum norm solution.

- If $d = 1$, $r = n, \Gamma$ nonsingular and $\sigma_n > \sigma_{n+1}$, (3.118) reduces to (2.26). The TLS solution is obtained from a simple scaling of the right singular vector v_{n+1} of $[A; b]$, associated with its minimal singular value σ_{n+1}.

- If $m > \frac{5}{3}(n + d)$, it is more efficient to first *triangularize* $[A; B]$, using a QR factorization [26], i.e.,

  (3.119) $[A; B] = Q\,R$ $\qquad$ R upper triangular and $Q^TQ = I_{n+d}$

  and then proceed to R. Indeed, $[A; B]$ and R have the same singular values and right singular vectors. Hence, the TLS solution $\widehat{X}$ of $AX \approx B$ can also be obtained from the SVD of R. This option has been incorporated in Algorithm 3.1 (see [176]). More details are given in Chapter 4.

- Observe from (3.117) that we only need to compute a *few* singular vectors, associated with the smallest singular values of $[A; B]$ to obtain the TLS solution $\widehat{X}$. Moreover, we only need to compute a *basis* of the singular subspace corresponding to the smallest $n + d - r$ singular values with r the numerical rank of $[A; B]$. Indeed, the TLS solution $\widehat{X}$ is *invariant* with respect to any nonsingular basis transformation P in its solution space $\begin{bmatrix} \widehat{X} \\ -I_d \end{bmatrix}$. With $\begin{bmatrix} Z \\ \Gamma \end{bmatrix}$ in (3.117) a basis of the TLS solution space, we have

  (3.120) $$\begin{bmatrix} Z \\ \Gamma \end{bmatrix} P = \begin{bmatrix} Z \cdot P \\ \Gamma \cdot P \end{bmatrix}$$

  and thus, $\widehat{X} = -(ZP)(\Gamma P)^{-1} = -ZPP^{-1}\Gamma^{-1} = -Z\Gamma^{-1}$ remains invariant.

  Based on these properties we were able to *improve the efficiency* of the TLS computations by computing the SVD of $[A; B]$ in Step 1 only "*partially.*" This results in the development of a new algorithm "partial total least squares (PTLS)." PTLS is about two times faster than the classical TLS Algorithm 3.1, while the same accuracy can be maintained. More details are given in Chapter 4.

- With respect to *rank determination*, the SVD is the only fully reliable way to treat *rank-deficient* problems. The TLS condition (3.2) requires that $\text{rank}([\widehat{A}; \widehat{B}]) \leq n$. Therefore, $r \leq n$. If the elements in $[A; B]$ are affected by independently and identically distributed zero-mean errors with variance σ_ν^2, an appropriate rank determinator $R_\nu = \sqrt{2\max\{m, n+d\}}\sigma_\nu$, which is implemented in [176], is derived as follows. The given data $[A; B]$ are approximated by an approximation $[\widehat{A}; \widehat{B}]$ of *maximal minimal signal-to-noise ratio* (maximal MSNR) but living in a subspace of dimension r [41], [156]. This dimension is chosen directly by the user (leading to a maximal MSNR for this dimension) or indirectly by specifying the standard deviation σ_ν of the error in the entries of $[A; B]$ and a required signal-to-noise ratio R_0. Using the properties of the SVD, it is proven that the MSNR is maximal when $R([\widehat{A}; \widehat{B}])$ coincides with the r-dimensional singular subspace of $[A; B]$, associated with $\sigma_1, \cdots, \sigma_r$. In this subspace, the expected MSNR $R \geq R_0$ is given by

$$R = \frac{(\sigma_r^0)^2}{\max\{m, n+d\}\sigma_\nu^2} \tag{3.121}$$

with σ_r^0 the rth singular value of the exact but unobservable data matrix $[A_0; B_0]$, measured by $[A; B]$. If the entries in $[A; B]$ are disturbed by independently and identically distributed zero-mean errors with variance σ_ν^2, the expected value $\mathcal{E}(\sigma_r^2)$ of the squared rth singular value σ_r^2 of $[A; B]$ is given by

$$\mathcal{E}(\sigma_r^2) = (\sigma_r^0)^2 + \max\{m, n+d\}\sigma_\nu^2. \tag{3.122}$$

Substituting $(\sigma_r^0)^2$ in (3.121) by (3.122) , we obtain from $R \geq R_0$

$$\sigma_r^2 - \max\{m, n+d\}\sigma_\nu^2 \geq \max\{m, n+d\}\sigma_\nu^2 R_0, \tag{3.123}$$

$$\text{or} \quad \sigma_r^2 \geq \max\{m, n+d\}(R_0+1)\sigma_\nu^2. \tag{3.124}$$

By choosing $R_0 = 1$, we obtain the rank determinator

$$R_\nu = \sqrt{2\max\{m, n+d\}}\sigma_\nu.$$

- In Step 3 of Algorithm 3.1, the *multiplicity* of σ_r must be determined. One approach is to regard all computed singular values in the interval $[\sigma_r, \sqrt{\sigma_r^2+\delta}]$ as being identical. $\delta > 0$ may be specified by the user or automatically computed from (3.122). In the latter case δ is given by $\max\{m, n+d\}\sigma_\nu^2$ with σ_ν the standard deviation of the error in each entry of the data matrix $[A; B]$.

- The upper triangular matrix Γ must be tested for *nonsingularity*. This can be done as follows. As long as the absolute values of all diagonal elements $|\Gamma_{ii}| > \varepsilon, i = 1, \cdots, d$ with ε a given small positive number, Γ is considered to be nonsingular. The reliability of this test can be improved by pivoting the rows in (3.117) when performing the Householder transformations [69, §5.4.1]. As soon as the pivot element is less than or equal to ε, Γ is considered to be singular. If pivoting is done, the inverse permutations must be performed in the last step of the algorithm. The choice of ε can be based on the numerical accuracy of the SVD computations [67] or, preferably, may depend on the perturbations of the data [204]. A generally applicable, reliable error bound ε in function of the standard deviation σ_ν of the perturbations can not be deduced. However, the user can get more insight in the choice of ε from the sensitivity analysis given in Chapter 7.

3.6.2. Scaling. In this book, we assume that the data $[A; B]$ have been *properly scaled* in advance. Therefore, it is interesting to give some *scaling* recommendations here. In general, the data are scaled by pre- and/or postmultiplying the matrix $[A; B]$ with nonsingular equilibration matrices D and C. The mathematically equivalent scaled version of the linear system $AX \approx B$ is given by

$$D[A; B]C\left(C^{-1}\begin{bmatrix} X \\ -I_d \end{bmatrix}\right) \approx 0. \tag{3.125}$$

The *basic* recommendation is that *scaling* of equations and unknowns must proceed on the basis of what the source problem proclaims about the significance of each entry in $[A; B]$. Most scaling strategies described in the literature, e.g., [15], [173], are intended to minimize the condition of the problem. However, these procedures may not be applied to the TLS problem of which the main concern is to equilibrate the measurement errors in the data matrix $[A; B]$. To attain the optimal TLS solution properties (see Chapter 8), we must assume a simple but revealing error model in which the entries of the error matrix $[\Delta A; \Delta B]$ must be *independent* random variables with zero mean and *equal* standard deviation σ_ν. This implies that the error covariance matrix is given by $\sigma_\nu^2 I_{n+d}$. The assumption of a common standard deviation has important implications for the scaling of $[A; B]$. From (3.125) it is evident that any scaling $D[A; B]C$ of $[A; B]$ automatically induces the same scaling $D[\Delta A; \Delta B]C$ of the error matrix $[\Delta A; \Delta B]$. Hence, the rows and columns of $[A; B]$ must be scaled so that all the elements of $[\Delta A; \Delta B]$ are roughly equal. The failure to observe this dictum can result in spurious indications of rank deficiency, a phenomenon sometimes referred to as artificial ill-conditioning [161]. Consequently, the following *scaling strategy* is recommended: estimate the absolute size of the errors in the matrix $[A; B]$ and then scale so that as nearly as possible all the estimates are equal. It is important to verify that the errors to be estimated include all initial errors (measurement, computational, and rounding errors). In many applications, rounding errors will be the least of these. If the error in each entry is proportional to the size of the element, the strategy reduces to scaling the original matrix $[A; B]$ so that all its elements are roughly equal. This is a frequently recommended strategy, but it is applicable only when the error matrix $[\Delta A; \Delta B]$ represents a uniform relative error in the entries of $[A; B]$. To obtain the proper scaling matrices D and C, "*equilibrium*" routines that scale a matrix so that its elements are roughly equal in size [15], [173] can be applied instead of the matrix of error estimates.

A commonly used special case is that in which the matrices D and C are diagonal, i.e., the entries of $[\Delta A; \Delta B]$ are assumed to be independent. In this case, left (respectively, right) multiplication of $[A; B]$ by D (respectively, C) can be interpreted as being a row (respectively, column) scaling operation in which row (respectively, column) i is multiplied by the diagonal element D_{ii} (respectively, C_{ii}). If we know the approximate size or standard deviation s_i of the uncertainty in the ith row (respectively, column) of $[A; B]$, it is usual to define $D_{ii} = 1/s_i$ (respectively, $C_{ii} = 1/s_i$) such that all elements of the modified matrix $D[A; B]C$ have errors with unit standard deviation. More generally, if we have sufficient statistical information about the uncertainties in $[A; B]$ to place numerical values on the correlation between the errors in different components of each column (respectively, row), we may express this information as an $m \times m$ (respectively, $(n + d) \times (n + d)$) positive-definite symmetric covariance matrix $\mathcal{D}$ (respectively, $\mathcal{C}$). The equilibration matrix D (respectively, C) is then given by the inverse square root (e.g., the Cholesky factor) of the error covariance matrix $\mathcal{D}$ (respectively, $\mathcal{C}$).

To avoid the explicit formation of the product $D[A; B]C$, a generalized TLS algorithm, based on the generalized SVD (GSVD) [179], [189], or a restricted TLS algorithm, based on the restricted SVD (RSVD) [194], can be used. The GSVD and RSVD avoid to transform the data A, B explicitly and are numerically more robust with respect to ill-conditioned covariance matrices than the explicit transformation procedures. It is of interest to note that if the data that lead to the determination of the a priori covariance matrix $\mathcal{C}$ were approximately the same as the present data $[A; B]$, then the matrix C will be approximately equal to R^{-1}, where R is the triangular matrix resulting from Householder triangularization of $[A; B]$. Then the condition number of $[A; B]C$ is likely to be fairly small [96, Chap. 25].

Finally, note that it is not always possible to equilibrate the error matrix appropriately by pre- and/or postmultiplying $[A; B]$ with a matrix, e.g., in the case of a nonlinear error distribution. Although in the latter case the optimal properties of the TLS solution are not obtained, they can be fairly well approximated and still motivate the use of TLS, as we could observe in some practical applications.

3.6.3. Mixed LS-TLS algorithm. If n_1 columns of the $m \times n$ data matrix A are *known exactly*, the solution of the mixed LS-TLS problem (3.107)–(3.111) is computed by the following algorithm.

ALGORITHM 3.2 (Mixed LS-TLS solution of $AX \approx B$).

Given :

An $m \times n$ data matrix A with n_1 exactly known linearly independent columns and $0 \leq n_1 \leq n$,
an $m \times d$ observation matrix B,
an n-dimensional vector t indicating whether the ith column of A is known exactly ($t_i = 0$) or not ($t_i = 1$), $i = 1, \cdots, n$.

Step 1. Column permutations in A

If $0 < n_1 < n$, permute the columns of A such that

$$A\,P = [\ \underset{n_1}{A_1}\ ;\ \underset{n_2}{A_2}\ ], \qquad n = n_1 + n_2, \tag{3.126}$$

where P is an $n \times n$ permutation matrix and A_1 contains the n_1 exactly known columns of A.

Step 2. QR factorization of $[A_1; A_2; B]$

If $n_1 > 0$ then begin
  compute the QR factorization:

$$[A_1; A_2; B] = Q \begin{bmatrix} R_{11} & R_{12} \\ 0 & R_{22} \end{bmatrix} \begin{matrix} n_1 \\ m - n_1 \end{matrix} \quad \text{(columns: } n_1,\ n_2 + d\text{)} \tag{3.127}$$

  with Q being orthogonal and R_{11} upper triangular.
  If $n_1 = n$ then begin {*LS solution* $\widehat{X}$}
    solve $R_{11}\widehat{X} = R_{12}$ by back-substitution
    stop
    end
  end
else $R_{22} \leftarrow [A; B]$

Step 3. SVD of R_{22}

$$U^T R_{22} V = \text{diag}(\sigma_1, \cdots, \sigma_s), \qquad s = \min\{m - n_1, n_2 + d\}$$

with U, $V = [v_1, \cdots, v_{n_2+d}]$ orthonormal, $\sigma_{i-1} \geq \sigma_i$ for $i = 2, \cdots, s$ and $\sigma_j = 0$ for $j > s$.

Step 4. Rank determination

If not user-determined, compute the numerical rank r ($\leq n_2$) of $[A_2; B]$:

$$\sigma_1 \geq \cdots \geq \sigma_r > R_\nu \geq \sigma_{r+1} \geq \cdots \geq \sigma_{n_2+d}$$

with R_ν an appropriate rank determinator.

Step 5. Solution space V_2

If $n_1 > 0$ then begin

$V_{22} \leftarrow [v_{r+1}, \cdots, v_{n_2+d}]$

solve $R_{11}V_{12} = -R_{12}V_{22}$ by back-substitution

$$V_2 \leftarrow \begin{bmatrix} V_{12} \\ V_{22} \end{bmatrix} \begin{matrix} n_1 \\ n_2 + d \end{matrix}$$

end

else $V_2 \leftarrow [v_{r+1}, \cdots, v_{n_2+d}]$

Step 6. TLS solution $\widehat{X}$

Compute with Householder transformations the orthogonal matrix Q such that

$$V_2 Q = \underset{\displaystyle\quad n_2 - r \qquad d}{\begin{bmatrix} Y & Z \\ 0 & \Gamma \end{bmatrix}} \begin{matrix} n \\ d \end{matrix} \quad \text{and } \Gamma \ d \times d \text{ upper triangular.}$$

If Γ singular then begin {TLS *solution is nongeneric*}

lower the rank r with the multiplicity of σ_r

go back to Step 5

end

else solve by forward elimination: $\widehat{X}\Gamma = -Z$

Step 7. Inverse row permutations in $\widehat{X}$

(3.128) If $0 < n_1 < n$, undo the permutations P: $\widehat{X} \leftarrow P\widehat{X}$

END

The following comments are in order:

- If $m - n_1 > \frac{5}{3}(n_2 + d)$, it is more efficient to compute the complete QR factorization of $[A_1; A_2; B]$ in Step 2 so that R_{22} is also upper triangular.
- If no columns of A are error free ($n_1 = n$), Algorithm 3.2 reduces to Algorithm 3.1.

- In computing the QR factorization of $[A_1; A_2; B]$ in Steps 2 and 3, *pivoting* [69, §5.4.1] may be done within three subsets of columns: the first n_1 columns A_1, the next n_2 columns A_2, and the last d columns B. Columns may not be pivoted with columns from another subset. Even for full rank problems, column pivoting seems to produce more accurate solutions [84]. If pivoting is done, the inverse permutations must be performed on the rows of V_2 in the beginning of Step 6.

- Mostly, rank $r = n_2$ is maximal. If $r < n_2$ (e.g., when the set of equations $AX \approx B$ is *underdetermined*), the TLS solution is no longer unique. In this case, Algorithm 3.2 singles out the *minimum norm* solution, as shown in the proof of Theorem 3.9.

- If A_1 does *not* have *full column rank*, i.e., $\text{rank}(A_1) = r_A < n_1$, we can always replace A_1 by a matrix having r_A independent columns selected from A_1, apply Algorithm 3.2 and set the coefficients of $\widehat{X}$ corresponding to the missing columns equal to zero without changing either the rank of the result or the norm of the difference (3.1). Note, however, that in this case the solution $\widehat{X}$ no longer has minimal norm.

- In particular, Algorithm 3.2 is able to compute the *Compensated Least Squares* (CLS) *estimate*, as derived by Stoica and Söderström [168]. Since Algorithm 3.2 first performs a QR factorization, it only needs to compute the SVD of a smaller submatrix R_{22} and hence is computationally more efficient than the computation procedure presented in [168].

3.7. Concluding Remarks

This chapter analyzes various extensions of the basic total least squares (TLS) problem and outlines the algorithms to solve these extended problems. First, it is shown how to solve the *multidimensional* TLS problem $AX \approx B$ in which B and X are matrices instead of vectors (§3.2) and how TLS singles out the minimum norm solution if the solution of the problem is *no longer unique* (§3.3).

However, when the set of equations $AX \approx B$ is highly incompatible or when the data matrix A is (nearly) rank-deficient, these TLS problems become very ill-conditioned or even lack a solution satisfying the TLS criteria, as proved by Golub and Van Loan. To make these problems solvable and well-conditioned, an extension of the generic TLS problem, called nongeneric TLS, is proposed. Nongeneric TLS problems satisfy the same TLS criteria as formulated in the generic TLS problem, but, in addition, impose constraints on the solution space. The TLS algorithm of Golub and Van Loan which fails to compute a finite solution in nongeneric TLS problems, is extended to solve both generic and nongeneric TLS problems (§3.4). It is shown that the proposed extension reduces the sensitivity of the TLS solution considerably. The additional constraints, which eliminate all nonpredictive multicollinearities from the TLS solution space, are scrutinized and compared with LS, latent root, and principal component regression.

Furthermore, mixed LS-TLS problems in which some of the columns of the data matrix A have no error (§3.5) are presented and solved and, finally, the TLS computations are *summarized* in one algorithm that includes all the extensions described above (§3.6). The Fortran 77 code is described in [178], [176] and is available through netlib [46].

Chapter 4

Direct Speed Improvement of the Total Least Squares Computations

4.1. Introduction

In this and the following chapter, we concentrate further on the computational aspects of the TLS problem: several algorithms, which *improve the efficiency* of the TLS computations as outlined in the classical TLS Algorithm 3.1 (§3.6.1), are presented and analyzed, and their efficiency in solving TLS problems is compared. Readers who are not interested in this algorithmic analysis may skip these chapters and move to Chapter 6 without loss of continuity.

Since the TLS solution of $AX \approx B$ is deduced from *only one* right singular vector or in general, *a basis of the right singular subspace* associated with the smallest singular values of the data $[A; B]$, a considerable saving in computation time is possible by calculating only those desired basis vectors. This can be done in a *direct way* by modifying the SVD algorithm, as described in this chapter, or in an *iterative fashion* by making use of start vectors, as discussed in the next chapter.

Because only a basis of the desired singular subspace is needed, the classical Singular Value Decomposition (SVD) algorithm can be modified in three ways. First, the Householder transformations of the bidiagonalization need only be applied to the basis vectors of the desired singular subspace. Second, the bidiagonal must be only partially diagonalized, and third, the convergence rate of the iterative diagonalization can be improved by an appropriate choice between QR and QL iteration steps. This improved algorithm, called "*partial* SVD (PSVD)," calculates in an efficient and reliable way a basis for the left and/or right singular subspace of a matrix associated with its smallest singular values. The dimension of the desired subspace may be fixed or depend on a given parameter θ [193]. It can be successfully used for the TLS computations and also for solving (non)homogeneous linear equations and for calculating the numerical null space of a perturbed matrix.

This chapter is organized as follows. In §4.2, the classical SVD algorithm is briefly described. Section 4.3 shows the modifications in the classical SVD algorithm that improve the computational efficiency in the cases mentioned above. In §4.4 both algorithms are compared. An analysis of the operation counts, as well as computational results, shows the relative efficiency of PSVD with respect to the classical SVD algorithm. By means of this algorithm PSVD, the TLS computations are sped up and summarized into an improved "*partial* TLS" algorithm PTLS, which is outlined and demonstrated in §4.5.

4.2. The classical SVD algorithm

Let C be an $m \times n$ matrix with $m \geq n$ and Singular Value Decomposition (SVD) given by

$$C_{m \times n} = U\Sigma V^T = [U_1; U_2] \begin{bmatrix} \Sigma_1 \\ 0 \end{bmatrix} V^T \tag{4.1}$$

with $U=[U_1;U_2], U_1=[u_1,\cdots,u_n]$ the left singular vectors,

$$U_2=[u_{n+1},\cdots,u_m], u_i \in \mathcal{R}^m, U^TU=I_m,$$

$V=[v_1,\cdots,v_n], v_i \in \mathcal{R}^n, V^TV = I_n$ the right singular vectors,
$\Sigma_{m \times n} = [\Sigma_1; 0]^T$, $\Sigma_1 = \mathrm{diag}(\sigma_1,\cdots,\sigma_n) \in \mathcal{R}^{n \times n}$, $\sigma_1 \geq \cdots \geq \sigma_n \geq 0$ the singular values.

If $\sigma_n \neq 0$, the columns of U_1 (respectively, V) constitute an orthonormal basis for the range (respectively, row space) of C. The columns of U_2 span the orthogonal complement of the range of C. The *rank* of C is immediately deduced from the number of singular values strictly greater than zero, i.e., rank $(C) = r$ if $\sigma_r > \sigma_{r+1} = \cdots = \sigma_n = 0$. There is no loss of generality in assuming that $m \geq n$. Indeed, if $m < n$ then we can instead compute the SVD of C^T.

The *classical* SVD algorithm consists of three phases. In the bidiagonalization phase, matrix C is made *bidiagonal*. We construct two finite sequences of Householder transformations [69, §5.1]:

$$\begin{aligned} P^{(k)}, k = 1,2,\cdots,p \quad &\text{with} \quad p = \min\{m-1,n\} \\ \text{and } Q^{(k)}, k = 1,2,\cdots,q \quad &\text{with} \quad q = \min\{n-2,m\} \end{aligned}$$

such that

$$J^{(0)} = P^{(p)} \cdots P^{(1)} C Q^{(1)} \cdots Q^{(q)} = \begin{bmatrix} q_1 & e_2 & 0 & \cdots & 0 \\ 0 & q_2 & e_3 & \cdots & \vdots \\ \vdots & 0 & \ddots & \ddots & 0 \\ 0 & \cdots & \cdots & \ddots & e_n \\ 0 & \cdots & \cdots & 0 & q_n \\ \hline & & 0_{(m-n) \times n} & & \end{bmatrix} \tag{4.2}$$

is an upper bidiagonal matrix. In particular, $P^{(i)}$ zeros out the subdiagonal elements in column i and $Q^{(j)}$ zeros out the appropriate elements in row j. Because all the transformations introduced are orthogonal, the singular values of $J^{(0)}$ are the same as those of C. Thus, if

$$J^{(0)} = G\Sigma H^T \tag{4.3}$$

is the SVD of $J^{(0)}$ then

$$C = PG\Sigma H^T Q^T \tag{4.4}$$

so that

$$\begin{aligned} U &= PG,\ V = QH \quad &(4.5)\\ \text{with } P &= P^{(1)} \cdots P^{(p)} \quad &(4.6)\\ \text{and } Q &= Q^{(1)} \cdots Q^{(q)} \quad &(4.7)\end{aligned}$$

In the process, the matrix C is destroyed when the $P^{(i)}$ (respectively, $Q^{(i)}$) are stored in factored form below (respectively, above) the main diagonal of C.

After the bidiagonalization, the orthogonal matrices P in (4.6) and/or Q in (4.7) must be computed explicitly during the *initialization* phase if U and/or V in (4.1) are required. This is done by accumulating the Householder transformations $P^{(i)}$ in (4.6) (respectively, $Q^{(i)}$ in (4.7)) in the space provided for U (respectively, V).

After this initialization phase, the bidiagonal matrix $J^{(0)}$ can be *diagonalized iteratively* by the QR method [69, §8.3] during the diagonalization phase. This diagonalization implies the construction of a sequence

$$J^{(0)} \to J^{(1)} \to \cdots \to \Sigma \text{ where } J^{(i+1)} = (S^{(i)})^T J^{(i)} T^{(i)}. \tag{4.8}$$

$S^{(i)}$ and $T^{(i)}$ are products of Givens transformations [69, §5.1] and are hence orthogonal. The matrices $T^{(i)}$ are chosen so that the sequence $J^{(i)T} J^{(i)}$ converges to a diagonal matrix while the matrices $S^{(i)}$ are chosen so that all $J^{(i)}$ have a bidiagonal form. The products of the $T^{(i)}$ and the $S^{(i)}$ are exactly the matrices H and G^T, respectively, in (4.5). It has been reported in [67] that the average number of iterations on $J^{(i)}$ in (4.8) is approximately $2n$, i.e., $J^{(2n)}$ is usually a good approximation to a diagonal matrix. To accumulate the transformations and obtain the final matrices U and/or V, the $S^{(i)}$ are applied to P from the right and $(T^{(i)})^T$ are applied to Q^T from the left, for each i.

If m is larger than $\frac{5}{3}n$, the computations are made more efficient by first transforming the matrix C into *upper triangular* form by Householder transformations on the left [26]:

(4.9)
$$L^{(n)} \cdots L^{(1)} C = L^T C = \begin{bmatrix} R \\ 0 \end{bmatrix}, \quad R_{n \times n} \text{ upper triangular and } L \text{ orthogonal.}$$

We then proceed to bidiagonalize R. The important difference here is that the matrix R is much smaller than C (if $n^2 << m \cdot n$). Hence, the work required to bidiagonalize R is much less than that originally done by the right transformations when $m >> n$. It can be easily verified that if the SVD of R is equal to $X\Sigma_1 Y^T$, then the SVD (4.1) of C is given by

$$C = U_1 \Sigma_1 V^T \text{ with } U_1 = L \begin{bmatrix} X \\ 0 \end{bmatrix} \text{ and } V = Y. \tag{4.10}$$

Observe that we require an extra product $L[X^T; 0]^T$. This can be obtained from

$$L \begin{bmatrix} X \\ 0 \end{bmatrix} = L \begin{bmatrix} I \\ 0 \end{bmatrix} X \tag{4.11}$$

by first accumulating $L[I; 0]^T$ in the space provided for U, and then multiplying by X. If U is not needed explicitly, e.g., in TLS problems, then we do not need to accumulate any left transformations. In that case, for $m \geq n$ it makes a substantial saving. This algorithm, which triangularizes C whenever $m \geq \frac{5}{3}n$, is called the *hybrid* SVD algorithm in [26].

4.3. The partial SVD algorithm PSVD

The original motivation is to improve the speed of the classical SVD algorithm when we require only a basis of the singular subspace of a matrix associated with its smallest singular values. In that case, three improvements are possible.

4.3.1. Delay initialization phase. During the second phase of the classical SVD algorithm, the matrices U and V are initialized by forming the products of applied Householder transformations (4.6) and (4.7) explicitly. Now, since only a few vectors in U and/or V are requested, we can save a great amount of work by *delaying the initialization* of U and V until the end of the diagonalization phase. If we then apply the Householder transformations $P^{(i)}$ (respectively, $Q^{(i)}$) only to those vectors in G (respectively, H) in (4.5) that are associated with the desired smallest singular values of the bidiagonal $J^{(0)}$, we obtain the desired basis vectors in U (respectively, V) with a reduced computational effort.

4.3.2. Partial diagonalization. During the diagonalization phase, the singular values and associated vectors are computed one after the other. If convergence has occurred to all desired singular values, then the associated singular vectors remain unchanged during the following QR iterations when the QR method is converging to the remaining singular values. Hence, we can stop the QR iterations immediately after all desired singular values have been computed. In the case of one singular vector, this means that the QR method only needs to converge to the associated singular value. No further

QR iterations are needed for the computation of the required singular vector. Thus, the iteration process may be *stopped* as soon as convergence has occurred to *all* desired singular values. Moreover, if only a basis for a singular subspace is required, we need not zero out all associated superdiagonal elements of the bidiagonal matrix $J^{(i)}$. We only have to partition the bidiagonal matrix into unreduced subbidiagonal matrices such that all singular values of each subbidiagonal matrix are either all greater than a given bound θ or all smaller than or equal to θ. The required basis then consists of all vectors associated with those subbidiagonal matrices whose singular values are all smaller than or equal to the bound θ.

To *partition a bidiagonal* matrix of the form (4.2) into *unreduced* subbidiagonal matrices, we must search the bidiagonal band for zeros. If $e_i = 0$ for some i, then the bidiagonal is immediately decoupled. If a diagonal element $q_i = 0$ for some i, then premultiplication by a sequence of Givens rotations $S^{(i)}$ can zero e_{i+1} and postmultiplication by a sequence of Givens rotations $T^{(i)}$ zeros out e_i. This canceling operation decouples the bidiagonal by isolating a 1 by 1 unreduced submatrix with zero singular value. Hence, its corresponding basis vector belongs to the desired singular subspace. Thus, we can partition the bidiagonal matrix whenever an $e_i, i = 2, \cdots, n$ or a $q_j, j = 1, \cdots, n$ is zero. See also [45, Chap. 11].

By choosing an appropriate shift for the QR iterations, the partial diagonalization proceeds as follows. We construct a sequence of bidiagonal matrices $J^{(i)}$, similar to (4.8) in the classical SVD algorithm. After each QR iteration step with appropriate shift (ith step), we obtain

$$
(4.12) \qquad J^{(i)} = \left[\begin{array}{c|c|c|c|c}
J_1^{(i)} & 0 & & & 0 \\
\hline
 & J_2^{(i)} & 0 & & \\
\hline
 & & \ddots & 0 & \\
\hline
0 & & & & J_l^{(i)}
\end{array}\right]
$$

containing l unreduced subbidiagonal matrices

$$J_j^{(i)} = \begin{bmatrix} x & x & \cdots & \cdots & 0 \\ 0 & x & x & & \vdots \\ \vdots & & \ddots & \ddots & 0 \\ \vdots & & & \ddots & x \\ 0 & \cdots & \cdots & 0 & x \end{bmatrix}, \quad j = 1, \cdots, l.$$

The subbidiagonal matrices $J_j^{(i)}$ are partitioned into three classes:

$C_1 = \{J_j | \text{all singular values of } J_j \text{ are } > \theta\}$.

$C_2 = \{J_j | \text{all singular values of } J_j \text{ are } \leq \theta\}$.

$C_3 = \{J_j | J_j \text{ has at least one singular value } \leq \theta \text{ and at least one singular value } > \theta\}$.

If C_3 is empty, then stop; otherwise, apply one QR iteration step to each element of C_3 and classify again.

To obtain a basis for the singular subspace of a matrix associated with its smallest singular values, *the shift* is set equal to the smallest (in absolute value) diagonal element of the considered subbidiagonal. This shift must be smaller than the given bound θ, otherwise it is set equal to zero. With this choice the iteration process is forced to converge first to the smallest singular values of the matrix and its associated vectors. This shift was experimentally found to give the best convergence results.

The classification of the l unreduced subbidiagonal matrices J_j (omit superscript (i) for simplicity) is based on applying Sylvester's Law of Inertia [69, Thm. 8.1.12], or, equivalently, Sturm sequences [119, p. 52].

SYLVESTER'S LAW OF INERTIA. *Let T be symmetric and W be nonsingular. Then T and WTW^T have the same number of positive, zero, and negative eigenvalues.*

In particular, suppose T is symmetric and tridiagonal, with diagonal entries $\alpha_1, \cdots, \alpha_t$ and offdiagonal entries $\beta_1, \cdots, \beta_{t-1}$. Then, via Gaussian elimination without pivoting, we can write $T - \theta I = LDL^T$, where L is unit lower triangular and bidiagonal and D is diagonal with entries d_i given by the recurrence [119, p. 47]:

$$d_1 = \alpha_1 - \theta \quad \text{and} \quad d_i = \alpha_i - \theta - \beta_{i-1}^2 / d_{i-1} \quad \text{for} \quad i > 1. \tag{4.13}$$

Observe that the recurrence restarts itself whenever a zero valued β_i occurs, i.e., when T is not unreduced. This recurrence will not break down ($d_i = 0$ for some $i < t$) as long as θ is not one of the $t(t-1)/2$ eigenvalues of leading submatrices of T. Then by Sylvester's Law of Inertia, the number of eigenvalues of T less than θ, equal to θ, and greater than θ is precisely the number of d_i which are negative, zero, and positive, respectively.

To apply this theory to the classification of the subbidiagonal matrices J_j, we must look for symmetric tridiagonal matrices associated with each J_j, of which the eigenvalues are related to the singular values of J_j. Here, the tridiagonal matrices $J_j^T J_j$ of which the eigenvalues are the squared singular values of J_j can be used [193], [63]. A numerically more accurate and equally efficient method is to identify the singular value problem of the bidiagonal matrix J_j with the eigenvalue problem of the following symmetric matrix:

$$J_j' = \begin{bmatrix} 0 & J_j^T \\ J_j & 0 \end{bmatrix}. \tag{4.14}$$

The eigenvalues of J_j' are the singular values of J_j and their negatives [69, p. 427]. Suppose that the bidiagonal matrix J_j is $k \times k$ with diagonal entries $q_1, \cdots, q_k$ and superdiagonal entries $e_2, \cdots, e_k$. Then by permuting the rows and columns of J_j' to appear in the new order $1, k+1, 2, k+2, \cdots, k, 2k$ we see that J_j' is orthogonally similar to the tridiagonal matrix T_j with zeros on its diagonal and offdiagonal entries $q_1, e_2, q_2, e_3, \cdots, e_k, q_k$ [37], [63]. Thus the singular values of J_j are the absolute values of the eigenvalues of the matrix T_j onto which Sylvester's Law of Inertia can be applied. Then the number of singular values of J_j less than or equal to θ is precisely the number of nonpositive d_i minus k, where the d_i are given by the recurrence (4.13), i.e.,

$$d_1 = -\theta \quad \text{and} \quad d_i = -\theta - \beta_{i-1}^2 / d_{i-1} \quad \text{for } i > 1 \tag{4.15}$$

with $\beta_1 = q_1, \beta_2 = e_2, \cdots, \beta_{2k-2} = e_k, \beta_{2k-1} = q_k$ the offdiagonal entries of T_j. If T_j has nonzero offdiagonal entries, it must have simple eigenvalues [119, p. 124] and $d_{2k} = 0$ if and only if θ is an eigenvalue of T_j.

This method is accurate, as proved by Kahan and Demmel in [37], [63]. If the computed number of nonpositive d_i is $k + s$, there must be at least s singular values of J_j less than or equal to $\theta/(1-(3k-1.5)\varepsilon)$ and no more than s singular values less than or equal to $\theta(1-(6k-2)\varepsilon)/(1-(3k-1.5)\varepsilon)$, where ε is the machine precision.

If no bound on the singular values can be given but instead, the *dimension* p of the *desired singular subspace* is known, then we can compute a bound θ with the *bisection method* [69, §8.4.1] such that the matrix has exactly p singular values smaller than or equal to θ. This is done at the end of the bidiagonalization phase with the following algorithm (# denotes the cardinality of the set).

ALGORITHM 4.1 (estimation of the bound θ).

Given :

— An $n \times n$ bidiagonal J, given by (4.2).

— The dimension p of the desired singular subspace of J.

— An initial estimate of θ ($\theta \leq \delta$ if no initial estimate is available).

— A stop criterion δ, defining the minimum subinterval length in the bisection method.

Step 1. Initialization of θ

if $\theta \le \delta$ then if $p = 1$ then $\theta \leftarrow \min_{i=1,\cdots,n} |q_i|$ else $\theta \leftarrow |q_{n-p+1}|$.

Step 2. Check the quality of the initial estimate of θ

$num \leftarrow \#\{\sigma_i(J) | \sigma_i \le \theta\}$
if $num = p$ return.

Step 3. Initialization of starting values for bisection method

if $num < p$ then begin $y \leftarrow \theta; z \leftarrow \max_{i=2,\cdots,n}(|q_i| + |e_i|, |q_{i-1}| + |e_i|)$ end
else begin $z \leftarrow \theta$; $y \leftarrow \delta$ end

Step 4. Bisection method

Do while $num \ne p$
  $\theta \leftarrow (y+z)/2$
  $num \leftarrow \#\{\sigma_i(J) | \sigma_i \le \theta\}$
  if $num < p$ then $y \leftarrow \theta$ else $z \leftarrow \theta$
  if $z - y < \delta$ then begin $\theta \leftarrow z$; $p \leftarrow \#\{\sigma_i(J) | \sigma_i \le \theta\}$; return end
end

END

Some comments are in order:

- In Step 1, using the properties of triangular matrices [96] we know that $\min |q_i|$ is always an upper bound for the minimal singular value of J and approaches it very closely, when the gap is large, i.e., $\sigma_{n-1}(J) >> \sigma_n(J)$. The initial estimate of θ in the case $p > 1$ was experimentally found to be suitable if the gap between $\sigma_{n-p}(J)$ and $\sigma_{n-p+1}(J)$ is large (this is the case for most TLS problems).

- In Steps 2 and 4, the computation of the number of singular values $\le \theta$ of $J^{(0)}$, given by (4.2), is based on applying Sylvester's Law of Inertia and the recurrence (4.13) to the tridiagonal matrix T with zero diagonal entries and offdiagonal entries $q_1, e_2, q_2, e_3, \cdots, e_n, q_n$, associated with $J^{(0)}$.

- In Step 3, the initial upper bound for the bisection method is derived from the *Gerschgorin Circle* Theorem [69, Thm. 7.2.1], applied to $J^{(0)\prime} = \begin{bmatrix} 0 & J^{(0)T} \\ J^{(0)} & 0 \end{bmatrix}$.

- In Step 4, the bisection method produces a sequence of subintervals $[y, z]$ that are repeatedly halved in length but such that

$$\#\{\sigma_i(J) | \sigma_i \le y\} < p < \#\{\sigma_i(J) | \sigma_i \le z\}.$$

We must find a bound θ such that $\sigma_{n-p}(J) > \theta \geq \sigma_{n-p+1}(J)$. Hence, if at a certain moment $z - y < \delta$ then at least two singular values of J lie in the interval $[y, z]$ within a distance $< \delta$ from each other. If δ is defined by the standard deviation of the noise on the data $[A; B]$ (like the bound (3.124), this means that σ_{n-p+1} may be considered as a singular value of multiplicity > 1. Hence, it makes no sense to continue the bisection further, trying to separate the singular values σ_{n-p} and σ_{n-p+1} of J. Instead, we consider the singular values of J in the interval $[y, z]$ as coinciding with σ_{n-p+1} and determine its *multiplicity* by the number of singular values of J in the interval $[y, z]$. Therefore, we should enlarge the dimension of the desired singular subspace so that it contains all basis vectors of the ρ-dimensional singular subspace associated with $\sigma_{n+p-1}(J)$ of multiplicity ρ. Observe, however, that this method only allows us to compute the multiplicity of a singular value rather arbitrarily and does not allow us to check the multiplicity of singular values as efficiently and reliably as the classical TLS Algorithm 3.1.

4.3.3. QL iterations. Usually, the smallest (in absolute value) diagonal element of each considered subbidiagonal matrix $J_j^{(i)}$ in (4.12) is the element in the lower right corner. This is surely the case if the gap between the singular values is large. Hence, QR iterations are appropriate. However, in some cases, e.g., when the gap is not so large, the smallest (in absolute value) diagonal element of $J_j^{(i)}$ may be in the *upper left corner*. In this case the number of iteration steps can be greatly reduced by using QL iterations. QL iterations diagonalize the bidiagonal matrix by zeroing the superdiagonal elements from top to bottom while QR iterations start the zeroing process at the bottom superdiagonal element and go on to the top superdiagonal element [21]. In other words, the QL method is identical to the QR method except that at each step the matrix $J_j^{(i)} -$ shift$\cdot I$ is factored into a product of an orthogonal matrix and a *lower* triangular matrix (with QR, we obtain an upper triangular matrix). For example, consider the 5×5 bidiagonal matrix with singular values 6, 5, 4, 3, and 0.9:

$$\begin{bmatrix} 1.18 & 2.47 & 0 & 0 & 0 \\ 0 & -3.05 & 3.05 & 0 & 0 \\ 0 & 0 & -4.73 & -2.25 & 0 \\ 0 & 0 & 0 & -4.00 & -1.98 \\ 0 & 0 & 0 & 0 & -4.75 \end{bmatrix}.$$

To compute its minimal singular value to a precision 10^{-15}, six QR iteration steps are needed to zero out the bottom superdiagonal element but only three QL iteration steps to zero out the top superdiagonal element. Observe that the smallest (in absolute value) diagonal element is in the upper left corner.

Therefore, if the upper left diagonal element of the considered subbidiagonal $J_j^{(i)}$ is larger in absolute value than its lower right diagonal element, QL iterations are performed; otherwise, QR iterations are used.

4.3.4. Outline of the PSVD algorithm. In essence, a delay of the initialization phase (§4.3.1) improves the efficiency of PSVD optimally if the number of desired vectors $<< n$. The partial diagonalization (§4.3.2) works best if the gap between the singular values associated with the desired and undesired vectors is large. This is usually the case when solving TLS problems and (non)homogeneous linear equations. In the case of a smaller gap, the convergence rate can be improved by using QL iterations (§4.3.3). All these improvements are incorporated in the following "*partial* SVD" *algorithm* PSVD.

ALGORITHM 4.2 (Partial SVD (PSVD)).

Given :

- — An $m \times n$ matrix C, $m \geq n$.
- — A bound θ on the singular values σ_i of C associated with the desired singular subspace or its dimension p with an initial estimate of θ.
- — Logical variables *wantu* and *wantv* indicating the need for a basis of a left or right singular subspace, respectively.
- — A stop criterion δ defining the minimum subinterval length in the bisection method (Step 4 of Algorithm 4.1).

Step 1. Bidiagonalization phase

1(a). If $m \geq \frac{5}{3}n$, transform C into upper triangular form R (see (4.9)) by Householder transformations $L^{(1)}, \cdots, L^{(n)}$.

1(b). Transform C (or R) into bidiagonal form $J^{(0)}$ (see (4.2)), using Householder transformations $P^{(1)}, \cdots, P^{(p)}$ and $Q^{(1)}, \cdots, Q^{(q)}$ with $p = \min\{m-1, n\}$ and $q = \min\{n-2, m\}$.

1(c). If *wantu* then $G \leftarrow I_m$; if *wantv* then $H \leftarrow I_n$.

Step 2. Partial diagonalization phase

If θ is not known, compute θ such that $\#\{\sigma_i(J^{(0)}) | \sigma_i \leq \theta\} = p$ using Algorithm 4.1.

For $i = 0, 1, \cdots$

classify the l unreduced subbidiagonal matrices $J_j^{(i)}, j = 1, \cdots, l$ of the bidiagonal matrix $J^{(i)}$ in (4.12), obtained in the ith iteration step, into three classes:

$C_1 = \{J_j | \sigma_i(J_j) > \theta, \forall i\}$,

$C_2 = \{J_j | \sigma_i(J_j) \leq \theta, \forall i\}$,

$C_3 = \{J_j | \exists i : \sigma_i(J_j) \leq \theta \text{ and } \exists t : \sigma_t(J_j) > \theta\}$;

if C_3 empty then go to Step 3
    else   begin
        for each $J_j^{(i)} \in C_3$ do
            $k \leftarrow$ dimension of $J_j^{(i)}$
            $q_{\min} \leftarrow \min_{s=1,\cdots,k}\{|q_s| \mid q_s = (s,s)\text{-th entry of } J_j^{(i)}\}$
            if $q_{\min} < \theta$ then $\zeta \leftarrow q_{\min}$ else $\zeta \leftarrow 0$
            perform one QR (if $|q_1| \geq |q_k|$) or QL (if $|q_1| < |q_k|$) iteration
            step with shift ζ, using Givens rotations $S^{(i)}$ and $T^{(i)}$:
                $J_j^{(i+1)} \leftarrow (S^{(i)})^T J_j^{(i)} T^{(i)}$
            if (*wantu*) then apply the Givens rotations $S^{(i)}$ to G
            if (*wantv*) then apply the Givens rotations $T^{(i)}$ to H
        end for
        end

end for

Step 3. Back transformation phase

3(a). Apply the Householder transformations $P^{(i)}$ (respectively, $Q^{(i)}$) to each column g_i of G (respectively, h_i of H) associated with the bidiagonal matrices of C_2:
    if (*wantu*) then $u_i \leftarrow P^{(1)} \cdots P^{(p)} g_i$ and $p = \min\{m-1, n\}$
    if (*wantv*) then $v_i \leftarrow Q^{(1)} \cdots Q^{(q)} h_i$ and $q = \min\{n-2, m\}$

3(b). If $m \geq \frac{5}{3}n$ and (*wantu*) then apply the Householder transformations $L^{(i)}$ to each vector u_i computed in Step 3(a).:
    $u_i \leftarrow L^{(1)} \cdots L^{(n)} u_i$.

END

This algorithm has been programmed in Fortran 77, fully documented and tested. A program listing is given in [176] and the code can be obtained through netlib [46].

Observe that only minor modifications in the algorithm are required to compute a basis of the singular subspace of a matrix, corresponding to its largest singular values ($> \theta$). Indeed, we must compute another shift to converge to the largest singular values, and the basis vectors associated with the subbidiagonal matrices of C_1 are now considered.

4.3.5. The case $m < n$. If $m < n$, we can apply the PSVD Algorithm 4.2 on C^T. Sometimes, it may be preferable to work on C. For those cases, the PSVD algorithm in §4.3.4 can also be used, but before the diagonalization phase one *preprocessing* step must be included to bring the bidiagonal matrix into its correct form. In fact, bidiagonalizing an $m \times n$ matrix $C, m < n$, produces an $m \times (m+1)$ bidiagonal matrix of the form.

$$(4.16) \qquad J^{(0)} = \begin{bmatrix} q_1 & e_2 & 0 & \cdots & 0 \\ 0 & q_2 & \ddots & & \vdots \\ \vdots & & \ddots & e_m & 0 \\ 0 & \cdots & 0 & q_m & e_{m+1} \end{bmatrix}.$$

As in [45, Chap. 11], the element e_{m+1} in the lower right corner is canceled by Givens rotations, which must be performed on the matrix V. If $J^{(0)}$ is unreduced, m Givens rotations are needed. To apply the algorithm to all possible m and n, an extra step is needed:

Step 1(d). if $m < n$ then begin

cancel e_{m+1} of $J^{(0)}$ in (4.16) by Givens rotations $F^{(i)}$:

$$J^{(0)} \leftarrow F^{(1)} \cdots F^{(r)} J^{(0)}, \quad r \leq m$$

if (*wantv*) then apply the Givens rotations $F^{(i)}$ to H:

$$H \leftarrow F^{(1)} \cdots F^{(r)} I_n, r \leq m$$

end

This step is included in the code given in [176].

4.3.6. Computational considerations. If $m > n$, the Givens rotations $S^{(i)}$, associated with each QR/QL iteration step in the partial diagonalization phase, need only be applied to the *first* n rows of G. This is immediately clear from the fact that the Givens rotations $S^{(i)}$ only affect the first n columns of G. Because the last $m - n$ components of the first n columns of G are set to zero during the initialization of G, they remain unaffected by the Givens rotations $S^{(i)}$. The same applies, if $m < n$, to the first m columns of H. However, in the preprocessing step, the $(n+1)$th row of H is affected by the Givens rotations $F^{(i)}$ and made nonzero. Hence, if $m < n$, the Givens rotations $T^{(i)}$ must be applied to the first $m+1$ rows of H.

The classical SVD algorithm accumulates the Householder transformations $P^{(i)}$ and $Q^{(i)}$ of the bidiagonalization phase, already before the diagonalization, in the space provided for U and V, respectively. Because all elements in U and V are nonzero, the Givens rotations $S^{(i)}$ and $T^{(i)}$ of the diagonalization phase must be applied to all components in the first n columns of U and V (if $m \geq n$). Thus, if $m \geq n$ and vectors from the left singular subspace U of C are wanted, then each Givens rotation in the diagonalization phase costs O(m) operations when using the classical SVD algorithm, while PSVD only requires O(n) operations (see §4.4.2).

Moreover, if $m > n$ and only a *basis for the orthogonal complement* of the range of a full rank matrix C (i.e., U_2 in (4.1)) is required, the computations with PSVD can be significantly reduced. First, the partial diagonalization phase can be entirely omitted. This is clear from the fact that the Givens rotations $S^{(i)}$ of each QR/QL iteration step only operate on the first n columns of U. The last $m-n$ columns, i.e., U_2, are unaffected. Second, if we first reduce C to upper triangular form R, we can omit the bidiagonalization phase of R entirely, if only the basis U_2 in (4.1) is required. This is clear from (4.10), which shows that the last $m-n$ columns of U, i.e., U_2, are unaffected by the bidiagonalization of R. Thus, if $m > n$, C of full rank and only the basis U_2 in (4.1) is desired, the PSVD algorithm reduces to the execution of the following steps.

ALGORITHM 4.3 (Orthogonal complement $R(U_2)$ of the range of a full rank matrix C).

Step 1. Transform $C_{m\times n}$ into upper triangular form R (see (4.9)), by Householder transformations $L^{(1)}\cdots L^{(n)}$ from the left.

Step 2. Apply the Householder transformations $L^{(i)}$ to the last $m-n$ columns of I_m:

$$U_2 = [u_{n+1}, \cdots, u_m] \leftarrow L^{(1)}\cdots L^{(n)} \begin{bmatrix} 0_{n\times(m-n)} \\ I_{m-n} \end{bmatrix}$$

END

The same applies to the case $m < n$ when only the basis vectors $v_{m+1}, \cdots, v_n$ of the null space of a full rank matrix C are required. In that case, C must be triangularized by Householder transformations from the right.

4.3.7. Solution of a set of nonhomogeneous equations. Using Algorithm 4.3, the computation time for solving nonhomogeneous linear equations $AX = B$ with SVD reduces significantly. Indeed, we can bring $AX = B$ with A $n\times n$, X and B $n\times d$ matrices, into the following form:

$$[X^T; -I][A; B]^T = 0 \tag{4.17}$$

with $[A; B]^T$ an $(n+d)\times n$ matrix of rank n. Taking the SVD (4.1) of $[A; B]^T$, $n+d$ left singular vectors u_i and n right singular vectors v_i can be computed. If $[A; B]$ has rank n, then its null space $N([A; B])$ contains exactly d basis vectors $u_{n+1}, \cdots, u_{n+d}$. Because the d solution vectors in (4.17) must be orthogonal with respect to $R([A; B]^T) = R(U_1)$, they must be within $R([u_{n+1}, \cdots, u_{n+d}]) = R(U_2)$, i.e.,

$$\exists \text{ nonsingular } S \in \mathcal{R}^{d\times d} : \begin{bmatrix} X \\ -I \end{bmatrix} = [u_{n+1}, \cdots, u_{n+d}]S = \begin{bmatrix} U_{21}S \\ U_{22}S \end{bmatrix} \begin{matrix} n \\ d \end{matrix} . \tag{4.18}$$

Hence, the solution matrix X is obtained from (4.18) by solving

$$XU_{22} = -U_{21}. \tag{4.19}$$

If the set of equations has only one right-hand side b, i.e., $d = 1$ in (4.17), (4.19) reduces to

$$\begin{bmatrix} x \\ -1 \end{bmatrix} = -u_{n+1}/u_{n+1,n+1} \tag{4.20}$$

with $u_{n+1,n+1}$ the $(n+1)$th component of the $(n+1)$th left singular vector u_{n+1}.

Thus, if a set of nonhomogeneous equations $AX = B$ must be solved and $\text{rank}(A) = n$, the solution is obtained by the following algorithm.

ALGORITHM 4.4 (Solution of $A_{n\times n}X = B_{n\times d}$ with A of full rank).

Step 1. Compute basis $U_2 = \underset{d}{\begin{bmatrix} U_{21} \\ U_{22} \end{bmatrix}} \begin{matrix} n \\ d \end{matrix}$ of $N([A; B])$, using Algorithm 4.3 with

$C = [A; B]^T$.

Step 2. Solve $XU_{22} = -U_{21}$.

END

Even if the rank r of $[A; B] < n$ and the minimum norm solution must be computed, a large reduction in computation time is obtained using the PSVD algorithm as only $n - r$ left singular vectors $u_{r+1}, \cdots, u_n$ of $[A; B]^T$ must be processed in the partial diagonalization phase (see §4.4.2). Indeed, if $[A; B]$ has rank r, then the minimum norm solution X is obtained by computing an orthonormal matrix Q such that

$$[u_{r+1}, \cdots, u_{n+d}]Q = \begin{array}{c} \begin{bmatrix} Y & Z \\ 0 & \Gamma \end{bmatrix} \begin{matrix} n \\ d \end{matrix} \\ \begin{matrix} n-r & \quad d \end{matrix} \end{array} . \tag{4.21}$$

X is obtained by solving

$$X\Gamma = -Z. \tag{4.22}$$

The solution X in (4.22) is the *minimum norm* TLS solution of the rank-deficient set $AX = B$ (see §3.3). If the $n - r$ smallest singular values of $[A; B]$ are exactly zero, X also equals the minimum norm LS solution.

4.4. Comparison between the classical SVD and the partial SVD algorithm

4.4.1. Storage requirements. If $m < \frac{5}{3}n$, the Householder transformations $P^{(i)}$ (respectively, $Q^{(i)}$) are stored below (respectively, above) the main diagonal of C. If $m \geq \frac{5}{3}n$ and U is needed, then the Householder transformations $L^{(i)}$ are stored in the lower part of C. The Householder transformations $Q^{(i)}$ applied to the rows of the triangular form R of C can be stored in the upper triangular part of C but the Householder transformations $P^{(i)}$ applied to the columns of R must be stored in $n(n+1)/2$ extra storage locations. The classical SVD algorithm accumulates the products of the Householder transformations $P^{(i)}$ and $Q^{(i)}$ immediately in the locations provided for U and V, respectively, during the initialization phase. Hence, they need not be stored and matrix C can be used to store U or V.

During the partial diagonalization phase of PSVD, the matrices G and H can be stored into the locations provided for U and V, i.e., the left and right singular vectors of C, respectively. No extra storage locations are needed. Thus by delaying the initialization phase, PSVD needs $m \cdot n$ *extra storage* locations if U (or U and V) are needed and n^2 extra storage locations, if only V is needed. However, the computations are made more efficient. In the partial diagonalization phase, no extra storage is required. Moreover, the computation time is greatly reduced, as shown below.

4.4.2. Operation counts. In §4.3, we indicated that PSVD should be more efficient than the classical SVD algorithm when only a few basis vectors of the left and/or right singular subspace of a matrix, associated with its smallest singular values, are needed. In this subsection we study the *relative efficiency* between the classical SVD and PSVD as a function of m and n. We do this by computing the number of multiplications (and divisions) for each algorithm. For the classical algorithm, the counts are based on the SVD algorithm DSVDC and the triangularization algorithm DQRDC of the LINPACK Library [45, Chaps. 9 and 11] (extra calculations, which deal with under- and overflow, are not taken into account). In all operation counts given in this section, we assume $m \geq n$ and only consider the third- and second-order terms in m and n. So the results are correct for moderate m and n. The average number of QR iteration steps for each singular value during the diagonalization in the classical SVD algorithm is denoted by s. The average number of QR/QL iteration steps needed for convergence to one basis vector in PSVD is denoted by z and p is the dimension of the desired singular subspace in PSVD.

1. **Bidiagonalization** (classical SVD and PSVD)

$$m < \tfrac{5}{3}n : 2mn^2 - \tfrac{2}{3}n^3 + 2n^2 \quad \text{mult.}$$

$$m \geq \tfrac{5}{3}n : \begin{cases} mn^2 - \frac{n^3}{3} + \frac{1}{2}n^2 + mn & \text{mult. (triangularization of } C) \\ \frac{4}{3}n^3 + 2n^2 & \text{mult. (bidiagonalization of } R). \end{cases}$$

2. **Initialization** (in classical SVD only)

$$V : \tfrac{2}{3}n^3 - \tfrac{n^2}{2} \quad \text{mult.}$$

$$\text{if } m < \tfrac{5}{3}n \quad \begin{cases} U_1 : mn^2 - \frac{1}{3}n^3 + \frac{3}{2}n^2 - nm & \text{mult.} \\ U \ : \frac{2}{3}m^3 - \frac{2}{3}(m-n)^3 + mn - \frac{1}{2}n^2 & \text{mult.} \end{cases}$$

$$\text{if } m \geq \tfrac{5}{3}n \quad \begin{cases} U_1 : 2mn^2 + \frac{1}{3}n^3 + 2n^2 - nm & \text{mult.} \\ U \ : \frac{4}{3}n^2 + 2m^2n - mn^2 + mn & \text{mult.} \end{cases}$$

3. **QR/QL iterations** (classical SVD and PSVD)
   classical SVD (U only) : $2sn^2m + 10sn^2 + (4-2s)nm$ mult.
   classical SVD (V only) : $2sn^3 + 8sn^2$ mult.
   classical SVD (U and V required) : $2sn^2m + 2sn^3 + 8sn^2 + (4-2s)nm$ mult.
   PSVD (p vectors u_i only) : $4zpn^2 + 2zpn(9-p)$ mult.
   PSVD (p vectors v_i only) : $4zpn^2 + 2zpn(9-p)$ mult.
   PSVD (p vectors u_i and v_i required) : $8zpn^2 + 4zpn(4-p)$ mult.

4. **Back transformation** (in PSVD only)
   for each vector $v_i \in V$: n^2 mult.
   if $m < \frac{5}{3}n$, for each vector $u_i \in U$: $2mn - n^2$ mult.
   if $m \geq \frac{5}{3}n$, for each vector $u_i \in U_1$: $2mn$ mult.
   for each vector $u_i \in U_2$: $2mn - n^2$ mult.

Some comments are in order:

- Usually, the average number s of QR iteration steps for convergence to each singular value (and corresponding vectors) during the diagonalization phase in the classical SVD is assumed to be 2 [67] and is experimentally found to be quite accurate.

- The average number z of QR/QL iteration steps needed for *convergence* to one basis vector in PSVD is usually slightly larger than s, depending on the desired *numerical accuracy* ε, the *dimension* p of the desired singular subspace and the *gap*, i.e., the distance between $\{\sigma_i(C)|\sigma_i(C) > \theta\}$ and $\{\sigma_i(C)|\sigma_i(C) \leq \theta\}$. First, if the elements of the matrix C are affected by noise, we need not reduce the superdiagonal elements of the bidiagonal matrix $J^{(i)}$ in the partial diagonalization phase of PSVD to machine precision. Instead, we could *stop* the iteration process as soon as the superdiagonal elements are *smaller* than a given error parameter ε, depending on the allowed approximation level. A value of ε that is substantially larger than the machine precision reduces the number of required QR/QL iterations and thus z. Second, it has been observed experimentally that the *computation time* does *not* increase linearly with the dimension p of the desired singular subspace. On the contrary, the computation time is rather independent of p. This follows from the fact that for large p, the bidiagonal matrix $J^{(i)}$ will preferably break down in subbidiagonal matrices belonging to the classes C_1 or C_2 (see §4.3.4). Hence, only a few QR/QL iterations are needed to process the remaining subbidiagonal matrices of C_3. Thus, for large p the average number z of QR/QL iterations will be smaller than for small p. Third, z depends on the gap. If the gap is large, i.e., larger than one order of magnitude, z is only slightly larger than s. Experimentally z is found to be approximately 3 in those cases for $p = 1$. But if the gap is small, z will be larger.

TABLE 4.1.
Total number of operations of the classical SVD *and* PSVD *for each of Cases* A–D.

Case	Classical SVD	PSVD
A. $m < \frac{5}{3}n$	$2mn^2 + 2sn^3 + (\frac{3}{2} + 8s)n^2$	$2mn^2 - \frac{2}{3}n^3 + (4zp + p + 2)n^2$
$m \geq \frac{5}{3}n$	$mn^2 + (\frac{5}{3} + 2s)n^3 + (2 + 8s)n^2 + mn$	$mn^2 + n^3 + (4zp + p + \frac{5}{2})n^2 + mn$
B. $m < \frac{5}{3}n$	$(3 + 2s)mn^2 - n^3 + (\frac{7}{2} + 10s)n^2 + (3 - 2s)nm$	$2mn^2 - \frac{2}{3}n^3 + (4zp - p + 2)n^2 + 2pmn$
$m \geq \frac{5}{3}n$	$3mn^2 + (\frac{4}{3} + 2s)n^3 + (\frac{9}{2} + 8s)n^2$	$mn^2 + n^3 + (4zp + \frac{5}{2})n^2 + (2p + 1)mn$
C. $m < \frac{5}{3}n$	$(3 + 2s)mn^2 + (2s - \frac{1}{3})n^3 + (3 + 8s)n^2 + (3 - 2s)nm$	$2mn^2 - \frac{2}{3}n^3 + (8zp + 2)n^2 + 2pmn$
$m \geq \frac{5}{3}n$	$3mn^2 + (2 + 4s)n^3 + (4 + 8s)n^2$	$mn^2 + n^3 + (8zp + p + \frac{5}{2})n^2 + (2p + 1)mn$
D. $m < \frac{5}{3}n$	$2m^2n + 2sn^2m + (\frac{3}{2} + 10s)n^2 + (5 - 2s)mn$	$2m^2n - mn^2 + \frac{1}{3}n^3 + (4zp - p - 1)n^2 - m^2 + (2p + 4)mn$
$m \geq \frac{5}{3}n$	$2m^2n + (\frac{7}{3} + 2s)n^3 + (\frac{5}{2} + 8s)n^2 + 2mn$	$2m^2n - 2mn^2 + 2n^3 + (4zp - \frac{1}{2})n^2 - m^2 + (2p + 5)mn$

However, due to the efficient choice between QR and QL iterations (see §4.3.3), z can be kept small.

- From these operation counts, we can compute in which cases the *delay* of the initialization phase in the classical SVD until the end, i.e., the back transformation phase in PSVD, is computationally *more efficient.* Therefore, the dimension p of the desired singular subspace must be smaller than the ratio of the operation counts needed for the initialization phase to the operation counts of the back transformation of one vector in PSVD, e.g.:
  If only vectors v_i are required, then $p \leq \frac{2}{3}n$ approximately.
  If $m < \frac{5}{3}n$ and only vectors $u_i \in U_1$ are required, then $p \leq \frac{n}{2}(\frac{m-n/3}{m-n/2})$ approximately.
  If $m < \frac{5}{3}n$ and vectors $u_i \in U_1$ and v_i are required, then $p \leq \frac{n}{2}(1 + \frac{n}{3m})$ approximately.

We now compare the efficiency in four different cases.

Case A. Only p vectors v_i are required explicitly.

Case B. Only p vectors $u_i \in U_1$ are required explicitly.

Case C. p vectors $u_i \in U_1$ and p vectors v_i are required explicitly.

Case D. p vectors $u_i \in U_1$ and the basis U_2 are required explicitly.

Case A arises in total least squares applications and in the solution of homogeneous linear equations.

Case B is Case A for the matrix C^T.

Case C arises in the computation of singular subspaces and Case D (and also Case A) arises in the computation of the null space of a perturbed matrix whose rank depends on a given perturbation bound θ. The total number of multiplications (and divisions) for each case is given in Table 4.1.

TABLE 4.2.
Ratio of operation counts of the classical SVD to that of PSVD for each of Cases A–D.

Case	Ratio ($\rho = \frac{m}{n}$)	$s = 2$
A. $m < \frac{5}{3}n$	$(\rho + s)/(\rho - \frac{1}{3} + 2z\frac{p}{n})$	$(\rho + 2)/(\rho - \frac{1}{3} + 2z\frac{p}{n})$
$m \geq \frac{5}{3}n$	$(\rho + 2s + \frac{5}{3})/(\rho + 1 + 4z\frac{p}{n})$	$(\rho + \frac{17}{3})/(\rho + 1 + 4z\frac{p}{n})$
B. $m < \frac{5}{3}n$	$(2s\rho + 3\rho - 1)/(2\rho - \frac{2}{3} + 4z\frac{p}{n})$	$(7\rho - 1)/(2\rho - \frac{2}{3} + 4z\frac{p}{n})$
$m \geq \frac{5}{3}n$	$(3\rho + \frac{4}{3} + 2s)/(\rho + 1 + 4z\frac{p}{n})$	$(3\rho + \frac{16}{3})/(\rho + 1 + 4z\frac{p}{n})$
C. $m < \frac{5}{3}n$	$(2s\rho + 3\rho + 2s - \frac{1}{3})/(2\rho - \frac{2}{3} + 8z\frac{p}{n})$	$(7\rho + \frac{11}{3})/(2\rho - \frac{2}{3} + 8z\frac{p}{n})$
$m \geq \frac{5}{3}n$	$(3\rho + 2 + 4s)/(\rho + 1 + 8z\frac{p}{n})$	$(3\rho + 10)/(\rho + 1 + 8z\frac{p}{n})$
D. $m < \frac{5}{3}n$	$(2\rho^2 + 2s\rho)/(2\rho^2 - \rho + \frac{1}{3} + 4z\frac{p}{n})$	$(2\rho^2 + 4\rho)/(2\rho^2 - \rho + \frac{1}{3} + 4z\frac{p}{n})$
$m \geq \frac{5}{3}n$	$(\rho^2 + s + \frac{7}{6})/(\rho^2 - \rho + 1 + 2z\frac{p}{n})$	$(\rho^2 + \frac{19}{6})/(\rho^2 - \rho + 1 + 2z\frac{p}{n})$

Using Table 4.1, we can compute approximations of the ratios of the operation counts of the classical SVD to that of PSVD in each of the four cases. The results are given in Table 4.2 where the ratio is expressed as a function of $\rho = m/n$. These ratios show the relative efficiency of PSVD in each of the four cases, mentioned above.

Observe that the relative *efficiency* of PSVD increases when the dimension p of the desired singular subspace is relatively *smaller* with respect to the dimension n, i.e., the ratio p/n decreases. If the average number z of QR/QL iteration steps for convergence to one basis vector in PSVD is comparable with the average number s of QR iteration steps for convergence to one singular vector in the classical SVD, PSVD is about 2.5 times faster than classical SVD in Cases A and D, three times faster in Case B and 3.5 times faster in Case C. The relative efficiency of PSVD with respect to classical SVD decreases with growing ratio m/n in Cases A and D, and increases in Case B. In Case C, the relative efficiency of PSVD is quasi-independent of the ratio m/n.

With respect to solving *nonhomogeneous equations* we can also compare the operation counts of Algorithm 4.4 with those of other currently used methods. For solving $AX = B$ with X and B n by d matrices and A of full rank, Algorithm 4.4 requires

$$\frac{2}{3}n^3 + \left(\frac{3}{2} + 2d\right) n^2 \text{ multiplications.}$$

These counts can be compared with (only third-order terms are taken):

Gaussian elimination method [69, §3.2] $\sim \frac{1}{3}n^3$ mult.

Householder triangularization [69, §5.2.1] $\sim \frac{2}{3}n^3$ mult.

SVD method (SVD of A) [69, §5.5], [145] $\sim 6n^3$ mult.

Classical TLS algorithm (see §3.6.1) $\sim 6n^3$ mult.

Moreover, if the rank of $[A; B] < n$, e.g., $n - 1$, the singular vector u_n corresponding to the smallest singular value of $[A; B]^T$ must be computed to obtain the minimum norm solution (see §4.3.7). This is easily done using the PSVD algorithm. The operation counts are $\sim \frac{4}{3}n^3$. This result is much better than the efficiency of the ordinary least squares method based on the SVD of A ($\sim 6n^3$) while the same numerical accuracy is maintained.

4.4.3. Computational results. The analysis based on the operation counts in §4.4.2 agrees well with our computational experiments for moderate n. All tests are run on the IBM 3030 of the Central Computer Center at the Katholieke Universiteit Leuven. Double precision is used throughout the calculation. The SVD routine DSVDC of the LINPACK Library [45, Chap. 11] is used as classical SVD routine. The matrices are randomly generated with specified singular value spectrum. In our examples given below, ε denotes the numerical accuracy. θ denotes the bound separating the singular values associated with the desired and undesired singular subspace. The CPU times are expressed in milliseconds.

Example 4.1. C is a 22×20 matrix with singular values 10, 9, 8, 7, 6, 5, 4, 3, 2, 1, .9, .8, .7, .6, .5, .4, .3, .2, .1, and .0001; $\varepsilon = 10^{-13}$ and $\theta = .01$. Hence, only the singular vector corresponding to $\sigma_{20} = .0001$ is desired. Two QR iteration steps are required for convergence. The CPU times obtained when Cases A–D (defined in §4.4.2) are considered are shown in Table 4.3.

Example 4.2. C is a 20×8 matrix with singular values 1000, 500, 100, 50, 10, 5, 1, and .01; $\varepsilon = 10^{-13}$ and $\theta = .5$. Hence, only one singular vector must be computed. Three QR iterations are required for convergence. If Cases A–D in §4.4.2 are considered, the CPU times obtained are those shown in Table 4.4.

Example 4.3. C is a 15×12 matrix with singular values 1000, 800, 400, 100, 80, 40, 20, 10, 8, 1, .1, .01. $\varepsilon = 10^{-13}$ and $\theta = 5$. Hence, three basis vectors are required. The total number of QR iteration steps is five. Because the diagonal element q_1 of the bidiagonal $J^{(0)}$ in (4.2) is larger than q_{12}, it is expected that QR iterations are most efficient. Indeed, the total number of QL iteration steps needed for convergence to three basis vectors is seven. Table 4.5 gives the CPU times obtained when Cases A–D (defined in §4.4.2)

TABLE 4.3.
Comparison of CPU *times* (*in* msec.) *for solving Example* 4.1.

	A		B		C		D	
	PSVD	DSVDC	PSVD	DSVDC	PSVD	DSVDC	PSVD	DSVDC
bidiagonalization	43	43	43	43	43	43	43	43
initialization U or V		17		19		38		24
diagonalization	7	80	7	84	12	129	7	85
back transformation	2		2		4		6	
Total	52	140	52	146	59	210	56	152

TABLE 4.4.
Comparison of CPU *times* (*in* msec.) *for solving Example* 4.2.

	A		B		C		D	
	PSVD	DSVDC	PSVD	DSVDC	PSVD	DSVDC	PSVD	DSVDC
triangularization	4	4	4	4	4	4	4	4
bidiagonalization of R	4	4	4	4	4	4	4	4
initialization U or V of R		2		2		3		2
diagonalization of R	3	9	3	9	4	12	3	9
back transformation	0	0	2	13	2	13	11	22
Total	11	19	13	32	14	36	22	41

TABLE 4.5.
Comparison of CPU times (*in* msec.) *for solving Example* 4.3.

	A		B		C		D	
	PSVD	DSVDC	PSVD	DSVDC	PSVD	DSVDC	PSVD	DSVDC
bidiagonalization	13	13	13	13	13	13	13	13
initialization U or V		4		6		10		9
diagonalization	7	20	7	20	12	29	8	21
back transformation	3		3		5		5	
Total	23	37	23	39	30	52	26	43

TABLE 4.6.
Comparison of CPU *times* (*in* msec.) *for solving Example* 4.4.

	A		B		C		D	
	PSVD	DSVDC	PSVD	DSVDC	PSVD	DSVDC	PSVD	DSVDC
triangularization	6	6	6	6	6	6	6	6
bidiagonalization of R	9	9	9	9	9	9	9	9
initialization U or V of R		3		3		6		3
diagonalization of R	6	17	6	17	9	23	6	17
back transformation	1	0	3	20	4	21	10	27
Total	22	35	24	55	28	65	31	62

are considered.

Example 4.4. C is an 18×11 matrix with singular values 10, 8, 4, 2, 1, .8, .4, .2, .1, .01, .001. $\varepsilon = 10^{-13}$ and $\theta = .05$. Hence a singular subspace of dimension $p = 2$ is requested. The total number of QR iterations is four. The CPU times shown in Table 4.6 have been obtained for the Cases A–D of §4.4.2.

The following examples illustrate the efficiency of solving a set of nonhomogeneous equations $Ax = b$ using Algorithm 4.4. They are compared with the CPU times needed to solve the set with the ordinary least squares method based on SVD, e.g., the procedure MINFIT in the EISPACK library [145].

Example 4.5. $[A; b]$ is a 10×11 matrix with singular values 10, 9, 8, 7, 6, 5, 4, 3, 2, 1. $\varepsilon = 10^{-13}$. Results are presented in Table 4.7.

Example 4.6. $[A; b]$ is a 15×16 matrix with singular values 10, 9, 8, 7, 6, 5, 4, 3, 2, 1, .9, .8, .7, .6, .5. $\varepsilon = 10^{-13}$. See Table 4.7 for results.

We also checked the accuracy of some of the computed results. The desired vectors of the matrices U and V in (4.1) agree to within a few units of the machine precision in almost all the cases but the signs of the corresponding vectors may be reversed. In fact, the SVD is only unique to within such a sign change.

TADLE 4.7.

Comparison of CPU *times* (*in* msec.) *for solving Examples* 4.5. *and* 4.6.

	Example 4.5		Example 4.6	
	Alg. 4.4	MINFIT	Alg. 4.4	MINFIT
triangularization of $[A;b]$	4		10	
bidiagonalization of A		7		20
initialization V of A		3		2
diagonalization of A		16		36
back transformation	1		2	
solution $x = V\Sigma^{-1}U^Tb$		1		2
Total	5	27	12	66

4.5. The partial TLS algorithm PTLS

4.5.1. Outline of the PTLS algorithm. The computational efficiency of the classical TLS Algorithm 3.1, presented in §3.6.1, can be improved using the "partial SVD" instead of the classical SVD algorithm in Step 1 [184]. Indeed, as shown in §3.6.1, the minimum norm TLS solution of $AX \approx B$ can be obtained from any orthogonal basis of the right singular subspace associated with the smallest $n+d-r$ singular values of the $m \times (n+d)$ matrix $[A; B]$. This improvement is incorporated in the following "partial TLS" algorithm PTLS.

ALGORITHM 4.5 (Partial TLS (PTLS) computation of the TLS solution $\hat{X}$ of $AX \approx B$).

Given :

- An $m \times n$ data matrix A and an $m \times d$ observation matrix B.
- A rank determinator θ on the singular values σ_i of $[A; B]$ (such that $\sigma_r([A; B]) > \theta \geq \sigma_{r+1}([A; B])$ or, if known, the numerical rank r of $[A; B]$ with an initial estimate of θ (the dimension of the desired singular subspace is then $n + d - r$).
- A stop criterion δ defining the minimum subinterval length in the bisection method (Step 4 in Algorithm 4.1).

wantu = .false.; *wantv* = .true.

Step 1. Bidiagonalization phase

Reduce $[A; B]$ to bidiagonal form J by performing Step 1 of the PSVD Algorithm 4.2.

Step 2. Partial diagonalization phase

Diagonalize J partially by performing Step 2 of the PSVD Algorithm

4.2.

Step 3. Back transformation phase

Apply the Householder transformations of the bidiagonalization (Step 1) to each basis vector v_i of the desired right singular subspace by performing Step 3 of the PSVD Algorithm 4.2.

Step 4. TLS solution $\widehat{X}$

Compute with Householder transformations the orthogonal matrix Q such that

$$[v_{r+1}, \cdots, v_{n+d}]Q = \begin{bmatrix} Y & Z \\ 0 & \Gamma \end{bmatrix} \begin{matrix} n \\ d \end{matrix} \quad \text{and } \Gamma \ d \times d \text{ upper triangular}$$
$$\qquad\qquad\qquad\qquad\quad n-r \quad\; d$$

(4.23)

with $v_{r+1}, \cdots, v_{n+d}$ the basis vectors of the desired right singular subspace of $[A; B]$.

If Γ singular then begin
 lower the rank r with one
 go back to Step 2
 end
else solve $\widehat{X}\Gamma = -Z$ by forward elimination

END

Some comments are in order:

- This algorithm has been programmed in Fortran, fully documented and tested. A program listing is given in [176] and the code can be obtained through netlib [46].

- If the TLS problem is nongeneric, i.e., Γ is singular, the rank must be lowered in Step 4 and some additional basis vectors must be computed in Step 2. The additional computations can be kept to a minimum. Indeed, we need only further diagonalize the partially reduced bidiagonal matrix by zeroing out some more superdiagonal elements (Step 2) and back transforming these newly computed basis vectors (Step 3). The basis vectors that were already computed remain unaffected.

- If the rank r is lowered in Step 4 but σ_r is of multiplicity > 1, the rank r will be further reduced with the multiplicity of σ_r in the beginning of Step 2 by the θ-estimation Algorithm 4.1. Here, the multiplicity of σ_r is determined by the number of singular values lying within an interval of length δ around σ_r. Observe, however, that this definition is rather arbitrary and does not allow to check the multiplicity of singular values as efficiently and reliably as the classical TLS algorithm, given in §3.6.1.

4.5.2. Comparison between the classical TLS and the partial TLS algorithm. Since the TLS computations are entirely dominated by the SVD, it is evident that a comparison between classical TLS and PTLS reduces mainly to a comparison between classical SVD and PSVD, Case A, discussed in §4.4.1. However, some additional comments are in order.

Concerning the *storage* requirements, PTLS needs *more memory space*, namely, $\frac{1}{2}(n+d)(n+d-1)$ (respectively, $m(n+d-(m+1)/2)$) extra storage locations in case $m \geq n+d$ (respectively, $m < n+d$). Indeed, the Householder transformations $Q^{(i)}$ performed onto the rows of $[A;B]$ during its bidiagonalization (see Step 1) must be stored until the end of the computations. The classical TLS algorithm accumulates the products of the Householder transformations $Q^{(i)}$ immediately into the right singular basis matrix V after the bidiagonalization. Hence, they need not be stored and matrix $[A;B]$ can be used to store V.

With respect to *the operation counts*, the following comments are worth mentioning. Once the desired basis vectors of $[A;B]$ are computed by classical SVD or PSVD, the TLS solution of $AX \approx B$ must be calculated. We first reduce those basis vectors with Householder transformations to the form (4.23) and then solve

$$\widehat{X}\Gamma = -Z \tag{4.24}$$

by forward elimination (Step 3 of Algorithm 3.1, or Step 4 of Algorithm 4.5). If the rank of $[A;B]$ equals n, d basis vectors have been calculated. In this case, the computation of Γ and the TLS solution (Step 4 of Algorithm 4.5) requires

$$\frac{1}{6}[(9n+4d+12)d^2+(15n+26)d]-3n-7 \text{ multiplications (and divisions).} \tag{4.25}$$

Observe that (4.25) reduces to n if $d=1$. This is indeed the number of multiplications required to scale the $(n+1)$th right singular vector of $[A;b]$ and obtain $\widehat{x}$ (see (2.26)).

The number of operations required for solving TLS problems $AX \approx B$ with A of full rank, using the classical TLS algorithm and the partial TLS algorithm are then immediately obtained. We add (4.25) to the operation counts of the classical SVD and PSVD, given in Table 4.1, Case A (and also in Table 5.2), and replace p by d and n by the number of columns $n+d$ in $[A;B]$.

Finally, we give some *computational results*. Observe that the comparison between classical SVD and PSVD largely agrees with the computational results of the classical SVD and PSVD. Indeed, when $d << n$, the additional computations (4.25) to obtain the TLS solution, are negligible with respect to the computations of the desired basis vectors by classical SVD or PSVD.

TABLE 4.8.

Comparison of CPU *times* (*in* msec.) *for solving Examples* 4.7. *and* 4.8.

	Example 4.7		Example 4.8	
	PTLS	classical TLS	PTLS	classical TLS
triangularization of $[A;B]$	21	21		
bidiagonalization	28	28	15	15
initialization V		11		5
diagonalization	8	47	9	32
back transformation	1		2	
solve $\widehat{X}\Gamma = -Z$	0	0	1	1
Total	58	107	27	53

The matrices $[A; B]$ are randomly generated with specified singular value spectrum. All tests are run on the IBM 3030 of the Central Computer Center at the Katholieke Universiteit Leuven. ε denotes the numerical accuracy. Double precision is used throughout the calculations. The CPU times are expressed in milliseconds.

Example 4.7. Consider the overdetermined set of equations $A_{30\times 16}x \approx b$ $(d = 1)$. The singular values of $[A; b]$ are 1000, 800, 400, 200, 100, 80, 40, 20, 10, 8, 4, 2, 1, .8, .4, .2, and .01 and the rank of $[\widehat{A}; \widehat{b}]$ is set to 16. Hence, only one basis vector must be computed. $\varepsilon = 10^{-13}$. Three QR iterations are required for convergence. Table 4.8 shows the improvement in CPU time for the different steps in each algorithm.

Example 4.8. Consider the overdetermined set of equations $A_{15\times 10}X \approx B_{15\times 3}$ $(d = 3)$. The singular values of $[A; B]$ are 100, 50, 28, 26, 24, 22, 20, 18, 15, 12, .3, .2, and .1 and the rank of $[\widehat{A}; \widehat{B}]$ is set to 10. Hence, three basis vectors must be computed. $\varepsilon = 10^{-13}$. Five QR iterations are required for convergence. The CPU times obtained are those shown in Table 4.8.

4.6. Concluding remarks

Since the TLS solution of $AX \approx B$ is deduced from only one right singular vector or in general, a basis of the right singular subspace associated with the smallest singular values of the data matrix $[A; B]$, its computational cost can be considerably reduced by computing its SVD only *partially*.

An improved algorithm PSVD is presented in this chapter, which computes the singular subspace of a matrix, associated with its smallest singular values. There are three reasons for its high efficiency with respect to the classical SVD algorithm. First, the Householder transformations of the bidiagonalization are only applied to the basis vectors of the desired singular subspace. Second, the bidiagonal is only partially diagonalized, and third, an appropriate choice is made between QR and QL iteration steps (§4.3). By analyzing the operation counts, the relative efficiency of PSVD with respect to classical SVD is demonstrated and confirmed with computational results (§4.4). Depending on the gap, the desired numerical accuracy and the dimension of the desired subspace, PSVD can be *three times faster* than the classical SVD algorithm while the *same accuracy* can be maintained. Using the new algorithm, a large reduction in computation time can be obtained in total least squares applications, in the computation of the null space of a matrix and in solving (non)homogeneous linear equations. Based on PSVD, an efficient and reliable algorithm is also derived for solving nonhomogeneous linear equations (§4.3.7). From a practical point of view, we may observe that in general the classical SVD algorithm is not likely to be improved in speed but is used extensively in many applications where the above-mentioned problems occur. Hence, the improved speed of PSVD should prove to be quite valuable for all these applications.

Incorporating the "partial SVD" algorithm into the TLS computations, results in an *improved "partial* TLS" *algorithm* PTLS. Its higher efficiency with respect to the classical TLS algorithm is demonstrated and confirmed with computational results (§4.5). Typically, PTLS reduces the computation time with a factor 2.

An alternative method for computing the basis of a left and/or right singular subspace of a matrix associated with its smallest singular values, appeared recently [28], [29]. As PSVD, this method does not need to compute the complete SVD but instead uses a *Rank Revealing* QR *factorization* (RRQR), i.e., a QR factorization of a matrix C in the particular form:

(4.26)
$$C_{m\times n}\Pi = QR = Q\begin{bmatrix} R_{11} & R_{12} \\ 0 & R_{22} \end{bmatrix} \begin{matrix} r \\ n-r \end{matrix} \qquad \text{with } r \text{ the numerical rank of } C$$

and $\|R_{22}\|_2$ small such that any numerical rank deficiency in C is revealed. Chan [27] has described an algorithm for choosing the permutation Π that guarantees $\|R_{22}\|_2$ to be small. Using this RRQR, upper and lower bounds for the small singular values are easily obtained and used to compute the numerical rank r of C. The RRQR also yields approximations to the corresponding $n-r$ singular vectors, which are then further improved by means of inverse subspace iteration. Finally, these vectors can be used to compute the TLS solution $\widehat{X}$ of $AX \approx B$, $C = [A; B]$, by performing Step 4 of Algorithm 4.5. As long as the matrix C has a well-determined numerical rank r, i.e., the gap between σ_r and σ_{r+1} is large, and the nullity is small compared to the dimensions of the matrix, the computational complexity of the RRQR algorithm is only slightly larger than that of the ordinary QR algorithm. Moreover, Bischof and Hansen [17] have recently demonstrated how the RRQR factorization of a sparse matrix can be computed efficiently without destroying the sparsity pattern of the matrix. Since PSVD carries out a complete reduction to bidiagonal form, it is not well suited for general sparse matrices. Based on this analysis, Chan and Hansen concluded that the RRQR based TLS algorithm described in [28], [29] is actually less demanding than PTLS, mainly because a reduction to bidiagonal form is avoided. It should be noted that their conclusions primarily hold when the nullity $n - r$, i.e., the dimension of the numerical null space of $[A; B]_{m \times n}$, is small, the gap between σ_r and σ_{r+1} is large and the TLS problem is not close-to-nongeneric, i.e., V_γ in (1.23) is well-conditioned.

However, when the gap is not so large, the RRQR-based TLS algorithm is still able to give good results but at the cost of a larger computational effort involved in the large number of inverse subspace iterations. Furthermore, the computation time of this algorithm increases linearly with increasing nullity. Moreover, in contrast to PSVD, this method assumes the rank r of matrix C to be known, or else, if only an upper bound θ on the smallest singular values is available, one additional singular value $> \theta$ and corresponding vector must be computed to assure convergence to all singular values $\leq \theta$. It is clear that these operations make the RRQR based TLS method less efficient. However, a more detailed comparative study is needed to reveal the relative merits of each algorithm.

Finally, we mention that, based on the concept of the RRQR factorization, Stewart [164] recently proposed a promising algorithm for *updating* the null space of a matrix whose rows are samples of a signal. At each time instant, a new sample arrives and is added to the data matrix such that the null space must be recomputed. Hereto, a so-called *Rank Revealing* URV decomposition is updated by means of Givens rotations and is defined as follows. Let $C_{m \times n}$ be of numerical rank r such that its singular values satisfy $\sigma_1 \geq \cdots \geq \sigma_r > \sigma_{r+1} \geq \cdots \geq \sigma_n$ where σ_r is large compared to σ_{r+1}. Then, C can be decomposed into the form:

$$C = U \begin{bmatrix} R & F \\ 0 & G \end{bmatrix} V^T,$$

where U, V are orthogonal, R and G are upper triangular, $\sigma_{\min}(R) \approx \sigma_r$, and $\|F\|_F^2 + \|G\|_F^2 \approx \sigma_{r+1}^2 + \cdots + \sigma_n^2$. It is shown that this decomposition is equally effective as the SVD in exhibiting the null space and can be updated in $\mathcal{O}(n^2)$ time. The updating technique can be run on a linear array of n processors in $\mathcal{O}(n)$ time.

Chapter 5

Iterative Speed Improvement for Solving Slowly Varying Total Least Squares Problems

5.1. Introduction

In the previous chapter, a speed improvement of the TLS computations is obtained in a direct way by computing only that part of the complete SVD that is needed. If, however, a priori *information* is available, the computation time can be further reduced by using an *iterative* algorithm. This happens, for instance, in the estimation of parameters of nonstationary systems that slowly vary with time, space, or frequency. In these problems, slowly varying sets of equations must be solved at each time instant and the TLS solution at step t is usually a good initial guess for the solution at step $t+1$. Therefore, *iterative algorithms* are indicated since these methods include this a priori information in the choice of their start vector or matrix. Two types of *varying* data sequences are possible. The changes in the data matrix at each step can be of rank *one* (or two), e.g., when a new column or row is added or deleted. In these cases, the computation time can be sped up considerably by using efficient rank-one updating algorithms [69, §12.6]. In other situations, e.g., block-processing, the changes are of small norm but still of *full* rank, e.g., when all elements of the data matrix change slowly from step to step.

There are still other reasons for preferring iterations in high speed computing. Each step of an iteration routine gives a new and better estimate of the solution. This allows us to control the level of convergence depending on the perturbations of the data. Furthermore, iterative algorithms are easy to code. Finally, an iterative algorithm enables us to use the given matrix over and over again without modifying it during the computation. This is most important if the matrix itself is highly structured or sparse.

It is the aim of this chapter to compare some iterative algorithms, in particular, inverse iteration and (inverse) Chebyshev iteration, for solving slowly varying TLS problems $AX \approx B$ of small or *moderate* size that require at each time instant the computation of the singular subspace of the data matrix $[A; B]$ associated with its smallest singular values.

This chapter is organized as follows. First, in §5.2, general convergence formulas are derived that allow us to compare the convergence behaviour of the iteration methods. Sections 5.3, 5.4.1, and 5.4.2 describe, respectively, the inverse iteration method, the ordinary Chebyshev iteration, and the inverse Chebyshev iteration method. The algorithms are outlined and their convergence properties are analyzed. Section 5.5 discusses some possible improvements in efficiency and implementation details. Based on the convergence rate and the operation counts per iteration step, the efficiency of these iterative algorithms is compared with that of the direct computation methods in §5.6. This evaluation shows for which class of problems each method is computationally most efficient. Simulated examples of slowly varying total least squares problems, presented in §5.7, confirm the theoretical analysis. In particular, the efficiency of the inverse iteration method, the PTLS and classical TLS algorithm, is illustrated in a practical real-life problem encountered in experimental modal analysis, namely multiple input estimation of frequency response functions. Finally, §5.8 describes other iterative methods, in particular, Rayleigh quotient iteration and Lanczos methods. Different algorithms are outlined and their convergence properties are analyzed, showing in which particular TLS problems these methods are computationally more efficient than the previously discussed methods.

5.2. Matrix functions

It is well known [56] that we can define functions on matrices, called *matrix functions*, certainly if those functions are analytic. For those functions, we can prove the following. Consider a real $n \times n$ symmetric matrix S with eigenvalue decomposition of the form

$$S = V \Lambda V^T. \tag{5.1}$$

V contains the n orthogonal eigenvectors and $\Lambda = \operatorname{diag}(\lambda_1, \cdots, \lambda_n)$. Then the matrix function f_k satisfies

$$f_k(S) = V f_k(\Lambda) V^T \quad \text{and} \quad f_k(\Lambda) = \operatorname{diag}(f_k(\lambda_1), \cdots, f_k(\lambda_n)). \tag{5.2}$$

If S is positive definite, the eigenvalue decomposition equals the SVD. Of course, expression (5.2) also applies to an arbitrary $m \times n$ rectangular matrix C. Indeed, since the matrices $C^T C$ and CC^T are real symmetric, we obtain from the relation between their eigenvalues and the SVD (4.1) of C

$$f_k(C^T C) = V f_k(\Sigma^T \Sigma) V^T \quad \text{and} \quad f_k(CC^T) = U f_k(\Sigma \Sigma^T) U^T,$$

where $\lambda_i(CC^T) = \lambda_i(C^T C) = \sigma_i^2(C)$ for $i = 1, \cdots, \min\{m, n\}$.
Now consider an arbitrary n-dimensional unit vector q_0 that has the following coordinates in the eigenvector basis V, called *internal coordinates*:

$$V^T q_0 = [g_1, \cdots, g_n]^T \quad \text{and} \quad g_i = v_i^T q_0 \quad i = 1, \cdots, n. \tag{5.3}$$

It is not difficult to see that the internal coordinates of the vector $f_k(S)q_0$ in the eigenvector basis V are given by

$$V^T f_k(S)q_0 = [f_k(\lambda_1)g_1, \cdots, f_k(\lambda_n)g_n]^T. \tag{5.4}$$

Now suppose that we want to compute the ith eigenvector $v_i, i \leq n$, with internal coordinates $g_j = v_j^T v_i = 1$ if $i = j$ else 0 for $j = 1, \cdots, n$, to a precision ε. For example, in one-dimensional TLS problems $Ax \approx b$, the $(n+1)$th eigenvector of the matrix $[A;b]^T[A;b]$ must be computed. Then the *sequence of functions* f_k must be chosen in such a way that $f_k(S)q_0$ *converges* to the *desired eigenvector* up to a *precision* ε. This implies that there must exist a finite K such that the internal coordinates of $V^T f_K(S)q_0$ satisfy

$$\frac{\max_{j\neq i} |f_K(\lambda_j)v_j^T q_0|}{|f_K(\lambda_i)v_i^T q_0|} \leq \varepsilon. \tag{5.5}$$

Taking φ as the angle between v_i and q_0, the convergence requirement is more elegantly expressed by

$$\frac{\max_{j\neq i} |f_K(\lambda_j)||\tan\varphi|}{|f_K(\lambda_i)|} \leq \varepsilon. \tag{5.6}$$

Since $\cos\varphi = |v_i^T q_0|$ and $\sin\varphi = \sqrt{1-\cos^2\varphi} \geq \max_{j\neq i} |v_j^T q_0|$, (5.6) is more severe than (5.5).

For example, *inverse iteration* uses the functions $f_k(\lambda) = (\lambda - \lambda_0)^{-k}$ with λ_0 an appropriately chosen shift, to converge to the eigenvector associated with the eigenvalue that approaches λ_0 most closely. By applying the orthogonal Chebyshev polynomials $T_k^{yz}(\lambda)$ defined over an appropriate interval $[y,z]$ containing all undesired eigenvalues of S, the *Chebyshev iteration* method allows the convergence to an eigenvector of S associated with an eigenvalue outside $[y,z]$.

If the ratio $|f_k(\lambda_j)/f_k(\lambda_i)|$ can be written as $|f_1(\lambda_j)/f_1(\lambda_i)|^k$ then, by taking the logarithm (with base 10) of (5.5), we can derive the *number of iteration steps* K required for *convergence* to the desired ith eigenvector v_i, up to a precision ε:

$$K \geq \frac{\log\varepsilon^{-1} + \log\left|\frac{v_j^T q_0}{v_i^T q_0}\right|}{\log\left|\frac{f_1(\lambda_i)}{f_1(\lambda_j)}\right|} \quad \text{with} \quad |f_K(\lambda_j)v_j^T q_0| = \max_{k\neq i} |f_K(\lambda_k)v_k^T q_0| \tag{5.7}$$

or analogously, from (5.6):

$$K \geq \frac{\log\varepsilon^{-1} + \log|\tan\varphi|}{\log\frac{|f_1(\lambda_i)|}{\max_{j\neq i}|f_1(\lambda_j)|}} \qquad 1 \leq i \leq n, 1 \leq j \leq n. \tag{5.8}$$

The derived formulas are interesting since they clearly show the three different *parameters* that influence the number of iteration steps:

1. $\log|f_1(\lambda_i)/f_1(\lambda_j)|$: the larger the *gap* between the value of the function $f_1(\lambda)$ in the desired eigenvalue λ_i with respect to that of the remaining eigenvalues λ_j, the faster the convergence. If $|f_1(\lambda_i)/f_1(\lambda_j)|$ approaches one, convergence will be very slow.

2. $\log\left|v_j^T q_0/v_i^T q_0\right|$ or $\log|\tan\varphi|$: the better the *quality of the initial guess* q_0, the smaller the number of iterations. The internal coordinates (5.3) of a very good initial vector q_0 satisfy $g_i >> g_j$, for all $j \neq i$.

3. $\log\varepsilon^{-1}$: this quantity measures the *number of desired correct digits* in the iteration vector converging to the desired eigenvector v_i. If, for instance, t correct digits are necessary, we set $\varepsilon = 10^{-t}$.

Hence, the numerator of (5.7) measures the number of digits of the initial guess vector q_0 that must be modified to obtain the desired eigenvector v_i, up to a precision ε. Observe that the *numerator* of (5.7) describes the *initial state information*, while the *denominator* determines the *convergence rate*, i.e., the number of correct decimal digits that can be gained in each iteration step.

Expression (5.8) can be extended to *subspace* iteration methods, as proved below.

THEOREM 5.1 (Convergence of subspace iteration methods). *Let* (5.1) *be the eigenvalue decomposition of a given symmetric matrix* $S \in \mathcal{R}^{n\times n}$. *Suppose that we want to compute* $R(V_n)$, $V_n = [v_{n-p+1},\cdots,v_n]$ *for* $p \geq 0$ *and that a sequence of functions* f_k *has been chosen in such a way that the iteration matrices* $Q_k = f_k(S)Q_0$, *with* Q_0 *a given start matrix, converge to a basis of* $R(V_n)$, *i.e., there exists a* K_0 *such that* $\operatorname{dist}(R(Q_K), R(V_n)) \leq \varepsilon$ *for all* $K \geq K_0$.
If $|\lambda_{n-p}| > |\lambda_{n-p+1}|$, $\sin\varphi = \operatorname{dist}(R(Q_0), R(V_n)) < 1$ *and* $f_k = f_1^k$ *for all* k, *then the number of iteration steps* K *at which* Q_K *has converged to a basis of* $R(V_n)$, *up to a precision* ε, *satisfies*

$$(5.9)\quad K \geq \frac{\log\varepsilon - \log|\tan\varphi|}{\log\|f_1(\Lambda_1)\|_2\|f_1^{-1}(\Lambda_n)\|_2} = \frac{\log\varepsilon^{-1} + \log|\tan\varphi|}{\log\left(\|f_1(\Lambda_1)\|_2\|f_1^{-1}(\Lambda_n)\|_2\right)^{-1}},$$

where $\Lambda_1 = \operatorname{diag}(\lambda_1,\cdots,\lambda_{n-p})$ *and* $\Lambda_n = \operatorname{diag}(\lambda_{n-p+1},\cdots,\lambda_n)$.

Proof. Consider the kth iteration matrix $f_k(S)Q_0$, orthonormalized as follows:

$$(5.10)\qquad f_k(S)Q_0 = Q_k R_k \qquad R_k \text{ upper triangular, } Q_k^T Q_k = I_p.$$

Substituting $S = V f_k(\Lambda) V^T$ in (5.10) transforms the problem into the eigenvector basis V:

$$(5.11)\qquad f_k(\Lambda)V^T Q_0 = V^T Q_k R_k.$$

Now partition $f_k(\Lambda) = \begin{bmatrix} f_k(\Lambda_1) & 0 \\ 0 & f_k(\Lambda_n) \end{bmatrix} \begin{matrix} n-p \\ p \end{matrix}$ and $V = \underset{n-p \quad\; p}{[V_1 \;\; ; V_n]}$,
$\qquad\qquad\qquad\qquad n-p \qquad\quad p$

where V_n is the desired p-dimensional eigenvector basis. Then

$$V^T Q_0 = \begin{bmatrix} V_1^T Q_0 \\ V_n^T Q_0 \end{bmatrix} = \begin{bmatrix} G_1^{(0)} \\ G_n^{(0)} \end{bmatrix} \begin{matrix} n-p \\ p \end{matrix} \quad \text{and}$$

$$V^T Q_k = \begin{bmatrix} V_1^T Q_k \\ V_n^T Q_k \end{bmatrix} = \begin{bmatrix} G_1^{(k)} \\ G_n^{(k)} \end{bmatrix} \begin{matrix} n-p \\ p \end{matrix}$$

represent, respectively, the internal coordinates of the orthonormalized start matrix and the orthonormalized kth iteration matrix. Using this partitioning in (5.11) yields

$$\begin{bmatrix} f_k(\Lambda_1) & 0 \\ 0 & f_k(\Lambda_n) \end{bmatrix} \begin{bmatrix} G_1^{(0)} \\ G_n^{(0)} \end{bmatrix} = \begin{bmatrix} G_1^{(k)} \\ G_n^{(k)} \end{bmatrix} R_k. \tag{5.12}$$

Since $\sin\varphi < 1$, it follows that $G_n^{(0)}$ is nonsingular. Indeed, $\sigma_{\min}(G_n^{(0)}) = \sigma_{\min}(V_n^T Q_0) = \sqrt{1 - \text{dist}^2(R(Q_0), R(V_n))} = \cos\varphi > 0$. Then, (5.12) can be solved for $G_1^{(k)}$ as follows:

$$G_1^{(k)} = f_k(\Lambda_1) G_1^{(0)} G_n^{(0)\,-1} f_k^{-1}(\Lambda_n) G_n^{(k)}.$$

By taking norms in this expression, we have

$$\|G_1^{(k)}\|_2 \leq \|f_k(\Lambda_1)\|_2 \|G_1^{(0)}\|_2 \|G_n^{(0)\,-1}\|_2 \|f_k^{-1}(\Lambda_n)\|_2 \|G_n^{(k)}\|_2. \tag{5.13}$$

According to [69, Cor. 2.6.2], $\|G_1^{(k)}\|_2 = \|V_1^T Q_k\|_2 = \text{dist}(R(Q_k), R(V_n))$. Convergence has occurred to the desired eigenvector space $R(V_n)$, up to a precision ε, if $\|G_1^{(k)}\|_2 \leq \varepsilon$. Also according to [69, Cor. 2.6.2], we have that

$$\|G_1^{(0)}\|_2 = \|V_1^T Q_0\|_2 = \text{dist}(R(Q_0), R(V_n)) = \sin\varphi$$

and $\sigma_{\min}(G_n^{(0)}) = \sqrt{1 - \text{dist}^2(R(Q_0), R(V_n))} = \sqrt{1 - \sin^2\varphi} = \cos\varphi$ so that $\|G_n^{(0)\,-1}\|_2 = 1/\cos\varphi$. Substituting these expressions in (5.13) and using $\|G_n^{(k)}\|_2 = \|V_n^T Q_k\|_2 \leq 1$ yields

$$\text{dist}(R(Q_k), R(V_n)) \leq |\tan\varphi| \|f_k(\Lambda_1)\|_2 \|f_k^{-1}(\Lambda_n)\|_2 \leq \varepsilon. \tag{5.14}$$

Substituting $f_k(\Lambda_1) = f_1^k(\Lambda_1)$ and $f_k(\Lambda_n) = f_1^k(\Lambda_n)$ in (5.14) and taking the logarithm (with base 10) yields (5.9) for $k = K$. $\square$

φ is the largest canonical angle between $R(Q_0)$ and $R(V_n)$ [69, pp. 584–585]. The condition $\sin\varphi < 1$ ensures that Q_0 is not deficient in the desired eigendirections. If convergence to one vector v_i is required, i.e., $V_n = v_i$, then expression (5.9) reduces to (5.8). For example, inverse subspace iteration with zero shift applies the functions $f_k(\lambda) = \lambda^{-k}$ to a matrix S. If S is given by (5.1), then $\|f_1(\Lambda_1)\|_2 = 1/\lambda_{n-p}$ and $\|f_1^{-1}(\Lambda_n)\|_2 = \lambda_{n-p+1}$.

5.3. Inverse iteration

5.3.1. Algorithms. As explained before, the TLS solution of (1.17), with r the numerical rank of $[A; B]$, is entirely determined by a *basis* of the right *singular subspace* of $[A; B]$ associated with its $n+d-r$ smallest singular values, i.e., a basis of the subspace spanned by the associated right singular vectors $v_{r+1}, \cdots, v_{n+d}$. The corresponding singular values $\sigma_{r+1}, \cdots, \sigma_{n+d}$ are typically *clustered.* Hence, *subspace* iteration methods are recommended if $d > 1$ [186].

The idea of inverse iteration is already quite old. Many authors studied this iterative method in (non)symmetric eigenvalue problems. See [119] and [69, §7.6.1] for references. According to Wilkinson [208], the inverse iteration method is the most powerful and most accurate method for computing eigenvectors.

By applying the *inverse power functions* $f_k(\lambda) = (\lambda - \lambda_0)^{-k}$ with λ_0 an appropriately chosen shift to the $n \times n$ matrix $S = C^T C$ with the SVD of C given by (4.1), we obtain the inverse of $S - \lambda_0 I$ and its powers:

$$f_k(S) = (S - \lambda_0 I)^{-k} = V(\Sigma^T\Sigma - \lambda_0 I)^{-k} V^T = VDV^T \tag{5.15}$$

with eigenvalues $d_j = f_k(\sigma_j^2) = (\sigma_j^2 - \lambda_0)^{-k}$ for $j = 1, \cdots, n$. Clearly, because the functions $(\lambda - \lambda_0)^{-k}$ are hyperbolic with a vertical asymptote in λ_0, the function values $f_k(\lambda_i)$ of the eigenvalues λ_i close to λ_0 will *dominate* the other. Hence, the iteration matrix Q_k,

$$Q_k = f_k(S)Q_0 = (S - \lambda_0 I)^{-1} Q_{k-1} = (S - \lambda_0 I)^{-k} Q_0, \tag{5.16}$$

with Q_0 a start matrix, becomes increasingly strong in the direction of the eigensubspace associated with the eigenvalues λ_i close to the shift λ_0. This is illustrated geometrically in Fig. 5.1. Observe from (5.15) that the singular values of the matrix functions $f_k(S)$ are changed but the singular vectors are conserved.

Of course, we will not compute the subsequent iteration matrices Q_k by computing the inverse in (5.16) but instead solve a *set of linear equations* of the following form:

$$(S - \lambda_0 I) Q_k = Q_{k-1} \tag{5.17}$$

in each iteration step. The usual way to solve (5.17) is by first factorizing the matrix $S - \lambda_0 I$. This factorization is then used to solve the set of linear

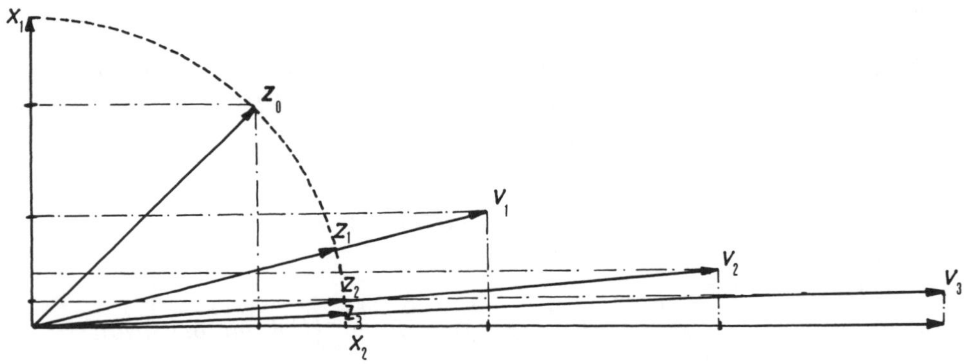

FIG. 5.1. *Geometric illustration of the inverse iteration process, applied to a* 2×2 *matrix with eigenvalues* $\lambda_1 = 2$ *and* $\lambda_2 = 0.5$ *and associated eigenvectors* x_1 *and* x_2. *The shift approaching* λ_2, *is set to zero, and the start vector* z_0 *is given by* $(x_1+x_2)/2$. *The figure shows the three first iteration vectors* v_1, v_2, *and* v_3 *and their corresponding unit vectors* z_1, z_2, *and* z_3, *converging to the eigenvector* x_2.

equations. We can take advantage of the *symmetry* of $S - \lambda_0 I$ to choose one of the well-known factorizations of symmetric matrices. The eigenvalue decomposition (5.1) is too costly and has no sense in an inverse iteration algorithm. Since $S - \lambda_0 I$ can be semidefinite or negative definite depending on the chosen shift λ_0, the symmetric triangular factorization $C = UDU^T$ with U unit upper triangular and D diagonal [208] and also the Cholesky factorization $C = R^T R$ with R upper triangular [69, §4.2.3] may not exist or may be unstable [45, Chap. 5]. The triangular factorization $C = UTU^T$ with U a product of elementary unit upper triangular and permutation matrices and T symmetric and tridiagonal [1], [69, §4.4.2] is stable but more costly than the stable *diagonal pivoting factorization* $C = UDU^T$, where U is a product of elementary unit upper triangular and permutation matrices and D is symmetric block diagonal with blocks of order 1 or 2 [22]–[24]. This factorization is a generalization of Lagrange's method for reducing quadratic forms to diagonal form and corresponds to block symmetric Gaussian elimination with partial pivoting. It is the best choice for solving the set of equations (5.17) in the inverse iteration algorithm. A comparison of various algorithms for solving symmetric indefinite systems of equations can be found in [14]. Given the factorization $S - \lambda_0 I = UDU^T$, we then solve (5.17) in each iteration step by computing $DU^T Q_k = U^{-1} Q_{k-1}$ implicitly. Using this method, the following inverse subspace iteration algorithm will converge, up to a precision ε, to a p-dimensional basis Q_k of the eigensubspace of a symmetric matrix S associated with the p smallest eigenvalues (in absolute value) of $S - \lambda_0 I$.

ALGORITHM 5.1 (Inverse subspace iteration using a diagonal pivoting factorization).

Given: an $n \times n$ symmetric nonsingular matrix $S - \lambda_0 I$ with shift λ_0,
the dimension p of the desired eigensubspace of S,
the desired numerical accuracy ε,
an $n \times p$ start matrix Q_0, $Q_0^T Q_0 = I_p$.

$UDU^T = S - \lambda_0 I$ {*diagonal pivoting factorization*}.
For $k = 1, 2, \cdots$
  solve $DU^T Z = U^{-1} Q_{k-1}$
  orthonormalize $Z : Z = Q_k R_z$, $Q_k^T Q_k = I_p$ {QR *factorization*}
  if $\text{dist}(R(Q_k), R(Q_{k-1})) \leq \varepsilon$, stop
end for
END

Some comments are in order:

- This method is also called *simultaneous* or *orthogonal* inverse iteration [69, §7.3.2] because the columns of the iteration matrix Q_k are orthonormalized in each step to converge to an orthonormal basis of the desired subspace.

- $\text{dist}(R(Q_k), R(Q_{k-1}))$ denotes the distance between the spaces $R(Q_k)$ and $R(Q_{k-1})$. More details of its computation are given in §5.5.

- For the computation of the eigensubspace of a matrix S associated with its smallest eigenvalues λ_i, e.g., in TLS problems, we take *shift* λ_0 zero or a better estimate, if available, satisfying

$$|\lambda_{n-p+i} - \lambda_0| < |\lambda_{n-p} - \lambda_0| \qquad i = 1, \cdots, p. \tag{5.18}$$

- Observe that (5.17) is very *ill-conditioned* if the shift λ_0 is very close to some eigenvalue λ_i of S. However, this fact does *not* affect the *accuracy* of the solution, as proved by Peters and Wilkinson [122] and Parlett [119].

- Note also that inverse iteration destroys the structure of the matrix S.

- Instead of factorizing $S - \lambda_0 I$, we could first *tridiagonalize* it using Householder transformations Q, as was done in the inverse iteration subroutine TINVIT of the EISPACK library [145]. Although the computation of the eigenvector of the tridiagonal matrix requires minimal cost, the necessary transformations are costly and make this algorithm less efficient for solving TLS problems [177, §10.4.1], certainly if the number of iteration steps $\leq n/2$ and the data matrix $S - \lambda_0 I$ and its factorization cannot be updated easily at each time step, e.g., via a lower rank modification.

However, when Algorithm 5.1 is used for solving TLS problems, the data $C = [A; B]$ must be available in their *cross-product* form $C^T C$, otherwise, this matrix must be computed explicitly before using Algorithm 5.1. This cross-product matrix can only be safely used when the underlying eigenvector problem is *not ill-conditioned*, i.e., when the eigenvector v_i associated with the eigenvalue σ_i^2 of $C^T C$ is not too sensitive. Consider the $m \times n$ matrix $C = C_0 + \Delta C$, where ΔC represents the perturbations of the true matrix C_0. Let (4.1) be the SVD of C and $C_0 = \sum_{i=1}^n \sigma_i^0 u_i^0 v_i^{0T}$. Then, the condition of the singular vector v_i of C associated with its ith singular value σ_i is given by [204] (see also Chap. 7)

(5.19)
$$\text{dist}(v_i, v_i^0) \leq \sqrt{2}\frac{\varepsilon_1}{\delta_1} \quad \text{with} \quad \varepsilon_1 = \max(\|\Delta C v_i\|_2, \|\Delta C^T u_i\|_2)$$
$$\text{and} \quad \delta_1 = \min_{i \neq j} |\sigma_j^0 - \sigma_i|.$$

The condition of the eigenvector of $C^T C$ associated with its ith eigenvalue σ_i^2 is given by

(5.20)
$$\text{dist}(v_i, v_i^0) \leq \sqrt{2}\tfrac{\varepsilon_2}{\delta_2} \quad \text{with } \varepsilon_2 = \|E v_i\|_2 \text{ and } \delta_2 = \min_{i \neq j} |\sigma_j^{02} - \sigma_i^2|,$$
$$E = C^T C - C_0^T C_0 + F = \Delta C^T \Delta C + C_0^T \Delta C + \Delta C^T C_0 + F.$$

F is the error matrix introduced by storing the cross-product matrix $C^T C$ explicitly in memory. Its elements are rounded at the same relative finite data storage precision ε_m as the elements of the original C. Since $\|C^T C\|_2 = \sigma_1^2$, it follows that $\|F\|_2 \approx \sigma_1^2 \varepsilon_m$. The corresponding perturbation in the eigenvalue σ_i^2 of $C^T C$ will be of order of magnitude $\|F\|_2$ and can be unacceptably large for small σ_i. This well-known problem has been treated by many authors, see, e.g., [161]. To compute the associated eigenvector v_i from $C^T C$ as accurately as the corresponding singular vector v_i from C, we must have

$$\frac{\varepsilon_2}{\delta_2} \approx \frac{\varepsilon_1}{\delta_1} \quad \text{or} \quad \frac{\|(E - F)v_i\|_2}{\delta_2} + \frac{\sigma_1^2 \varepsilon_m}{\delta_2} \approx \frac{\|\Delta C v_i\|_2}{\delta_1}. \tag{5.21}$$

When (5.21) is satisfied, Algorithm 5.1 may be safely used to compute the desired singular vector (or subspace) of C without loss of accuracy. This implies that the problem is not too ill-conditioned, i.e., δ_2 is sufficiently larger than $\|F\|_2 = \sigma_1^2 \varepsilon_m$, or that the perturbations $\|\Delta C\|_2$ are large enough so that the rounding error matrix F is negligible with respect to the perturbations on the data.

TABLE 5.1.

Comparison in accuracy between the singular vector v_6 *computed from* C, *and the corresponding eigenvector* e_6 *computed from* C^TC, *as described in Example* 5.1.

σ_1^0	$\|\Delta C\|_2$	dist(v_6, v_6^0)	dist(e_6, v_6^0)
10^4	10^{-9}	1.4 x 10^{-8}	1.6 x 10^{-7}
10^6	10^{-9}	3.1 x 10^{-9}	2.7 x 10^{-3}
10^4	10^{-6}	2.2 x 10^{-6}	2.2 x 10^{-6}
10^6	10^{-6}	2.2 x 10^{-6}	2.9 x 10^{-3}
$\leq 10^5$	10^{-3}	2.2 x 10^{-3}	2.2 x 10^{-3}
10^6	10^{-3}	2.2 x 10^{-3}	3.5 x 10^{-3}

However, if (5.21) is not satisfied, the cross-product matrix must be used implicitly in the algorithm (see Algorithm 5.2 below). In this case, the condition of v_i is given by (5.19). Experimental results confirm these theoretical considerations.

Example 5.1. Consider a 10×6 matrix C_0 with singular values $\sigma_i^0, i = 2, \cdots, 6$ given by 100, 80 , 50, 0.1, and 10^{-11}, respectively. σ_1^0 is variable. The data are stored in double precision, i.e., $\varepsilon_m = 10^{-16}$. Now perturb C_0 10 times by independent and equally sized random errors ΔC. Call $C = C_0 + \Delta C$. Denote by v_6 the singular vector of C associated with σ_6, and by e_6 the corresponding eigenvector of C^TC associated with σ_6^2. Table 5.1 shows the obtained average results.

In any case, it is always recommended to use C^TC *implicitly* in the algorithm whenever possible. For example, in the following algorithm inverse iteration is performed implicitly on R^TR after a preceding QR factorization of C [69, §5.2].

ALGORITHM 5.2 (Inverse subspace iteration applied implicitly to R^TR).

Given : an $m \times n$ full rank matrix C, $m \geq n$,
the dimension p of the desired right singular subspace of C,
the desired numerical accuracy ε,
an $n \times p$ start matrix Q_0, $Q_0^TQ_0 = I_p$.

$C = Q_CR_C, Q_C^TQ_C = I_n, R_C$ upper triangular {QR *factorization*}
For $k = 1, 2, \cdots$
solve $R_C^TY = Q_{k-1}$
solve $R_CZ = Y$
orthonormalize $Z : Z = Q_kR_z$, $Q_k^TQ_k = I_p$ {QR *factorization*}
if dist$(R(Q_k), R(Q_{k-1})) \leq \varepsilon$, stop
end for
END

The following comments are in order:

- No back transformation is needed since R_C and C have the same right singular vectors. Therefore, this algorithm is nearly as efficient as computing C^TC and then applying Algorithm 5.1 (see Table 5.2).

- Note, however, that this algorithm does not allow a nonzero shift. For most TLS problems this is not a restriction. Indeed, if r is the numerical rank of $[A; B]$ and the ratio σ_r/σ_{r+1} of the rth and $(r+1)$th singular values of [A;B] is high, the efficiency can only be marginally improved by choosing an appropriate nonzero shift.

In summary, it is justified to use Algorithm 5.2 based on computing the cross-product form C^TC explicitly:

- When the data are immediately acquired in the form $[A; B]^T[A; B]$, e.g., auto- and crosscorrelation values in signal processing, auto- and crosspower spectra. Hence, the use of Algorithm 5.1 for measuring frequency response functions (see §5.5) is adequate here because the data are only available in cross-product form. Indeed, the auto- and crosspowers are immediately computed during data collection.

- In well-conditioned eigenvector problems, i.e., (5.21) is satisfied.

- When the algorithm can improve the efficiency, e.g., by using an appropriately chosen shift λ_0, and the implicit use of $C^TC - \lambda_0 I$ is not possible: the increase in efficiency is considered to be more valuable than a possible loss in accuracy.

5.3.2. Convergence properties. By substituting (5.15) into (5.9) (or (5.7) and (5.8) if only one singular vector is desired), the number K_i of iteration steps necessary for numerical convergence of the inverse iteration Algorithms 5.1 and 5.2, up to $t = -\log \varepsilon$ decimal digits, can be estimated as given in Table 5.2. *Fast convergence* occurs when the convergence rate depending on the ratio:

$$\frac{|\sigma_{n-p}^2 - \lambda_0|}{|\sigma_{n-p+1}^2 - \lambda_0|} \quad \text{or} \quad \frac{|\lambda_{n-p} - \lambda_0|}{|\lambda_{n-p+1} - \lambda_0|} \tag{5.22}$$

is high. λ_{n-p} and λ_{n-p+1} are the eigenvalues of the matrix S in Algorithm 5.1 or the squared singular values σ_{n-p}^2 and σ_{n-p+1}^2 of matrix C in Algorithm 5.2. This is the case if the *shift* λ_0 approaches σ_{n-p+1}^2 or λ_{n-p+1} and the *gap* between σ_{n-p}^2 and σ_{n-p+1}^2 is *large*. The last requirement is especially important and is satisfied for many TLS problems. For example, if $\sigma_{n-p}/\sigma_{n-p+1}$ has order $\mathcal{O}(10^k)$, the convergence rate is of order $\mathcal{O}(10^{2k})$. Hence, inverse iteration can be very efficient in solving TLS problems of known rank, as illustrated in §5.7. Even if the range of the start matrix Q_0 is orthogonal to the desired singular subspace $R(V_n)$ of C, convergence is slowed down but still occurs because of roundoff errors.

5.4. Chebyshev iteration

5.4.1. Ordinary Chebyshev iteration.

Algorithm. The usefulness of the inverse iteration method for convergence to a singular subspace associated with the p smallest singular values of a matrix C depends on (5.22) since this ratio dictates the rate of convergence. In other words, convergence is slowed down if the gap between $|\sigma_{n-p}^2 - \lambda_0|$ and $|\sigma_{n-p+1}^2 - \lambda_0|$ is not sufficiently large. Especially in these cases, convergence can be accelerated by applying the Chebyshev polynomials $T_k^{yz}(x)$ (defined below) instead of the inverse power functions.

The application of Chebyshev polynomials to accelerate iterative eigenvalue algorithms is certainly not new, see, e.g., [119], [137], [208], [42], and [64]. Chebyshev polynomials $T_k(x)$ are orthogonal over an interval [-1,1] with respect to the density function $1/\sqrt{1-x^2}$. They can be defined in several ways [128], but here we only need some important, well-known properties. As all orthogonal polynomials, the Chebyshev polynomials satisfy a three-term recurrence relation:

$$(5.23) \qquad T_{k+1}(x) = 2xT_k(x) - T_{k-1}(x) \text{ with } T_0(x) = 1, T_1(x) = x.$$

They are defined by

$$\begin{aligned}
(5.24)\ T_k(x) &= \cos(k \arccos x) = \cos k\varphi \quad |x| \le 1,\\
(5.25)\ |T_k(x)| &\le 1 \text{ (resp., } \ge 1) \text{ for } |x| \le 1 \text{ (resp., } \ge 1) \text{ and } T_k(1) = 1\ \forall k,\\
(5.26)\ T_k(x) &= 0.5\left[\left(x+\sqrt{x^2-1}\right)^k + \left(x+\sqrt{x^2-1}\right)^{-k}\right] \quad |x| > 1,\\
(5.27)\qquad &\approx 0.5\left(x+\sqrt{x^2-1}\right)^k \quad |x| > 1,\ k \text{ large}.
\end{aligned}$$

The *monic* Chebyshev polynomial $2^{1-k}T_k(x)$ has the *steepest* slope outside $[-1,1]$ with respect to all monic polynomials of degree k [128]. This *rapid growth* of $T_k(x)$ for x ouside $[-1,1]$ and large k makes the Chebyshev polynomials attractive for computing extreme singular values.

To get the analogous results for an arbitrary interval $[y,z]$ we can use the *adapted* Chebyshev polynomials with similar properties over $[y,z]$:

$$(5.28) \quad T_k^{yz}(x) = T_k(f(x)) \text{ with } f(x) = t = \frac{2x-y-z}{y-z} = 1 + 2\frac{x-y}{y-z}.$$

Observe that now every $x \in [y,z]$ is projected onto a $t \in [-1,1]$ by the linear spectrum compression function f. Now, by choosing the interval $[y,z]$ as small as possible so that it contains *all undesired* eigenvalues of a matrix S, the Chebyshev iteration method will converge to a basis of the eigensubspace associated with the remaining eigenvalues outside $[y,z]$. In particular, if the singular subspace associated with the p *smallest* singular values of an $m \times n$ matrix C, $m \geq n$, is desired, e.g., in multidimensional TLS problems, then $[y, z]$ must contain the $n-p$ largest eigenvalues of the cross-product matrix $S = C^T C$. If (4.1) is the SVD of C, this implies that

$$z \geq \sigma_1^2 \text{ and } \sigma_{n-p+1}^2 < y \leq \sigma_{n-p}^2. \tag{5.29}$$

By applying the adapted Chebyshev polynomials (5.28) satisfying (5.29) onto the matrix $S = C^T C$, we obtain the sequence of matrices:

$$\begin{aligned} T_k^{yz}(S) &= T_k(f(C^T C)) = V\,T_k\left(\frac{2}{y-z}\Sigma^T\Sigma - \frac{y+z}{y-z}I\right)V^T \\ &= V \operatorname{diag}(T_k(d_1), \cdots, T_k(d_n))\,V^T. \end{aligned}$$

The eigenvalues of $T_k^{yz}(S)$ are the diagonal elements $T_k(d_i)$, with $d_i = f(\sigma_i^2) = (2\sigma_i^2 - y - z)/(y-z)$ and satisfy the conditions (5.25). Indeed, $|d_i| \leq 1$ for $i = 1, \cdots, n-p$ and $d_j > 1$ for $j = n-p+1, \cdots, n$. Only the singular values (eigenvalues of S) are changed but the singular vectors are conserved. This means that the following Chebyshev iteration algorithm is able to converge to a *basis* Q_k of the *right singular subspace* of an $m \times n$ matrix C, associated with its p *smallest* singular values (and hence allows to solve the multidimensional TLS problem if C is replaced by the $m \times (n+d)$ matrix $[A; B]$).

ALGORITHM 5.3 (Ordinary Chebyshev iteration).

Given: an $m \times n$ matrix C,
the dimension p of the desired right singular subspace of C,
bounds y, z satisfying (5.29),
the desired numerical accuracy ε,
an $n \times p$ start matrix Q_0, $Q_0^T Q_0 = I_p$.

$Q_1 \leftarrow \frac{2}{y-z} C^T(CQ_0) - \frac{y+z}{y-z} Q_0$
$k \leftarrow 1$
do while $\operatorname{dist}(R(Q_k), R(Q_{k-1})) > \varepsilon$
  $Q_{k+1} \leftarrow \frac{4}{y-z} C^T(CQ_k) - 2\frac{y+z}{y-z} Q_k - Q_{k-1}$
  $k \leftarrow k+1$
end while
orthonormalize Q_k
END

Some comments are in order:

- Computing $C^T C Q_k$ as $C^T(CQ_k)$ is numerically more accurate than $(C^T C)Q_k$ (as illustrated in Example 2.1). However, if only $S = C^T C$ is available, replace $C^T C$ by S in Algorithm 5.3 to converge to a basis of the p-dimensional eigensubspace of S associated with its p smallest eigenvalues. In this case, substitute $2m$ by n in Table 5.2 to obtain the required number of operations.

- Since the iteration matrices Q_k, used in the recurrence relation of Algorithm 5.3, cannot be orthonormalized, they are not well suited for numerical purposes, especially when $p > 1$. In particular, when $p > 1$, the iteration process should be restarted as soon as the columns of Q_k are no longer fully independent. See §5.5 for more details.

- Observe that the matrix C does not change during the computations of Algorithm 5.3. Hence, we can take advantage of the structure or sparsity pattern of the matrix C to compute the matrix products $C^T(CQ_k)$ very efficiently. However, if C is not structured, it may be more efficient to perform first a QR factorization of C, i.e.,

$$C = Q_C R_C, \qquad Q_C^T Q_C = I_n, \text{ and } R_C \in \mathcal{R}^{n\times n} \text{ upper triangular}$$

  and then proceed with R_C instead of C in the iteration loop. This implementation with preceding QR factorization requires as many operations per iteration step as inverse Chebyshev iteration (see Table 5.2) and is to be preferred in the following cases.

  First, if C is not structured, this latter implementation is more efficient than Algorithm 5.3 if

(5.30)
$$K_c \geq \frac{mn - \frac{n^2}{3} + \frac{n}{2} + m + \frac{5}{6} - \frac{4}{n}}{(2m-n-1)\,p}$$
$$\text{or, for large } m \text{ and } n, \; K_c \geq \frac{m - \frac{n}{3}}{(2\frac{m}{n} - 1)\,p}.$$

  This is the case for most unstructured TLS problems. Observe that this limit value increases with increasing m and n or when m approaches n or p approaches 1.

  Moreover, when solving slowly varying TLS problems, where the modifications at each time instant are of low rank, e.g., when a new row is appended successively, efficient rank-one updating algorithms for the QR factorization [69, §12.6] can be applied for computing the new factorization from the previous one and can speed up the computation time considerably. In this case, the implementation with preceding QR factorization is more efficient than Algorithm 5.3 from a much lower number K_c of iteration steps on than the bounds (5.30) and is recommended, certainly when combined with weighting the past data by means of an exponential forgetting factor, see, e.g., [107].

Convergence properties. The total number K_c of iteration steps necessary for numerical convergence of Algorithm 5.3, up to $t = -\log \varepsilon$ decimal digits can be estimated as follows. For sufficiently large K_c and $|d_{n-p+1}|$, $T_{K_c}^{yz}(\sigma_{n-p+1}^2) =$

$T_{K_c}(d_{n-p+1})$ with $d_{n-p+1} = f(\sigma_{n-p+1}^2)$ (see (5.28)) can be approximated from (5.27) by

$$(5.31) \qquad T_{K_c}(d_{n-p+1}) \approx 0.5 \left(d_{n-p+1} + \sqrt{d_{n-p+1}^2 - 1} \right)^{K_c}.$$

Substitute this into (5.14) (or (5.5) or (5.6) if $p = 1$), use the property that $\max_{j=1,\cdots,n-p} |T_{K_c}(d_j)| \approx 1$, and take the logarithm (with base 10), then we obtain the expression in Table 5.2. Convergence is optimal if $|d_{n-p+1} - 1|$ is as large as possible. It is not difficult to prove that this corresponds to optimal choices of y and z as

$$(5.32) \qquad y = \sigma_{n-p}^2 \text{ and } z = \sigma_1^2.$$

Observe from Table 5.2 that the convergence rate $\log(d_{n-p+1} + \sqrt{d_{n-p+1}^2 - 1})$ not only depends on the *gap* between σ_{n-p}^2 and σ_{n-p+1}^2, but also on the relative *spread* of the squared undesired singular value spectrum $(\sigma_1^2 - \sigma_{n-p}^2)/\sigma_{n-p}^2$. It is mainly this last parameter that considerably *slows* down the convergence of the ordinary Chebyshev iteration algorithm, even if the gap between σ_{n-p} and σ_{n-p+1} is quite large (see Fig. 5.3(a)). For most TLS problems $AX \approx B$ the relative spread of the squared singular value spectrum, given by $(\sigma_1^2 - \sigma_r^2)/\sigma_r^2$ with r the numerical rank of $[A; B]$, is not small so that the Chebyshev iteration Algorithm 5.3 is nearly always *less efficient* than the inverse iteration Algorithms 5.1 and 5.2 unless C is very sparse (see §§5.6 and 5.7.1).

5.4.2. Inverse Chebyshev iteration.

Algorithm. To accelerate convergence, the Chebyshev polynomials $T_k^{yz}(x)$ can be applied to the matrix $\tilde{S} = S^{-1} = (C^T C)^{-1}$ (instead of S). Choose the interval $[\tilde{z}, \tilde{y}]$ as small as possible such that it contains all undesired eigenvalues of $\tilde{S}$, i.e.,

$$(5.33) \qquad \frac{1}{\sigma_{n-p+1}^2} > \tilde{y} \geq \frac{1}{\sigma_{n-p}^2} \geq \cdots \geq \frac{1}{\sigma_1^2} \geq \tilde{z}.$$

Then we obtain the sequence of matrices:

$$\begin{aligned} T_k^{\tilde{y}\tilde{z}}(\tilde{S}) &= T_k(f((C^T C)^{-1})) = V\, T_k\Big(\frac{2}{\tilde{y} - \tilde{z}}(\Sigma^T \Sigma)^{-1} - \frac{\tilde{y} + \tilde{z}}{\tilde{y} - \tilde{z}} I\Big) V^T \\ &= V \operatorname{diag}(T_k(\tilde{d}_1), \cdots, T_k(\tilde{d}_n))\, V^T. \end{aligned}$$

The eigenvalues of $T_k^{\tilde{y}\tilde{z}}(\tilde{S})$ are the diagonal elements $T_k(\tilde{d}_i)$, with $\tilde{d}_i = f(\sigma_i^{-2}) = (2\sigma_i^{-2} - \tilde{y} - \tilde{z})/(\tilde{y} - \tilde{z})$, and satisfy conditions (5.25). Indeed, $|\tilde{d}_i| \leq 1$ for $i = 1, \cdots, n-p$ and $\tilde{d}_j > 1$ for $j = n-p+1, \cdots, n$. The singular values of C are changed but the singular vectors are conserved. This means that the following *inverse* Chebyshev iteration algorithm will converge, up to a precision ε, to a basis Q_k of the eigensubspace of $\tilde{S}$ associated with its p largest eigenvalues

$1/\sigma_n^2, \cdots, 1/\sigma_{n-p+1}^2$ or equivalently, to a basis of the right singular subspace of C associated with its p smallest singular values $\sigma_{n-p+1}, \cdots, \sigma_n$.

ALGORITHM 5.4 (Inverse Chebyshev iteration).

Given: an $m \times n$ full rank matrix C, $m \geq n$,
the dimension p of the desired right singular subspace of C,
bounds $\tilde{y}, \tilde{z}$ satisfying (5.33),
the desired numerical accuracy ε,
an $n \times p$ starting matrix Q_0, $Q_0^T Q_0 = I_p$.

$C = Q_C R_C$, $\quad Q_C^T Q_C = I_n$, R_C upper triangular {QR *factorization*}
solve $R_C^T G = Q_0$
solve $R_C H = G$
$Q_1 \leftarrow \frac{2}{\tilde{y}-\tilde{z}} H - \frac{\tilde{y}+\tilde{z}}{\tilde{y}-\tilde{z}} Q_0$
$k \leftarrow 1$

do while dist$(R(Q_k), R(Q_{k-1})) > \varepsilon$
  solve $R_C^T G = Q_k$
  solve $R_C H = G$
  $Q_{k+1} \leftarrow \frac{4}{\tilde{y}-\tilde{z}} H - 2\frac{\tilde{y}+\tilde{z}}{\tilde{y}-\tilde{z}} Q_k - Q_{k-1}$
  $k \leftarrow k+1$
end while
orthonormalize Q_k
END

Analogously to Algorithm 5.3, Algorithm 5.4 must be used with special care when $p > 1$. Since the columns of Q_k cannot be orthonormalized in each step, the iteration process should be restarted as soon as the columns of Q_k are no longer fully independent. See §5.5 for more details.

Convergence properties. The total number K_{ic} of iteration steps required for convergence of the inverse Chebyshev iteration Algorithm 5.4, up to $t = -\log \varepsilon$ decimal digits, can be estimated analogously as for ordinary Chebyshev iteration. For sufficiently large K_{ic} and $|\tilde{d}_{n-p+1}|$, $T_{K_{ic}}(\tilde{d}_{n-p+1}) = T_{K_{ic}}(f(\sigma_{n-p+1}^{-2}))$ can be approximated by (5.27) with $x = \tilde{d}_{n-p+1}$. Substitute this into (5.14) (or (5.5) or (5.6) if $p = 1$), use the property that $\max_{j=1,\cdots,n-p} |T_{K_{ic}}(\tilde{d}_j)| \approx 1$, and take the logarithm, then we obtain the expression in Table 5.2. These expressions do not depend on the number of times the iteration process was restarted. In the case of restarting, the total number of performed iteration steps until convergence must be considered. Optimal convergence conditions occur for

$$\tilde{z} = \frac{1}{\sigma_1^2} \quad \text{and} \quad \tilde{y} = \frac{1}{\sigma_{n-p}^2}. \tag{5.34}$$

In practice, the lower bound $\tilde{z}$ will be close to zero since σ_1 is usually large enough. Setting $\tilde{z}$ equal to zero hardly influences the convergence rate, as experimentally verified in §5.6.2.

Finally, it is interesting to note that we can estimate d_n or $\tilde{d}_n$ from the norm of two consecutive iteration vectors. Indeed, denote d_n or $\tilde{d}_n$ by $\bar{d}_n$ and let $q_1^{(k)}$ be the first column of the iteration matrix Q_k after k iteration steps, then

$$\|q_1^{(k)}\|_2^2 = T_k^2(\bar{d}_n)g_n^2 + \sum_{j=1}^{n-1} T_k^2(\bar{d}_j)g_j^2, \tag{5.35}$$

where $g_i = q_1^{(0)T} v_i$ denotes the ith internal coordinate and v_i the ith right singular vector of C. Provided that $\bar{d}_n$ is sufficiently larger than $\bar{d}_{n-1}$ and k is large enough, the sum in (5.35) becomes negligible and hence we have

$$\frac{\|q_1^{(k+1)}\|_2}{\|q_1^{(k)}\|_2} \approx \frac{T_{k+1}(\bar{d}_n)}{T_k(\bar{d}_n)} \approx \bar{d}_n + \sqrt{\bar{d}_n^2 - 1} \quad \text{for } k \text{ large.} \tag{5.36}$$

Using (5.36), the estimate of $\bar{d}_n$ is given by

$$\bar{d}_n = \frac{t^2+1}{2t} \quad \text{with } t = \frac{\|q_1^{(k+1)}\|_2}{\|q_1^{(k)}\|_2}.$$

This quantity estimates d_n or $\tilde{d}_n$ quite accurately, as experimentally verified, once a certain number of iteration steps have been performed [52].

5.5. Implementation aspects.

This section describes how to implement the iterative algorithms outlined before and also discusses some possible improvements in efficiency.

To orthonormalize the columns of the iteration matrix Q_k, a Householder approach [69, §5.2.1] is used here to get Q_k in factored form and then compute the first p columns of Q_k explicitly. The modified Gram-Schmidt method (MGS) [69, §5.2.8], which is about twice as efficient as Householder orthogonalization, could be used as well.

The outlined iteration algorithms are mainly used to converge to a subspace associated with clustered singular values, e.g., in multidimensional TLS problems. If only *one* singular vector is required, e.g., in one-dimensional TLS problems $Ax \approx b$ with a unique solution ($p = 1$), the orthonormalization is simply a reduction of the computed iteration vector Z to unit length, i.e.,

$$Q_k \leftarrow \frac{Z}{\|Z\|_2}. \tag{5.37}$$

The distance function $\text{dist}(R(Q_k), R(Q_{k-1}))$ describes the closeness of the two subspaces $R(Q_k)$ and $R(Q_{k-1})$ and is defined by $\sin\varphi$ with φ the largest canonical angle between these subspaces [69, §2.6.3 and pp. 584–585]. To compute this quantity, note that $\sin\varphi$ is given by [28]:

$$\sin\varphi = \text{dist}(R(Q_k), R(Q_{k-1})) = \|(I - Q_k Q_k^T)Q_{k-1}\|_2 = \|P_k Q_{k-1}\|_2 \tag{5.38}$$

in which Q_k and Q_{k-1} are the orthonormalized bases and P_k is the projection matrix for orthogonal projection onto the orthogonal complement of $R(Q_k)$. The matrix $P_k Q_{k-1}$ can be computed simply by orthonormalizing the columns of Q_{k-1} with respect to the columns of Q_k (e.g., by MGS). The quantity $\text{dist}(R(Q_k), R(Q_{k-1}))$ can then be estimated by a simple estimate of $\|P_k Q_{k-1}\|_2$ and compared with the desired accuracy ε, e.g., by $\|P_k Q_{k-1}\|_F$, as done in §5.7. In particular, when the p smallest singular values of C do not coincide so that the columns of Q_k converge at different rates to the desired basis vectors, $\|P_k Q_{k-1}\|_F$ estimates $\|P_k Q_{k-1}\|_2$ very accurately, as experimentally verified [52].

If only one basis vector is required, the classical convergence test can be used, i.e., after orthonormalization of the iteration vectors Q_{k-1} and Q_k, the iteration process is stopped when

$$\nu = \|Q_{k-1} - \text{sign}(Q_k^T Q_{k-1})Q_k\|_2 \leq \varepsilon. \tag{5.39}$$

Since singular vectors are only defined up to a scaling factor, we must take care in (5.39) that the angle φ between the subsequent iteration vectors remains smaller than 90°. Therefore, Q_k is multiplied with the sign of $Q_k^T Q_{k-1}$. This test requires as many multiplications and divisions per iteration step as the computation of the distance function (5.38) for $p = 1$, namely $3n$ with n the dimension of Q_k. This is $2n$ operations more than the computation of the classical convergence test $\||Q_{k-1}| - |Q_k|\|_2 \leq \varepsilon$ proposed in [186]. However, this latter test can sometimes erroneously stop the iteration process, e.g., when $Q_{k-1} = (4n-3)^{-1/2}[-1, 2, \cdots, 2]^T$ and $Q_k = 1/(4n-3)^{-1/2}[1, 2, \cdots, 2]^T$. Note that for $\varepsilon < 10^{-1}$ both tests (5.38) and (5.39) are equivalent in the sense that they stop the iteration process at the same number of iteration steps. Indeed, since $\nu^2 = 2(1 - \cos\varphi)$ with $\cos\varphi = |Q_k^T Q_{k-1}|$, it is immediately proved that $\text{dist}(R(Q_k), R(Q_{k-1})) = \sin\varphi = \nu\sqrt{1 - \frac{\nu^2}{4}} \approx \nu$ for $\nu < 10^{-1}$, as also experimentally verified [52]. Although the tests $\cos\varphi \geq \sqrt{1-\varepsilon^2}$ and $\cos\varphi \geq 1 - \varepsilon^2/2$ are mathematically equivalent with the distance test $\sin\varphi \leq \varepsilon$ and the classical convergence test $\nu \leq \varepsilon$ respectively, they may not be implemented as such since they suffer from numerical errors whenever $\varepsilon \leq \sqrt{\varepsilon_m}$, where ε_m is the machine precision used.

In contrast to the orthonormal iteration matrices Q_k produced by the inverse iteration Algorithms 5.1 and 5.2, the iteration matrices Q_k used in the recurrence relation of Algorithms 5.3 and 5.4 are not well suited for numerical purposes since they may not be orthonormalized. Hence, Algorithms 5.3 and

5.4 can only be used with special care, especially when $p > 1$, while Algorithms 5.1 and 5.2 are nearly fullproof. Indeed, the Chebyshev recurrence relation tends to make the columns of Q_k more and more parallel as $k \leftarrow \infty$. To guarantee the same accuracy ε on the final result, the columns of Q_k must be kept fully independent during iteration. If this is no longer the case, the iteration process should be stopped after, say, μ iteration steps and Q_μ must be orthonormalized. If the basis thus obtained does not yet approximate the final result within the accuracy ε, then restart the iteration process using as start matrix the lastly computed and properly orthonormalized iteration matrix Q_μ. As long as

$$\left(\frac{\bar{d}_n + \sqrt{\bar{d}_n^2 - 1}}{\bar{d}_{n-p+1} + \sqrt{\bar{d}_{n-p+1}^2 - 1}}\right)^\mu < 10^t$$

with $\bar{d}_n = d_n$ or $\tilde{d}_n$ and $\bar{d}_{n-p+1} = d_{n-p+1}$ or $\tilde{d}_{n-p+1}$, parallelization of the columns of Q_μ will not have gone further than that at most t digits are canceled out when these columns are orthonormalized. Different choices of μ are given in [135] and [136]. If the convergence rate is high so that the number K of required iteration steps is small, say $K \leq 20$ for $\varepsilon = 10^{-14}$, $\mu = 5$ gives satisfactory results in terms of efficiency and accuracy. For slower convergence rates, a better efficiency is obtained for larger values of μ, e.g., $\mu = 9$ (see Table 5.5). Alternatively, we could control the independence of the columns of Q_k by estimating its condition number κ. Perform the QR factorization of Q_k with column pivoting, i.e., $Q_k\Pi = Q_q R_q$, and compute the ratio of the first to the last diagonal element of the triangular factor R_q : $\kappa = r_{11}/r_{pp}$. It was experimentally found that restarting the iteration process as soon as κ exceeds 10^5 also gives satisfactory results in terms of efficiency and accuracy, provided convergence is fast enough and the p smallest singular values $\sigma_{n-p+1}, \cdots, \sigma_n$ of C do not coincide or are not intrinsically close to σ_{n-p}. Even if all smallest σ_i are equal, better convergence properties are obtained by restarting the iteration process regularly because of the improved condition of the start matrix with orthonormal columns. These facts have been experimentally investigated in [52].

The work load per iteration step can be considerably reduced by orthonormalizing the iteration matrices Q_k and performing the convergence test only from time to time. Note, however, that then the iteration matrices Q_k of the inverse iteration Algorithms 5.2 and 5.1 are given by $Q_k = (R_C^T R_C)^{-1} Q_{k-1}$ and $Q_k = S^{-1} Q_{k-1}$, respectively. Since these are no longer orthonormal, special care is needed as in the Algorithms 5.3 and 5.4. In particular, Q_k should be orthonormalized as soon as its columns are no longer independent. More details are given in [135].

The efficiency is further improved as follows. If the p smallest singular values of the given matrix C do not coincide, the columns of the iteration matrix Q_k converge at different rates to the desired basis vectors. Since a converged column no longer changes (up to a scaling factor) in the next iteration steps, it may be "frozen," i.e., this column is no longer subject to iteration and orthogonalization but is still used to orthogonalize the unconverged columns. Having thus computed v_n in the first column of Q_k, the iteration is—with obvious modifications of the testing process—continued to find v_{n-1}, etc.

An alternative method for computing the p desired basis vectors, $p > 1$, is to compute the basis vectors consecutively by applying the iterative algorithms for the case $p = 1$ and preventing the algorithm from recomputing the results already found. This is done by using deflation techniques or by orthogonalizing in each step the iteration vector to the basis vectors already found [119]. This procedure may be easier to apply.

Furthermore, precautions against under- and overflow are not included in the iterative algorithms presented here. To avoid overflow, compare the norm of each column of Q_k with a certain limit depending on the largest possible number of the computer used. If this limit is exceeded, then scale Q_k (and also Q_{k-1} in Algorithms 5.3 and 5.4) appropriately.

Finally, note that the (inverse) Chebyshev iteration algorithms require the *estimation* of *good bounds* y, z or $\tilde{y}, \tilde{z}$. The upper bound z of ordinary Chebyshev iteration can be estimated by $\|C\|_F$. If σ_1 of C is one order of magnitude larger than the other singular values, this estimate approaches σ_1 very closely. The lower bound $\tilde{z}$ of inverse Chebyshev iteration can be estimated by $1/z$ or, more simply, can be set to zero. The lower bounds $y = 1/\tilde{y}$ can be estimated by an appropriate rank determination method. For instance, in TLS problems $A_{m\times n}X \approx B_{m\times d}$ where the elements in A and B are affected by independently and identically distributed zero-mean errors of equal variance σ_ν^2, an appropriate rank determinator R_ν is given by $R_\nu^2 = 2\max\{m, n+d\}\sigma_\nu^2$ (see §3.6.1). By setting $y = 1/\tilde{y} = R_\nu^2$, convergence can occur to a basis of the right singular subspace of C, associated with its singular values $\leq R_\nu$, from which a TLS solution can be computed. The efficiency of this bound depends on the closeness of R_ν to the undesired singular value spectrum ($> R_\nu$). If the dimension p of the desired singular subspace is not known, then Sylvester's Law of Inertia [69, Thm. 8.1.12] or, equivalently, the Sturm sequence property [69, Thm. 8.4.1] can be applied to the matrix C after bidiagonalization to compute how many singular values of C are smaller than the given bound. Such a procedure is described in §4.3.2. If no bounds on the singular values can be given but instead, the dimension p of the desired singular subspace is known, then we can compute bounds $y = 1/\tilde{y}, z = 1/\tilde{z}$, sufficiently close to the optimal ones, by using the bisection method [69, §8.4.1] such that the matrix C has exactly p singular values smaller than or equal to the bound y. The matrix C needs to be bidiagonalized first. Such procedures are outlined in [69,

§8.4.1] and §4.3.2. Efficient estimation procedures that dynamically estimate the bounds y, z are not yet analyzed and need further investigation.

Of course, these estimation strategies increase the work load. Alternatively, if the bounds y, z or $\tilde{y}, \tilde{z}$ cannot be obtained efficiently and if the gap between the desired and undesired singular value spectrum is small, a (block) Lanczos method can be used (see §5.8.2). This method converges at least as fast as the Chebyshev iteration method with optimal bounds. Moreover, no bounds on the singular value spectrum need to be known.

5.6. Comparison in efficiency

5.6.1. Iterative versus direct computation methods. Since the TLS computations are entirely dominated by the calculation of the desired singular subspace of the data [A;B], a comparison between the iterative TLS methods and the direct computation methods classical TLS (§3.6.1) and PTLS (§4.5) reduces mainly to a comparison between the iterative algorithms discussed here, classical SVD and PSVD. The additional computations, namely, Step 4 of Algorithm 4.5 (excluding nongeneric TLS problems), are the same for all TLS algorithms discussed here. Therefore, the computational efficiency of these algorithms can be evaluated by comparing their operation counts required for computing the desired basis vectors. These results are summarized in Table 5.2. The total computational cost of each iterative algorithm α is obtained by multiplying the operation counts in each iteration step with the required number K_α of iteration steps. Note that the computed operation counts also depend on the implementation of the algorithms. The convergence test for subspace iteration is based on (5.38). By using the modified Gram–Schmidt method and computing the Frobenius norm, the desired quantity is estimated. For vector iteration ($p = 1$), the test (5.39) can be used as well and requires as many operations. As a classical SVD routine, DSVDC of the LINPACK library [45, Chap.11] is used. PSVD is given in [176]. Extra calculations that deal with under- and overflow are not included.

By taking the *ratio* of the total number of operations in Table 5.2, the computational efficiency of the methods discussed here can be compared in function of the dimensions m, n of the original matrix C, the dimension p of the required singular subspace and the number of required iteration steps. This is done in Table 5.3. As long as the number K_α of iteration steps determined by the convergence rate remains smaller than the bound given in Table 5.3, the iterative method α is computationally more efficient than the direct computation method used in the comparison. Numerical values of these bounds are given in Table 5.4 for specific choices of m, n, and p.

TABLE 5.2.

Computational efficiency of classical SVD, PSVD, *inverse iteration, ordinary Chebyshev and inverse Chebyshev iteration, determined by the operation counts (second column) and the convergence rate (third column). An $m \times n$ matric C with singular values $\sigma_1, \cdots, \sigma_n, \sigma_1 \geq \cdots \geq \sigma_n$, and corresponding right singular vectors $v_1, \cdots, v_n$ is considered; p basis vectors of the singular subspace $R(V_n)$, $V_n = [v_{n-p+1}, \ldots, v_n]$, of C are desired. ε denotes the desired convergence accuracy. Q_0 is the $n \times p$ start matrix, $Q_0^T Q_0 = I_p$ and φ is the largest canonical angle defined by* $\sin\varphi = \mathrm{dist}(R(Q_0), R(V_n))$ *(or, if $p = 1$, by* $\cos\varphi = |v_n^T Q_0|$*). Before using Algorithm* 5.1, *the matrix $S = C^T C$ must be available, else $mn(n+1)/2$ extra multiplications are required for its explicit computation.*

Method	Total number of multiplications (and divisions)	Number of required iteration steps
Classical SVD	$m < 5n/3$: $2mn^2 + 2sn^3 + (\frac{3}{2} + 8s)n^2$ $+(\frac{5}{2} - s)n - 2m - 10 - 9s$ $m \geq 5n/3$: $mn^2 + (\frac{5}{3} + 2s)n^3 + (2 + 8s)n^2 + mn$ $+(\frac{4}{3} - s)n - 17 - 9s$ s = average number of required QR iteration steps for convergence to one singular vector v_i.	convergence ultimately cubic
Partial SVD	$m < 5n/3$: $2mn^2 - \frac{2}{3}n^3 + [2 + (4s+1)p]n^2$ $+[\frac{8}{3} + 2sp(11-p)]n - 2m - 12sp^2 - (10s+4)p - 7$ $m \geq 5n/3$: $mn^2 + n^3 + [\frac{5}{2} + (4s+1)p]n^2 + mn$ $+[\frac{3}{2} + 2sp(11-p)]n - 12sp^2 - (10s+4)p - 14$ s = average number of required QR/QL iteration steps for convergence to one basis vector of $R(V_n)$.	convergence ultimately cubic
Inverse iteration	Algorithm 5.1 for $p = 1$: $\frac{n^3}{6} + \frac{3}{4}n^2 + \frac{7}{3}n + n(n+7)K_i$ Algorithm 5.1 for $p > 1$: $\frac{n^3}{6} + \frac{3}{4}n^2 + \frac{7}{3}n$ $+[pn^2 + (3+4p)pn - \frac{2}{3}p^3 + 2p^2 - \frac{p}{3} - 4]K_i$ Algorithm 5.2 for $p = 1$: $mn^2 - \frac{n^3}{3} + \frac{n^2}{2} + mn + \frac{5}{6}n - 4 + n(n+6)K_i$ Algorithm 5.2 for $p > 1$: $mn^2 - \frac{n^3}{3} + \frac{n^2}{2} + mn + \frac{5}{6}n - 4$ $+[pn^2 + (2+4p)pn - \frac{2}{3}p^3 + 2p^2 - \frac{p}{3} - 4]K_i$ K_i = number of inverse iteration steps	$K_i \approx \frac{\log \varepsilon^{-1} + \log M}{\log(\lvert\sigma_{n-p}^2 - \lambda_0\rvert / \lvert\sigma_{n-p+1}^2 - \lambda_0\rvert)}$ with $M = \lvert\tan\varphi\rvert$ (or, if $p = 1$, $M = \max_{j \neq n} \lvert v_j^T Q_0\rvert / \lvert v_n^T Q_0\rvert$) λ_0 is the shift ($\lambda_0 = 0$ in Algor.5.2)
Ordinary Chebyshev iteration	Algorithm 5.3 for $p = 1$: $(2m+7)nK_c$ Algorithm 5.3 for $p > 1$: $[(2m + 4p + 3)np - \frac{2}{3}p^3 + 2p^2 - \frac{p}{3} - 4]K_c$ Algorithm 5.3 with preceding QR factorization: *see* inverse Chebyshev iteration with $K_c = K_{ic}$ K_c = number of ordinary Chebyshev iteration steps	$K_c \approx \frac{\log(2\varepsilon^{-1}) + \log M}{\log(d_{n-p+1} + \sqrt{d_{n-p+1}^2 - 1})}$ with $M = \lvert\tan\varphi\rvert$ (or, if $p = 1$, $M = \max_{j \neq n} \lvert v_j^T Q_0\rvert / \lvert v_n^T Q_0\rvert$), $d_{n-p+1} = \frac{2\sigma_{n-p+1}^2 - y - z}{y - z}$, y, z = bounds satisfying (5.29)
Inverse Chebyshev iteration	Algorithm 5.4 for $p = 1$: $mn^2 - \frac{n^3}{3} + \frac{n^2}{2} + mn + \frac{5}{6}n - 4 + n(n+8)K_{ic}$ Algorithm 5.4 for $p > 1$: $mn^2 - \frac{n^3}{3} + \frac{n^2}{2} + mn + \frac{5}{6}n - 4$ $+[pn^2 + 4p(p+1)n - \frac{2}{3}p^3 + 2p^2 - \frac{p}{3} - 4]K_{ic}$ K_{ic} = number of inverse Chebyshev iteration steps	$K_{ic} \approx \frac{\log(2\varepsilon^{-1}) + \log M}{\log(\bar{d}_{n-p+1} + \sqrt{\bar{d}_{n-p+1}^2 - 1})}$ with $M = \lvert\tan\varphi\rvert$ (or, if $p = 1$, $M = \max_{j \neq n} \lvert v_j^T Q_0\rvert / \lvert v_n^T Q_0\rvert$), $\bar{d}_{n-p+1} = \frac{2/\sigma_{n-p+1}^2 - \bar{y} - \bar{z}}{\bar{y} - \bar{z}}$, $\bar{y}, \bar{z}$ = bounds satisfying (5.33)

TABLE 5.3.

Upper bounds on the number K_α of iteration steps for method α (first column) such that α is computationally more efficient, i.e. requires less operation counts, than method β (second column). p basis vectors of the singular subspace of an $m \times n$ matrix C, associated with its p smallest singular values, are desired and s is the average number of required QR/QL *iteration steps for convergence to one basis vector with the considered direct computation method β. It is assumed that the matrix $S = C^TC$ in Algorithm* 5.1 *must be computed explicitly, unless stated otherwise.*

	Method α	Method β	α more efficient than β if K_α is smaller than:
1	Inverse iteration (Algor.5.1),$p = 1$	Classical SVD	$m < 5n/3 : [18m + (12s - 1)2n + 96s + 9 - 6m/n]n/(12n + 84)$ $m \geq 5n/3 : [2m + (3 + 4s)2n + 32s + 5 + 2m/n]n/(4n + 28)$ S available: $[(11 + 12s)2n + 96s + 9]n/(12n + 84)$
2	Inverse iteration (Algor.5.1),$p = 1$	Partial SVD, $p = 1$	$m < 5n/3 : [18m - 10n + 48s + 27 - 6m/n]n/(12n + 84)$ $m \geq 5n/3 : [6m + 10n + 48s + 33 + 6m/n]n/(12n + 84)$ S available: $[14n + 48s + 27]n/(12n + 84)$
3	Inverse iteration (Algor.5.2), $p = 1$	Classical SVD	$m < 5n/3 : [3m + (6s + 1)n + 24s + 3 - 3m/n]n/(3n + 18)$ $m \geq 5n/3 : [(1 + s)4n + 16s + 3]n/(2n + 12)$
4	Inverse iteration (Algor.5.2),$p = 1$	Partial SVD, $p = 1$	$m < 5n/3 : [6m - 2n + 24s + 15 - 6m/n]n/(6n + 36)$ $m \geq 5n/3 : [4n + 12s + 9]n/(3n + 18)$
5	Ordinary Chebyshev iteration (Algor.5.3), $p = 1$	Classical SVD	$m < 5n/3 : [2m + 2sn + 8s + 3/2]n/(2m + 7)$ $m \geq 5n/3 : [3m + (5 + 6s)n + 24s + 6 + 3m/n]n/(6m + 21)$
6	Ordinary Chebyshev iteration (Algor.5.3), $p = 1$	Partial SVD, $p = 1$	$m < 5n/3 : [6m - 2n + 12s + 9]n/(6m + 21)$ $m \geq 5n/3 : [2m + 2n + 8s + 7 + 2m/n]n/(4m + 14)$
7	Inverse Chebyshev iteration (Algor.5.4), $p = 1$	Classical SVD	$m < 5n/3 : [3m + (6s + 1)n + 24s + 3 - 3m/n]n/(3n + 24)$ $m \geq 5n/3 : [(1 + s)4n + 16s + 3]n/(2n + 16)$
8	Inverse Chebyshev iteration (Algor.5.4), $p = 1$	Partial SVD, $p = 1$	$m < 5n/3 : [6m - 2n + 24s + 15 - 6m/n]n/(6n + 48)$ $m \geq 5n/3 : [4n + 12s + 9]n/(3n + 24)$
9	Inverse iteration (Algor.5.1), $p > 1$	Classical SVD	$m < 5n/3 : [18m + (12s - 1)2n + 96s + 9 - 6m/n]n/[12p(n + 4p + 3)]$ $m \geq 5n/3 : [2m + (3 + 4s)2n + 32s + 5 + 2m/n]n/[4p(n + 4p + 3)]$ S available: $[(11 + 12s)2n + 96s + 9]n/[12p(n + 4p + 3)]$ (if $p << m$ and $p << n$)
10	Inverse iteration (Algor.5.1), $p > 1$	Partial SVD	$m < 5n/3 : [18m - 10n + (4s + 1)12p + 15 - 6m/n]n/[12p(n + 4p + 3)]$ $m \geq 5n/3 : [6m + 10n + (4s + 1)12p + 21 + 6m/n]n/[12p(n + 4p + 3)]$ S available: $[(14n + (4s + 1)12p + 15]n/[12p(n + 4p + 3)]$ (if $p << m$ and $p << n$)
11	Inverse iteration (Algor.5.2), $p > 1$	Classical SVD	$m < 5n/3 : [3m + (6s + 1)n + 24s + 3 - 3m/n]n/[3p(n + 4p + 2)]$ $m \geq 5n/3 : [4(1 + s)n + 16s + 3]n/[2p(n + 4p + 2)]$ (if $p << m$ and $p << n$)
12	Inverse iteration (Algor.5.2), $p > 1$	Partial SVD	$m < 5n/3 : [6m - 2n + (4s + 1)6p + 9 - 6m/n]n/[6p(n + 4p + 2)]$ $m \geq 5n/3 : [4n + (4s + 1)3p + 6]n/[3p(n + 4p + 2)]$ (if $p << m$ and $p << n$)

TABLE 5.4.

Numerical values of the upper bounds on the number K_α of iteration steps for method α such that α requires fewer operation counts than method β for specific values of m, n, p, and s. p basis vectors of the singular subspace of an $m \times n$ matrix C, associated with its p smallest singular values, are desired. s is the average number of required QR/QL *iteration steps for convergence to one basis vector using the considered direct computation method β. If m equals n, $S = C^TC$ is considered to be available and need not be computed.*

matrix size				pairs α, β of computation methods $1, \cdots, 12$ refer to rows of Table 5.3.											
m	n	p	s	1	2	3	4	5	6	7	8	9	10	11	12
10	8	1	2	33.0	9.6	34.5	9.5	20.6	7.6	30.2	8.3				
10	10	1	2	44.2	12.9	43.3	10.1	28.7	9.0	38.5	9.0				
20	18	1	2	82.9	17.8	85.4	17.5	49.6	14.9	78.8	16.2				
30	28	1	2	134.8	25.1	137.7	24.8	79.2	21.9	130.1	23.4				
30	30	1	2	155.5	36.7	146.7	24.6	88.4	22.8	138.9	23.3				
18	10	1	2	48.3	17.0	48.4	15.2	22.0	9.6	43.1	13.5				
40	20	1	2	109.8	35.9	105.8	29.0	39.8	16.9	98.2	26.9				
50	30	1	2	168.7	49.9	164.6	42.5	67.2	26.1	155.9	40.3				
100	50	1	2	301.1	90.7	283.5	69.3	97.4	39.5	273.7	67.0				
10	8	2	2									13.0	5.7	13.4	5.7
10	10	2	2									17.9	7.4	17.3	6.3
20	18	2	2									35.8	10.5	36.6	10.4
30	30	3	2									42.6	14.1	40.0	10.8
18	10	2	2									19.6	9.0	19.4	8.3
50	30	3	2									46.3	17.7	44.9	15.7

With these comments in mind, the following *conclusions* can be drawn from the Tables 5.2, 5.3, and 5.4:

- The efficiency of *iterative* methods in general depends on *many* parameters. The most important ones are those that influence the *convergence rate.* This factor is first determined by the *quality of the initial information*, such as knowledge of a good start matrix, an appropriate shift, or tight bounds. If these parameters cannot be obtained sufficiently accurately, then iterative methods can *fail* to converge to the desired p-dimensional singular subspace $R(V_n)$ associated with the smallest singular values of matrix $[A;B]$. Indeed, Chebyshev iteration methods fail to converge to the desired singular subspace associated with the p smallest singular values of a matrix C, given by (4.1), if the bounds y, z or $\tilde{y}, \tilde{z}$ are not well defined, e.g., when $y = 1/\tilde{y} > \sigma_{n-p}^2$ or $z = 1/\tilde{z} < \sigma_1^2$. If the shift λ_0 makes the ratio (5.22) smaller than one, then inverse iteration does not converge to the desired singular subspace $R(V_n)$. Note, however, that a zero shift always ensures convergence to $R(V_n)$. Second, the convergence rate of an iterative method depends on *the matrix characteristics* such as its *dimensions* m and n and the *gap* between the singular values associated with the desired and undesired singular subspace. Third, the desired *accuracy* influences the convergence rate. Moreover, the efficiency of an iterative method is determined by the *dimension* p of the desired *singular subspace.*

- On the other hand, the efficiency of the *direct* methods is essentially determined by only a *few* parameters, namely, the *dimensions* m and n of the considered matrix. In the PSVD case, the efficiency also depends on the desired accuracy ε, the dimension p of the desired singular subspace, and the gap between the corresponding singular values and the remaining ones. Moreover, the direct methods classical SVD and PSVD *always* converge to the desired vectors for any matrix.

- The relative efficiency of the iterative methods with respect to the direct computation methods is most pronounced when only *one singular vector* is required, e.g., in one-dimensional TLS problems with a unique solution. In multidimensional TLS problems, additional calculations are required in each step to make the iteration matrix Q_k orthonormal or to compute an orthonormal basis of $R(Q_k)$. This increases the operation counts in each iteration step and reduces the efficiency of the iterative method.

- Moreover, when using iterative methods, the *dimension* p of the desired singular *subspace* must always be *known*. The direct methods on the contrary also allow us to compute the basis vectors of a singular subspace of a matrix associated with its singular values smaller than or equal to a given bound θ. By using deflation techniques or by orthogonalizing the iteration vector to the singular vectors already found in each step [119], however, it is possible to compute subsequent singular triplets with an iterative method until all singular vectors associated to singular values $\leq \theta$ are computed. But the number of operations in each iteration step is increased, and one additional singular triplet corresponding to a singular value $> \theta$ must also be computed to assure convergence to all singular values $\leq \theta$. It is clear that these operations make the iterative methods less efficient. This is the case for TLS problems of which the numerical rank of the matrix [A;B] is determined by an error bound θ, but also nongeneric TLS problems require the computation of an additional basis of the singular subspace corresponding to the next larger singular value. Moreover, in the latter TLS problems the start vectors used for calculating the additional basis vectors with an iterative method are in general of bad quality, so convergence to the desired TLS solution is slowed down.

We can thus conclude that iterative methods are efficient in solving d-dimensional TLS problems (1.17) with SVD given by (1.22), provided
— the start matrix is good and the problem is generic;
— the desired accuracy is low;
— the numerical rank $r(\leq n)$ of $[A; B]$ is known;
— the dimension d of the problem is low;
— the matrix dimensions m and n are moderate;
— the convergence rate is sufficiently high.

The last requirement is especially important and essentially determines the usefulness of a particular iterative method in solving TLS problems. The next section compares the convergence rate of the iterative methods presented here and shows for which class of problems each method is computationally most efficient.

5.6.2. Inverse iteration versus Chebyshev iteration. The computational efficiency of the different iterative algorithms presented in this chapter depends on the operation counts per iteration step, as well as on the convergence behavior of the method used. The total computational cost of each algorithm is obtained by multiplying the operation counts in each iteration step with the number of iteration steps.

First, observe from Table 5.2 that ordinary Chebyshev iteration with preceding QR factorization requires as many operations per iteration step as inverse Chebyshev iteration. The inverse Chebyshev iteration Algorithm 5.4 coincides with the inverse iteration Algorithm 5.2, except for the computation of the recurrence. Hence, by adding $2np$ to the operation counts of Algorithm 5.2, given in Table 5.2, we obtain the operation counts of Algorithm 5.4. Thus, inverse iteration is as efficient as inverse Chebyshev iteration (respectively, ordinary Chebyshev iteration with preceding QR factorization) if

$$K_i = \gamma K_{ic} \ \text{(respectively, } K_i = \gamma K_c\text{)}, \tag{5.40}$$

where

$$\gamma = \begin{cases} \frac{n+8}{n+6} & \text{for } p = 1 \\ \frac{\beta+2pn}{\beta} & \text{with } \beta = pn^2 + (1+2p)2pn - \frac{2}{3}p^3 + 2p^2 - \frac{p}{3} - 4 \ \text{ for } p > 1. \end{cases}$$

γ approaches 1 for increasing n and p so that this difference in operation counts per iteration step becomes negligible. Hence, for unstructured or dense matrices with n or p not too small, we can conclude that differences in efficiency between the iterative algorithms discussed here are mainly due to differences in convergence behavior reflected in the number of required iteration steps. Estimates are given in Table 5.2 and agree well with the experimentally obtained numbers of iteration steps, as shown in Table 5.5. Note that these

TABLE 5.5.

Comparison between the estimated (est.) numbers (as predicted by Table 5.2) and the experimentally (exp.) obtained numbers of iteration steps averaged over 20 test cases, required by the inverse iteration (K_i), ordinary Chebyshev iteration (K_c), and inverse Chebyshev iteration (K_{ic}) algorithm for convergence, up to a precision ε, to a p-dimensional right singular subspace of an $m \times n$ matrix C associated with its p smallest singular values $\sigma_{n-p+1}, \cdots, \sigma_n$. $m = 18, n = 11$ and the start matrix contains t correct digits. σ_1 (respectively σ_{n-p}) is the largest (respectively $(n-p)$th) singular value of C and the intermediate ones are equally spaced between these two values. For (inverse) Chebyshev iteration the optimal bounds $y = \frac{1}{\tilde{y}} = \sigma_{n-p}^2$ and $z = \frac{1}{\tilde{z}} = \sigma_1^2$ are taken and, if $p > 1$, iteration is restarted each μ steps : $\mu = 9$ is used for values of p marked with an asterisk, else $\mu = 5$.

problem specifications						K_i		K_c		K_{ic}	
p	ε	t	σ_1	σ_{n-p}	$\leq \sigma_{n-p+1}$	est.	exp.	est.	exp.	est.	exp.
1	10^{-14}	0	2.0	1.0	0.5	23.25	24.35	34.22	38.75	11.41	13.75
1	10^{-12}	0	2.0	1.0	0.5	19.93	21.25	29.43	33.75	9.81	12.20
1	10^{-10}	0	2.0	1.0	0.5	16.61	17.65	24.65	28.70	8.22	10.50
1	10^{-5}	0	2.0	1.0	0.5	8.30	9.50	12.68	16.85	4.23	6.60
1	10^{-14}	1	2.0	1.0	0.5	21.59	19.60	31.82	31.95	10.61	11.40
1	10^{-14}	3	2.0	1.0	0.5	18.27	16.50	27.04	27.00	9.01	10.00
1	10^{-14}	5	2.0	1.0	0.5	14.95	13.35	22.25	22.00	7.42	8.00
1	10^{-14}	7	2.0	1.0	0.5	11.63	10.10	17.47	17.35	5.82	7.00
1	10^{-14}	10	2.0	1.0	0.5	6.64	5.25	10.29	10.05	3.43	4.00
1	10^{-14}	0	5.0	1.0	0.5	23.25	24.40	93.62	103.80	12.34	14.70
1	10^{-14}	0	10.0	1.0	0.5	23.25	24.40	189.40	206.10	12.46	14.90
1	10^{-14}	0	1.2	1.0	0.5	23.25	24.95	15.22	17.95	8.85	11.00
2	10^{-14}	0	2.0	1.0	0.5,0.5	23.25	25.30	34.22	44.90	11.41	14.30
2	10^{-14}	0	2.0	1.0	0.6,0.3	31.55	32.70	36.85	47.75	13.53	15.95
2	10^{-14}	0	2.0	1.0	0.9,0.4	152.98	147.45	66.10	99.15	30.86	39.45
2	10^{-14}	0	2.0	1.0	0.3,0.1	13.39	15.10	31.29	39.80	8.18	10.00
2*	10^{-14}	0	2.0	1.0	0.5,0.5	23.25	25.30	34.22	41.95	11.41	14.15
2*	10^{-14}	0	2.0	1.0	0.6,0.3	31.55	32.70	36.85	44.20	13.53	16.25
2*	10^{-14}	0	2.0	1.0	0.9,0.4	152.98	147.45	66.10	85.60	30.86	36.85
2*	10^{-14}	0	2.0	1.0	0.3,0.1	13.39	15.10	31.29	37.20	8.18	14.95

estimates cannot be implemented in practice since the singular values of the matrix C are not known a priori. Therefore, different but practically feasible convergence tests must be used. That is the reason why the experimentally obtained numbers of iteration steps performed by Algorithms 5.2, 5.3, and 5.4 slightly differ from the estimates given in Table 5.2. Note also that the estimated numbers of (inverse) Chebyshev iteration steps agree well with the experimentally obtained numbers of iteration steps as long as the applied approximation (5.27) with $x = d_{n-p+1}$ or $x = \tilde{d}_{n-p+1}$ is sufficiently accurate. For small K_c or K_{ic} or values d_{n-p+1} or $\tilde{d}_{n-p+1}$ very close to 1, the estimates largely underestimate the real numbers of iteration steps.

The estimates of Table 5.2 are also important for our theoretical analysis since they clearly reveal the parameters that influence the convergence and hence the efficiency of the iterative algorithms under study.

First, it is clear that each iterative algorithm requires *less* iteration steps if the desired *accuracy* ε *decreases* or the *start matrix* is of *better quality*. This is illustrated in Figs. 5.2(a) and (b), respectively. As predicted by Table 5.2, the number of required iteration steps *linearly* increases (respectively, decreases) with larger number of desired correct digits $t = -\log\varepsilon$ (respectively, larger number of correct digits t in the start matrix).

The differences between the presented iterative algorithms are entirely determined by the differences in *convergence rate*, given by the denominator of the estimates in Table 5.2. Let us now analyze those influencing parameters σ_1, σ_{n-p}, σ_{n-p+1}, and the quality of the bounds y, z, $\tilde{y}$, and $\tilde{z}$.

First, the convergence rate of ordinary Chebyshev iteration strongly depends on the *spread of the squared singular value spectrum*. This is clear from Fig. 5.3(a). Observe that the convergence rate decreases considerably, causing an increase in the number of iteration steps, when the ratio σ_1/σ_{n-p} increases, i.e., when the singular value spectrum enlarges. Moreover, this number of required iteration steps further increases with decreasing quality of the upper bound $z \geq \sigma_1^2$.

In contrast, the convergence rate of inverse Chebyshev iteration is nearly independent of the spread of the squared singular value spectrum, as well as of the quality of the lower bound $\tilde{z} \leq 1/\sigma_1^2$, provided σ_1 is not too close to σ_{n-p}, say $\sigma_1/\sigma_{n-p} \geq 2$, as illustrated in Fig. 5.3(b). Note that the convergence rate of inverse iteration is not at all influenced by the value of σ_1, as evidenced by Table 5.2.

Second, the *gap* between σ_{n-p} and σ_{n-p+1} and additionally the *quality of the bounds* y and $\tilde{y}$ in case of (inverse) Chebyshev iteration, strongly influences the convergence rate of each iterative algorithm, although in a different way. This difference in behavior is best illustrated by comparing the ratios of the number of iteration steps required by each algorithm as a function of two important parameters: $r = \sigma_{n-p}/\sigma_{n-p+1}$ and α defined by $\alpha = y/\sigma_{n-p}^2$ or $\alpha = \tilde{y}\sigma_{n-p}^2$ (see Figs. 5.4–5.6). The singular value ratio $r \geq 1$ expresses the influence of the gap between σ_{n-p} and σ_{n-p+1}. The smaller the gap, the closer r is to 1. α expresses the influence of the quality of the bounds $y = \alpha\sigma_{n-p}^2$ or $\tilde{y} = \alpha/\sigma_{n-p}^2$. The closer the bounds are to the optimal ones, the closer α is to 1. Since $\sigma_{n-p+1}^2 < y \leq \sigma_{n-p}^2$ (respectively, $1/\sigma_{n-p+1}^2 > \tilde{y} \geq 1/\sigma_{n-p}^2$), α must satisfy $1 \geq \alpha > 1/r^2$ (respectively, $1 \leq \alpha < r^2$).

Let us first compare *inverse iteration* with *inverse Chebyshev iteration*. For identical start matrices and desired accuracy ε, the ratio of the number K_{ic} of inverse Chebyshev iteration steps to the number K_i of inverse iteration steps is approximately given by

$$(5.41) \qquad S = \frac{K_{ic}}{K_i} \approx \frac{\log(\sigma_{n-p}^2/\sigma_{n-p+1}^2)}{\log(\tilde{d}_{n-p+1} + \sqrt{\tilde{d}_{n-p+1}^2 - 1})},$$

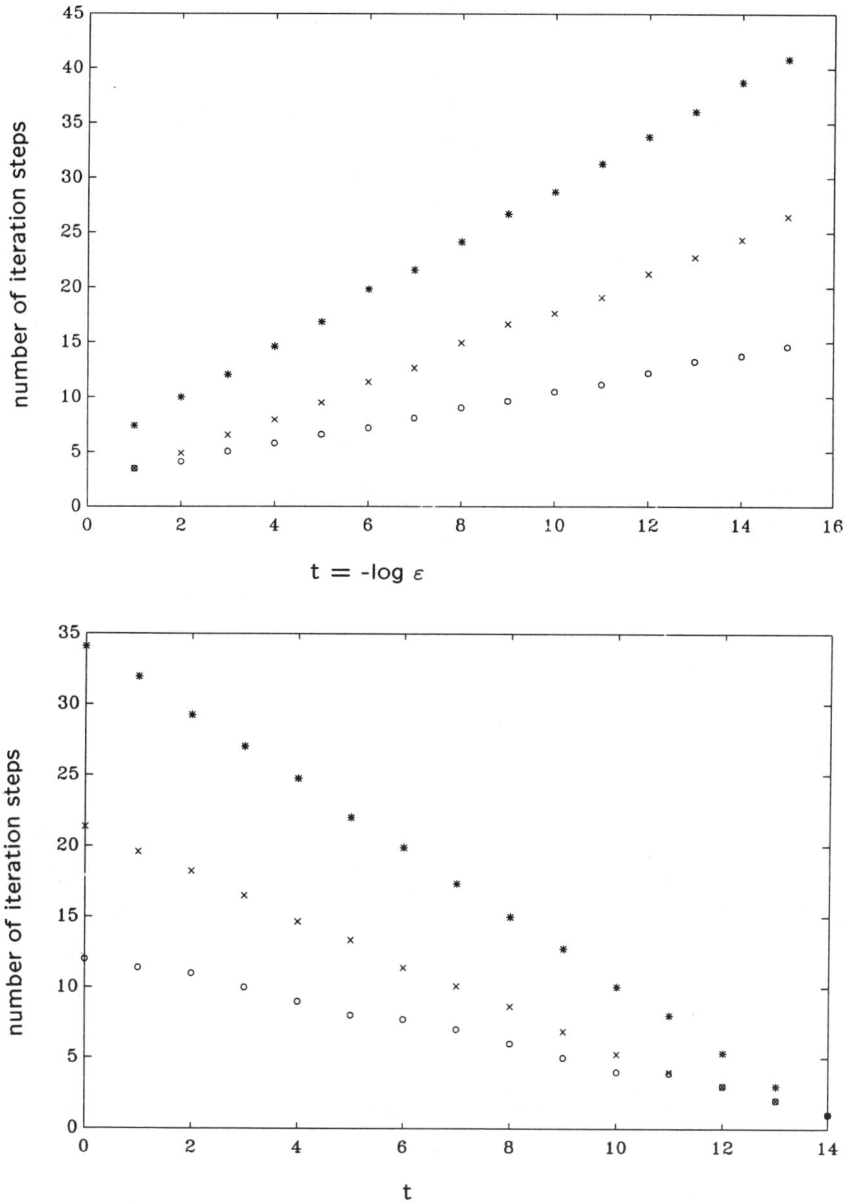

(a)

(b)

FIG. 5.2. *Comparison of the experimentally obtained average numbers of iteration steps required by the inverse iteration* ($\times$), *ordinary Chebyshev iteration* ($*$), *and inverse Chebyshev iteration* ($\circ$) *algorithm for convergence, up to a precision* ε, *to the* 11*th right singular vector of an* 18×11 *matrix* C *in function of* (a) *the desired number of correct digits* $t = -\log \varepsilon$ *in the result* (b) *the number of correct digits* t *in the start vector with* $\varepsilon = 10^{-14}$. *The singular values of* C *are* $\sigma_1 = 2, \sigma_{10} = 1,\ \sigma_{11} = 0.5$ *and the remaining ones equally spaced between* σ_1 *and* σ_{10}. *For Chebyshev iteration optimal bounds* $y = \frac{1}{\bar{y}} = 1$ *and* $z = \frac{1}{\bar{z}} = 4$ *are taken. Each experiment is repeated* 20 *times using different randomly chosen start vectors.*

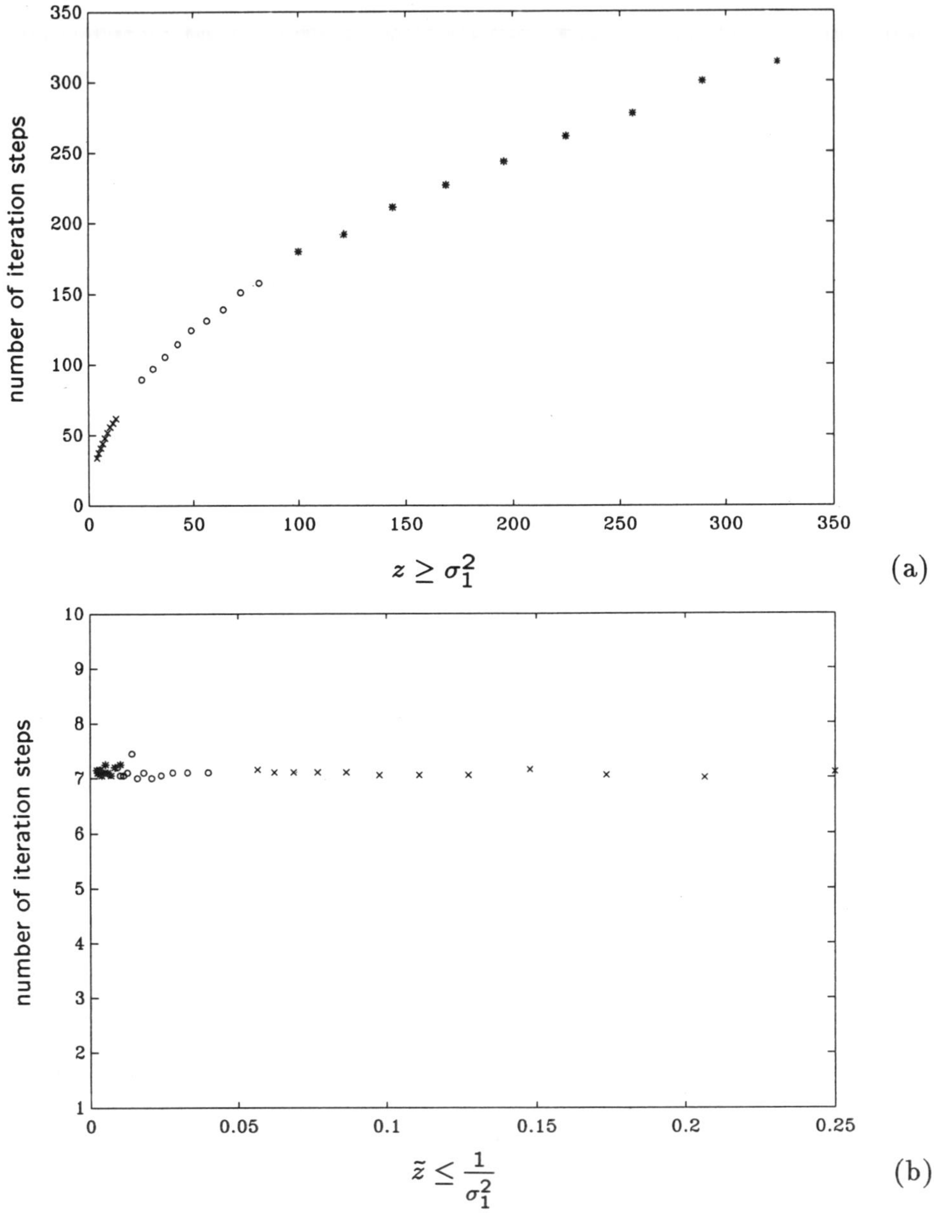

FIG. 5.3. *Comparison of the experimentally obtained average numbers of iteration steps required by the* (a) *ordinary and* (b) *inverse Chebyshev iteration algorithm for convergence, up to a precision* 10^{-14}, *to the* 11*th right singular vector of an* 18×11 *matrix* C *in function of the quality of the given* (a) *upper bound* $z \geq \sigma_1^2$ *and* (b) *lower bound* $\tilde{z} \leq \frac{1}{\sigma_1^2}$. *The singular values of* C *are* σ_1 *variable,* $\sigma_{10} = 1$, $\sigma_{11} = 0.5$ *and the remaining ones equally spaced between* σ_1 *and* σ_{10}. *The symbols* $\times$, $\circ$ *and* $*$ *correspond with the values* $\sigma_1 = 2, 5$ *and* 10, *respectively, and indicate the corresponding* (a) *upper bounds* z *chosen in the interval* $[\sigma_1^2, 3.24\sigma_1^2]$ *and* (b) *lower bounds* $\tilde{z}$ *chosen in the interval* $[0, \frac{1}{\sigma_1^2}]$. *The bound* $y = \tilde{y} = 1$ *is optimal and each experiment is repeated* 20 *times using different randomly chosen start vectors.*

where

$$\tilde{d}_{n-p+1} = \frac{2\sigma_{n-p+1}^{-2} - \tilde{y} - \tilde{z}}{\tilde{y} - \tilde{z}} \quad \text{with} \quad \frac{1}{\sigma_{n-p+1}^2} > \tilde{y} \geq \frac{1}{\sigma_{n-p}^2} \quad \text{and} \quad \frac{1}{\sigma_1^2} \geq \tilde{z}.$$

When $S < 1$, the inverse Chebyshev iteration method converges faster than the inverse iteration method. Since the lower bound $\tilde{z}$ hardly influences the convergence rate of inverse Chebyshev iteration whenever σ_1/σ_{n-p} is not too small (≥ 2), we can take $\tilde{z} = 0$. Then, using the parameters r and α, defined by $\tilde{y} = \alpha/\sigma_{n-p}^2$, (5.41) reduces to

$$S \approx \frac{\log r^2}{\log\left(\frac{2r^2}{\alpha} - 1 + \sqrt{(\frac{2r^2}{\alpha} - 1)^2 - 1}\right)}. \tag{5.42}$$

This convergence rate ratio is plotted in Fig. 5.4(a) as a function of α for different values of r. The *smaller* the *gap* between the desired singular values ($\leq \sigma_{n-p+1}$) and the undesired ones ($\geq \sigma_{n-p}$), i.e., $r \to 1$, the *larger* the *gain in speed* of the inverse Chebyshev iteration method in comparison with the inverse iteration method, but, of course, the more difficult it is to estimate $\tilde{y}$. In other words, a smaller gap increases the number of iteration steps required by inverse iteration at a faster rate than that required by inverse Chebyshev iteration. Whenever r is large, typically $r > 10$, the convergence rate of inverse iteration is high enough to ensure convergence to the desired basis in less than ten iteration steps. Although in these cases the ratio S is still slightly smaller than 1, it does not pay to use inverse Chebyshev iteration since we can win at most one iteration step. Moreover, good bounds $\tilde{y}$ must be estimated. For smaller gaps, say $r \leq 10$, the gain in speed may be quite large, depending on the quality of the estimate $\tilde{y}$, expressed by α. If $\tilde{y}$ approaches its optimal value $1/\sigma_{n-p}^2$ ($\alpha \to 1$), the gain in speed can be considerable but if $\tilde{y}$ is badly chosen and approaches $1/\sigma_{n-p+1}^2$, α tends to its limit value r^2 and induces such an increase in the number of iteration steps that the ratio S largely exceeds 1; i.e., inverse iteration is clearly the most efficient method now.

Using (5.42), it follows that, for $\tilde{z} = 0$, inverse Chebyshev iteration requires less iteration steps than inverse iteration if

$$S \leq 1 \;\Leftrightarrow\; 1 \leq \alpha \leq \left(\frac{2r^2}{r^2+1}\right)^2 \tag{5.43}$$

or, inversely,

$$S \leq 1 \;\Leftrightarrow\; 1 \leq r^2 \leq \frac{\sqrt{\alpha}}{2 - \sqrt{\alpha}}. \tag{5.44}$$

For example, take $r = \frac{10}{5}$. Then (5.43) tells us that the inverse iteration method should be preferred if the estimated bound $\tilde{y}$ is not better than $2.56/\sigma_{n-p}^2$. For

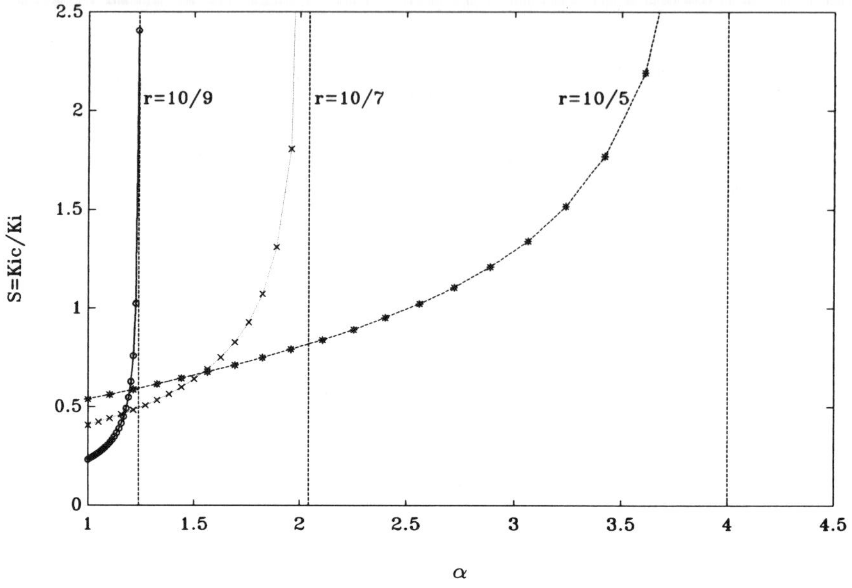

(a)

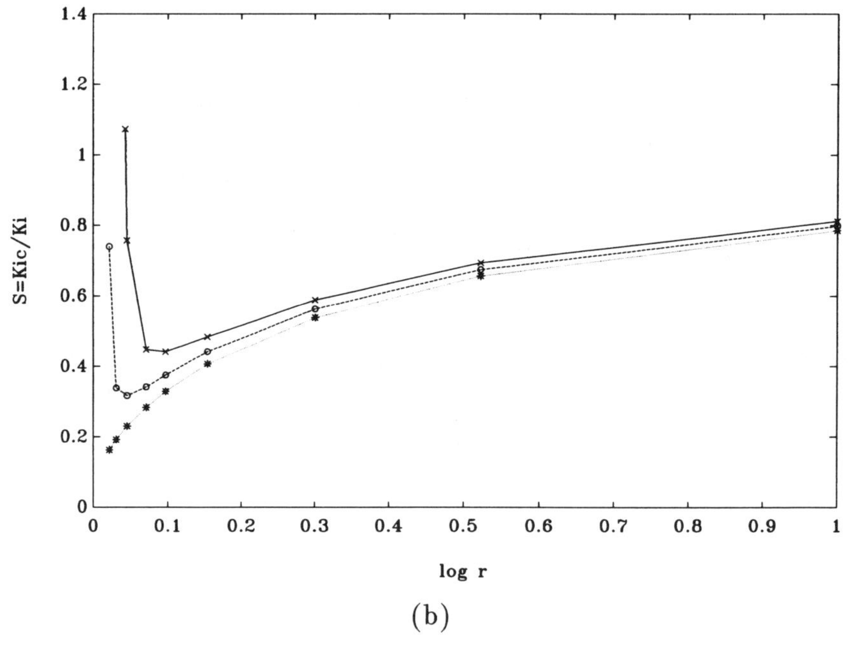

(b)

FIG. 5.4. *Comparison of the ratio of iteration steps* $S = \frac{K_{ic}}{K_i}$, *defined by* (5.42), *in function of* (a) *the quality* α *of the chosen upper bound* $\tilde{y} = \alpha \frac{1}{\sigma_{n-p}^2}$ *with the singular value ratio* $r = \frac{\sigma_{n-p}}{\sigma_{n-p+1}}$ *as parameter and* (b) *the ratio* r *with* α *as parameter. The symbols* $*$, $\circ$, *and* $\times$ *correspond with* (a) $r = \frac{10}{5}, \frac{10}{9}$ *and* $\frac{10}{7}$ *and* (b) $\alpha = 1$ (*optimal*), 1.05^2 *and* 1.1^2, *respectively. Vertical asymptotes occur at values* $\alpha = r^2$.

large r it follows that $S \leq 1$ whenever $\alpha \leq 4$. This means that, independently of the singular value spectrum, inverse iteration is always better than inverse Chebyshev iteration if the bound $\tilde{y}$ can not be sufficiently well estimated, i.e., $\tilde{y} > 4/\sigma_{n-p}^2$.

For correct and optimal estimation of $\tilde{y} = 1/\sigma_{n-p}^2$ corresponding to $\alpha = 1$, S is always ≤ 1 for any r, as shown in Fig. 5.4(b); i.e., inverse Chebyshev iteration converges faster than inverse iteration for all r. The smaller the gap is between σ_{n-p} and $\sigma_{n-p+1}(\log r \to 0)$, the better the convergence rate. As α increases, this gain in speed of inverse Chebyshev iteration becomes relatively smaller. Observe also the sudden increase in number of required inverse Chebyshev iteration steps when α approaches its limit value r^2. If $\alpha = r^2$, then $\tilde{y} = 1/\sigma_{n-p+1}^2$ and $S = \infty$, this means that the start matrix cannot converge to the desired basis because $|T_k(\tilde{d}_{n-p+1})| = |T_k(1)| = 1$ for all k, as opposed to inverse iteration, which still converges provided $r > 1$.

Finally, note that we assume $\tilde{z} = 0$ in the comparison. Choosing a better estimate $\tilde{z} \leq 1/\sigma_1^2$ may improve the efficiency of inverse Chebyshev iteration, in particular when the undesired singular value spectrum is quite dense, say $\sigma_1/\sigma_{n-p} < 2$, such that the better efficiency of inverse Chebyshev iteration is even more pronounced (see Table 5.5).

Everything said here about the convergence of inverse iteration and inverse Chebyshev iteration, respectively, also holds for the convergence of the *power* method [69, §7.3.1] and ordinary Chebyshev iteration, respectively, provided convergence to the singular subspace of the matrix C associated with its p *largest* singular values $\sigma_1, \cdots, \sigma_p$ is considered. This is because inverse iteration and inverse Chebyshev iteration can also be considered as iterative methods that converge to the singular subspace associated with the p largest singular values $1/\sigma_n, \cdots, 1/\sigma_{n-p+1}$ of a matrix, namely, the pseudo-inverse of C. Hence, by replacing $1/\sigma_{n-i+1}$ by σ_i in (5.41) and (5.42), we obtain the equivalent ratios of iteration steps required by the power method to that required by ordinary Chebyshev iteration. These expressions are given in [42].

We now compare *inverse iteration* with *ordinary Chebyshev iteration*. For identical start matrices and desired accuracy ε, the ratio of the number K_c of Chebyshev iteration steps to the number K_i of inverse iteration steps is approximately given by

$$S = \frac{K_c}{K_i} \approx \frac{\log(\sigma_{n-p}^2/\sigma_{n-p+1}^2)}{\log(d_{n-p+1} + \sqrt{d_{n-p+1}^2 - 1})}, \tag{5.45}$$

where

$$d_{n-p+1} = \frac{2\sigma_{n-p+1}^2 - y - z}{y - z} \quad \text{with} \quad \sigma_{n-p+1}^2 < y \leq \sigma_{n-p}^2. \quad \text{and} \quad \sigma_1^2 \leq z.$$

When $S < 1$ the ordinary Chebyshev iteration method converges faster than the inverse iteration method. Using the parameters $r = \sigma_{n-p}/\sigma_{n-p+1}$

and α, defined by $y = \alpha\sigma_{n-p}^2$, (5.45) reduces to

$$S \approx \frac{\log r^2}{\log(\beta + \sqrt{\beta^2 - 1})} \quad \text{with} \quad \beta = \frac{\left(\frac{2}{r^2} - \alpha\right)\sigma_{n-p}^2 - z}{\alpha\sigma_{n-p}^2 - z}. \tag{5.46}$$

This convergence rate ratio is plotted in Fig. 5.5(a) as a function of α for different values of r. Figure 5.5(b) shows the convergence rate ratio as a function of r with α as parameter. The *smaller* the *gap* between the singular values associated with the desired ($\leq \sigma_{n-p+1}$) and the undesired singular subspace ($\geq \sigma_{n-p}$), i.e., $r \to 1$, the *faster* the *convergence* of the Chebyshev iteration method in comparison with inverse iteration but the more difficult it is to estimate the bound y. Analogously to inverse Chebyshev iteration, we observe a sudden increase of the number of iteration steps required by ordinary Chebyshev iteration when α approaches its limiting value $1/r^2$. In this limit, $S = \infty$, i.e., the start matrix cannot converge to the desired basis since $d_{n-p+1} = f(\sigma_{n-p+1}^2) = 1$ and hence $|T_k(d_{n-p+1})| = 1$ for all k. Inverse iteration still converges as long as $r > 1$ and is clearly more efficient.

Using (5.46), we compute that ordinary Chebyshev iteration requires less iteration steps than inverse iteration if

$$S \leq 1 \quad \Leftrightarrow \quad 1 \leq r^2 \leq \beta + \sqrt{\beta^2 - 1}$$

$$\Leftrightarrow \quad 1 \leq r^2 \leq \frac{2\sigma_{n-p}\sqrt{(\alpha - 1)z + \alpha\sigma_{n-p}^2} + \alpha\sigma_{n-p}^2 + z}{z - \alpha\sigma_{n-p}^2}. \tag{5.47}$$

For example, take $\alpha = 1$, $z = 4$, and $\sigma_{n-p} = 1$; then (5.47) tells us that Chebyshev iteration converges faster than inverse iteration when $r^2 \leq \frac{7}{3}$. This agrees well with our experimental results shown in Fig. 5.5(b).

Comparing Figs. 5.4 and 5.5, we observe the larger values of the ratio $S = K_c/K_i$ showing that the convergence of *ordinary Chebyshev iteration* is slower, sometimes considerably, than that of *inverse Chebyshev iteration.* It is not difficult to prove that this is always true. Indeed, for identical start matrices and desired accuracy ε, the ratio of the number K_{ic} of inverse Chebyshev iteration steps to the number K_c of ordinary Chebyshev iteration steps is given by

$$S = \frac{K_{ic}}{K_c} \approx \frac{\log(d_{n-p+1} + \sqrt{d_{n-p+1}^2 - 1})}{\log(\tilde{d}_{n-p+1} + \sqrt{\tilde{d}_{n-p+1}^2 - 1})}. \tag{5.48}$$

Good bounds y, z for ordinary Chebyshev iteration always imply good bounds $\tilde{y}$, $\tilde{z}$ for inverse Chebyshev iteration since we can always take $\tilde{y} = 1/y$ and $\tilde{z} = 1/z$. For this choice, inverse Chebyshev iteration requires fewer iteration steps than ordinary Chebyshev iteration if

$$S \leq 1 \Leftrightarrow d_{n-p+1} \leq \tilde{d}_{n-p+1} \Leftrightarrow (\sigma_{n-p+1}^2 - y)(\sigma_{n-p+1}^2 - z) \geq 0,$$

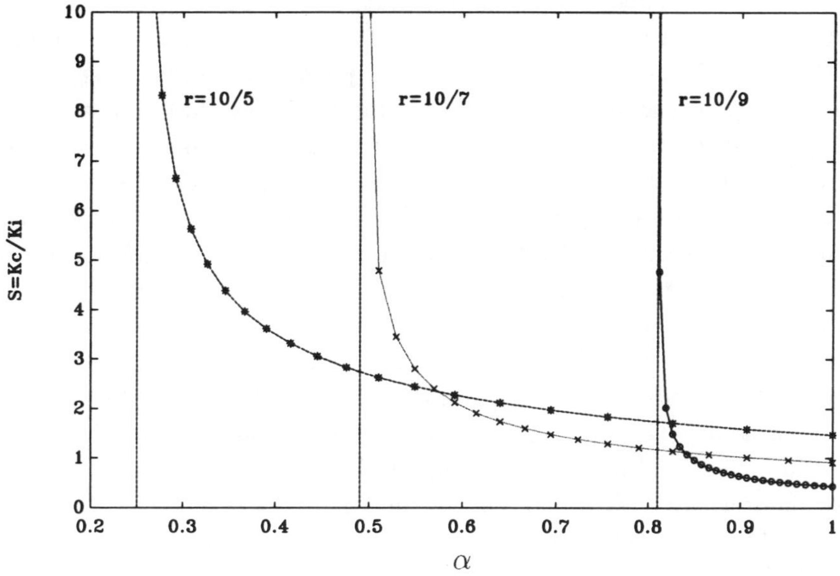

(a)

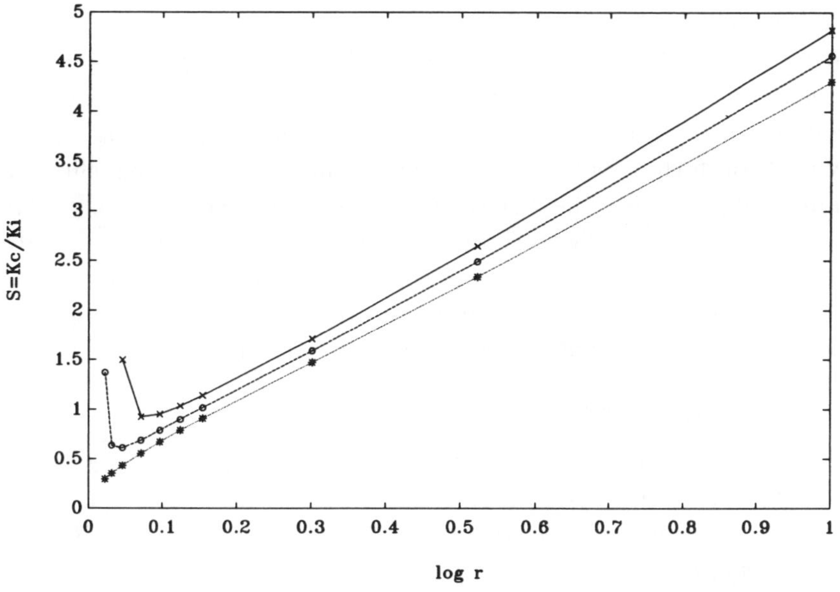

(b)

FIG. 5.5. *Comparison of the ratio of iteration steps* $S = \frac{K_c}{K_i}$, *defined by* (5.46), *in function of* (a) *the quality* α *of the chosen lower bound* $y = \alpha\sigma_{n-p}^2$ *with the singular value ratio* $r = \frac{\sigma_{n-p}}{\sigma_{n-p+1}}$ *as parameter and* (b) *the ratio* r *with* α *as parameter. The symbols* $*$, $\circ$, *and* $\times$ *correspond with* (a) $r = \frac{10}{5}$, $\frac{10}{9}$ *and* $\frac{10}{7}$ *and* (b) $\alpha = 1$ *(optimal)*, $\frac{1}{1.05^2}$ *and* $\frac{1}{1.1^2}$, *respectively. Vertical asymptotes occur at values* $\alpha = \frac{1}{r^2}$.

which is always satisfied for any given matrix C since $x \geq y \geq \sigma_{n-p+1}^2$. The convergence rate ratio $S = K_{ic}/K_c$ is plotted in Fig. 5.6(a) as a function of α, defined by $\tilde{y} = 1/y = \alpha/\sigma_{n-p}^2$, with r as parameter and in Fig. 5.6(b) as a function of r with α as parameter. These figures show that inverse Chebyshev iteration is *always more efficient* than ordinary Chebyshev iteration. The larger the gap between σ_{n-p} and σ_{n-p+1} ($r \to \infty$,) or the worse the quality of the estimated bounds $\tilde{y} = 1/y = \alpha/\sigma_{n-p}^2$ ($\alpha \to r^2$), the larger the gain in speed of the inverse Chebyshev iteration in comparison with the ordinary Chebyshev iteration. When α approaches its limiting value, the number of inverse Chebyshev iteration steps, as well as the number of Chebyshev iteration steps, increases considerably since $d_{n-p+1} \to 1$ and $\tilde{d}_{n-p+1} \to 1$. Observe from Fig. 5.6 that the increase is more pronounced for ordinary Chebyshev iteration as $K_c \to \infty$ and $K_{ic} \to \infty$. Since we cannot go beyond this limiting value $\alpha = r^2$, the curves of Fig. 5.6 are stopped abruptly.

The theoretically obtained curves in Figs. 5.4–5.6 agree well with the experimentally obtained curves given in [180]. Note also that the estimated bounds of this subsection, e.g., (5.43), (5.44), and (5.47), are deduced from theoretical estimates of the number of required iteration steps K_i, K_{ic}, K_c. Hence, they approximately hold in practice as long as the theoretical results are good approximations.

The inverse (Chebyshev) iteration Algorithms 5.1, 5.2, and 5.4 are not recommended for solving very large, sparse, or structured problems if no sparse QR factorization algorithm can be found that exploits the structure or sparsity pattern in the given matrix C. Since the ordinary Chebyshev iteration algorithm does not alter the matrix C, it may be recommended. Although its convergence rate is slower than that of the inverse Chebyshev iteration Algorithm 5.4, efficient matrix product computations $C^T(CQ_k)$ can reduce the number of operations per iteration step in such a way that Algorithm 5.3 now becomes the most efficient one, as illustrated below.

5.7. Slowly varying TLS problems

5.7.1. Simulation results. This section compares the efficiency of the iterative algorithms, discussed in this chapter, with that of the classical TLS Algorithm 3.1 in solving *slowly varying* TLS *problems* $A_{m\times n}X \approx B_{m\times d}$. These problems are simulated as follows.

Example 5.2. Consider the exact overdetermined sets of linear equations $A_0X_0(t_i) = B_0(t_i)$, $i = 1, \cdots, 100$ with A_0 of full rank. The $n \times d$ components $X_{kj}^0(t)$ of $X_0(t)$ are slowly varying sinusoidal functions. Perturb at each time instant the set $A_0X_0(t_i) = B_0(t_i)$ with zero-mean normally distributed noise $\Delta A(t_i)$, $\Delta B(t_i)$ of equal variance $\sigma_\nu^2 = 0.00001$, and take the solution of the previous set as initial guess for the next perturbed set. Compute the TLS solution of the sets $A(t_i)X(t_i) \approx B(t_i)$, $A(t_i) = A_0 + \Delta A(t_i)$, $B(t_i) =$

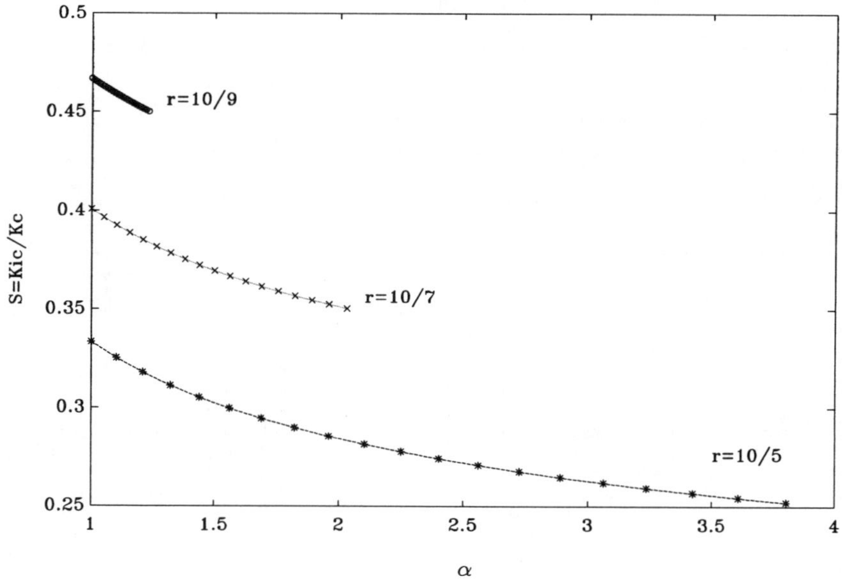

(a)

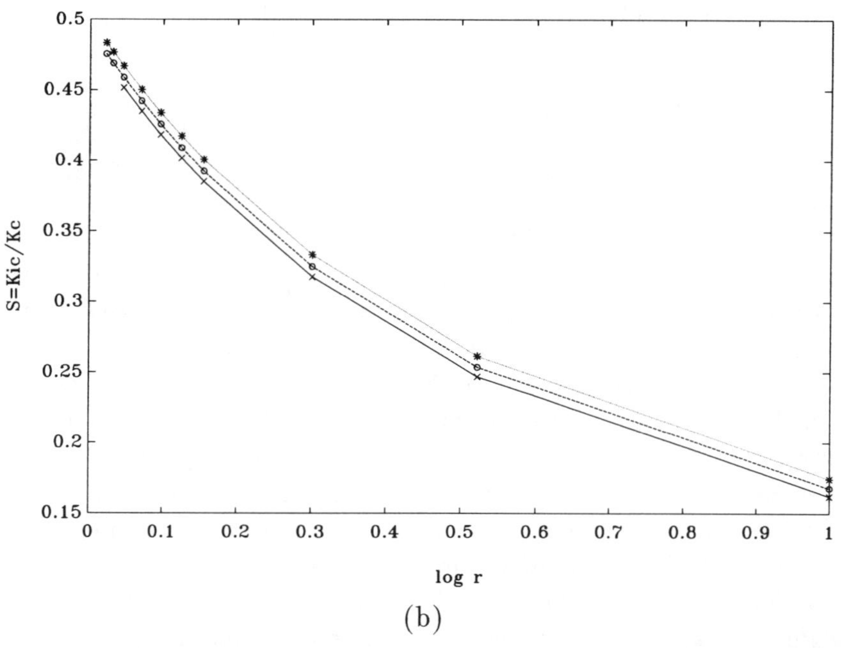

(b)

FIG. 5.6. *Comparison of the ratio of iteration steps* $S = \frac{K_{ic}}{K_c}$, *defined by* (5.48), *in function of* (a) *the quality* α *of the chosen upper bound* $\tilde{y} = \frac{1}{y} = \alpha \frac{1}{\sigma_{n-p}^2}$ *with the singular value ratio* $r = \frac{\sigma_{n-p}}{\sigma_{n-p+1}}$ *as parameter and* (b) *the ratio* r *with* α *as parameter. The symbols* $*$, $\circ$, *and* $\times$ *correspond with* (a) $r = \frac{10}{5}, \frac{10}{9}$ *and* $\frac{10}{7}$ *and* (b) $\alpha = 1$ (*optimal*), 1.05^2 *and* 1.1^2, *respectively.*

TABLE 5.6.

Comparison in efficiency of classical TLS, *inverse iteration, ordinary Chebyshev iteration with preceding* QR *factorization, and inverse Chebyshev iteration for solving Example* 5.2. A_0 *is a* 10×5 *matrix with singular values* $\sigma_1, 0.4, 0.3, 0.2, \sigma_n$. *The* $5 \times d$ *elements* $X^0_{kj}(t_i)$ *are given by* $0.1 \sin(\gamma t_i)$ *if* k *is even and* $0.1 \cos(\gamma t_i)$ *if* k *is odd, with* $\gamma = 0.0001((d+5)(j-1)+k+9), t_i = i$.

problem specifications						cl.TLS	inv.it.		ord.Cheb.it.		inv.Cheb.it.	
d	ε	σ_n	σ_1	y	z	flops	flops	# it.	flops	# it.	flops	# it.
1	10^{-12}	0.1	0.5	0.09^2	0.55^2	4714	1180	8.16	13267	78.76	1153	6.65
1	10^{-12}	0.1	0.5	0.09^2	0.80^2	4714	1180	8.16	18939	112.52	1155	6.66
1	10^{-12}	0.1	0.5	0.09^2	1.10^2	4714	1180	8.16	25692	152.72	1159	6.69
1	10^{-12}	0.1	0.5	0.06^2	0.55^2	4714	1180	8.16	20246	120.30	1343	7.78
1	10^{-12}	0.1	0.5	0.035^2	0.55^2	4714	1180	8.16	38322	227.89	1845	10.77
1	10^{-6}	0.1	0.5	0.09^2	0.55^2	4714	637	4.39	6121	36.22	694	3.92
1	10^{-6}	0.1	5.0	0.09^2	5.5^2	4356	633	4.36	46811	278.42	692	3.91
1	10^{-12}	0.05	0.5	0.04^2	0.55^2	4663	1757	12.16	32211	191.52	1664	9.69
1	10^{-12}	0.04	0.5	0.035^2	0.55^2	4679	2095	14.52	38401	228.36	1855	10.82
2	10^{-12}	0.1	0.5	0.09^2	0.55^2	8818	8107	8.88	108665	117.40	7014	7.18
2*	10^{-12}	0.1	0.5	0.09^2	0.55^2	8818	8107	8.88	88320	95.21	8862	9.17
2	10^{-12}	0.05	0.5	0.04^2	0.55^2	8791	12619	14.08	362375	392.52	10646	11.11
2*	10^{-12}	0.05	0.5	0.04^2	0.55^2	8791	12619	14.08	297425	321.65	11658	12.20
2	10^{-6}	0.05	0.5	0.04^2	0.55^2	8791	6697	7.26	146395	158.31	5909	5.98
2*	10^{-6}	0.05	0.5	0.04^2	0.55^2	8791	6697	7.26	119705	129.30	5956	6.02
2	10^{-12}	0.04	0.5	0.035^2	0.55^2	8601	15526	17.43	431545	467.52	12103	12.69
2*	10^{-12}	0.04	0.5	0.035^2	0.55^2	8601	15526	17.43	391675	423.72	12769	13.40

$B_0(t_i) + \Delta B(t_i), i = 1, \cdots, 100$ with the classical TLS Algorithm 3.1 and also by means of each iterative algorithm discussed here, as follows. First, the basis vectors $v_{n+1}, \cdots, v_{n+d}$ of the right singular subspace of $[A(t_i); B(t_i)]$ associated with its d smallest singular values are computed iteratively, up to a convergence accuracy ε, and then the generic TLS solution $\widehat{X}$ is obtained by performing Step 4 of Algorithm 4.5. The numerical rank of $[A(t_i); B(t_i)]$ is set equal to n. Compute per problem the number of floating point operations ($+$, $-$, $\times$, $/$ or any elementary function), performed in MATLAB [104] and denoted by "flops," as well as the number of required iteration steps, denoted by "# it.," as a measure of efficiency. Also, if more than one basis vector is to be computed, i.e., when $d > 1$, (inverse) Chebyshev iteration is restarted every μ steps: $\mu = 9$ is used for values of d marked with an asterisk, otherwise $\mu = 5$. Different values of ε, d, $\mu, \sigma_1(A_0)$, and $\sigma_n(A_0)$ are tried, as well as different estimates of the lower and upper bounds $y = 1/\tilde{y}$, $z = 1/\tilde{z}$ of (inverse) Chebyshev iteration.

Table 5.6 shows the obtained average numbers of performed flops and required iteration steps for solving one TLS problem. These results do not include the number of required flops for computing the QR factorization (i.e.,

773 flops for $d = 1$ and 981 flops for $d = 2$) for the following reasons. First, since the number of required Chebyshev iteration steps K_c is larger than the computed bound (5.30) ($K_c \geq 4.7$), the Chebyshev iteration algorithm with preceding QR factorization is chosen here. Hence, each considered algorithm (even classical TLS) first performs a QR factorization and, therefore, these calculations are not relevant for a relative comparison in efficiency. Second, the results of Table 5.6 are independent of the type of data sequences considered. Indeed, in Example 5.2 the modifications in the sequence of data matrices $[A(t_i); B(t_i)]$ from one time instant to the next one are of small norm but still of *full* rank, so that at each time instant the QR factorization of the data matrix $[A(t_i); B(t_i)]$, as required by all considered iterative and classical TLS algorithms, must be entirely recomputed. However, if the modifications at each time instant are of *low rank*, the QR factorization $Q_i R_i$ of $[A(t_i); B(t_i)]$ can be *updated* very efficiently to obtain the QR factorization $Q_{i+1} R_{i+1}$ of $[A(t_{i+1}); B(t_{i+1})]$, using [69, §12.6.3]. For instance, the modifications are of rank one when one row $c^T(t_{i+1})$ is appended so that

$$[A(t_{i+1}); B(t_{i+1})] = \begin{bmatrix} [A(t_i); B(t_i)] \\ c^T(t_{i+1}) \end{bmatrix}.$$

Mostly the factor R_i is weighted with an exponentially decaying factor $\tau < 1$ to forget past data. In these problems, each considered algorithm (including classical TLS) computes the QR factorization of the data matrices via *rank-one updating* of the previously computed QR factorization and then proceeds with the triangular factors R_i. This updating operation requires considerably fewer flops than the whole QR factorization. By excluding this number of flops, similar average results to those listed in Table 5.6 would have been obtained. Hence, similar conclusions to those given below hold for these examples.

Observe that these results agree with our theoretical analysis of §5.6. Mostly, the iterative algorithms (except ordinary Chebyshev iteration) are more efficient than the classical TLS algorithm. This difference in efficiency is more pronounced with decreasing convergence accuracy $t = -\log \varepsilon$. Observe that the number of flops, as well as the number of required iteration steps for $\varepsilon = 10^{-6}$, is about one half the corresponding numbers for $\varepsilon = 10^{-12}$, as predicted by the estimates in Table 5.1.

The singular value ratio $r = \sigma_n([A(t_i); B(t_i)])/\sigma_{n+1}([A(t_i); B(t_i)])$ can be estimated by $(\sigma_n^2([A_0; B_0(t_i)]) + m\sigma_\nu^2)^{1/2}/m^{1/2}\ \sigma_\nu$ (see §3.6.1) and is expected to be larger (respectively, smaller) than 10 if the smallest singular value $\sigma_n(A_0) \geq 0.1$ (respectively, ≤ 0.05). Therefore, we expect a small (respectively, large) reduction in iteration steps required by the inverse Chebyshev iteration method compared to the inverse iteration method provided the bound $y = 1/\tilde{y}$ is well estimated, e.g., $y = 0.09^2$ for $\sigma_n(A_0) = 0.1$. This is confirmed by Table 5.6. If the bound y can not be sufficiently well estimated, e.g., when

$y = 0.035^2$ for $\sigma_n(A_0) = 0.1$, the performance of the Chebyshev iteration algorithms becomes worse. This performance reduction is more pronounced for ordinary Chebyshev iteration. Moreover, the efficiency of this latter method depends strongly on the spread of the squared singular value spectrum $[\sigma_n^2, \sigma_1^2]$ and the quality of the upper bound $z = 1/\tilde{z}$, as opposed to inverse Chebyshev iteration. Observe also that the efficiency of the inverse iteration method is independent from σ_1, y, z, and μ.

Although the number of iteration steps required by inverse Chebyshev iteration is usually less than that of inverse iteration, this is not always true for the number of flops, especially when $d = 1$. This is because (inverse) Chebyshev iteration requires extra operations per iteration step for the computation of the Chebyshev recursion. This difference in number of flops is only significant for small n and d and becomes negligible for increasing n and d, as shown in (5.40) and illustrated in Table 5.6 for $d = 2$.

As discussed in §5.5, the Chebyshev iteration algorithms are restarted every μ steps when $d > 1$ to avoid numerical problems. If the convergence rate is high ($K_{ic} \leq 20$), $\mu = 5$ exhibits the best efficiency while for slower convergence rates ($K_c \geq 20$) the best efficiency is obtained at larger μ, e.g., $\mu = 9$ as shown in Table 5.6.

Because of the orthonormalization required for the computation of the distance function, the work load of each iterative algorithm increases considerably when $d > 1$. If the ratio $(n + d)/d$ is small, then classical TLS may even be more efficient than either iterative algorithm (see Table 5.6) as soon as a larger number of iterative steps up to convergence is required, e.g., when the desired accuracy ε is high, the convergence rate is slow, and the start matrix is of bad quality. In these cases, classical TLS is the best choice, while iterative algorithms are typically recommended when $(n + d)/d$ is large enough or the required accuracy ε is low.

Example 5.3. Now consider the same Example 5.2 but assume that each matrix $[A(t_i); B(t_i)]$ is 6×5 and *sparse* as follows:

$$
(5.49) \quad [A(t_i); B(t_i)] = \begin{bmatrix} a_{11}(t_i) & a_{12}(t_i) & 0 & 0 & B_1(t_i) \\ 0 & a_{22}(t_i) & 0 & a_{24}(t_i) & B_2(t_i) \\ 0 & 0 & a_{33}(t_i) & 0 & B_3(t_i) \\ 0 & a_{42}(t_i) & 0 & 0 & B_4(t_i) \\ a_{51}(t_i) & 0 & a_{53}(t_i) & 0 & B_5(t_i) \\ 0 & a_{62}(t_i) & 0 & 0 & B_6(t_i) \end{bmatrix}.
$$

In this example, 50 percent of the elements of $[A(t_i); B(t_i)]$ are zero. Since the QR factorization need not be computed, we can expect a considerable increase in efficiency of the ordinary Chebyshev iteration Algorithm 5.3 (expressed by the number of flops) provided the matrix products $C^T(CQ_k)$ are computed at least as efficiently as $R_C^T(R_C Q_k)$. Therefore, at least 50 percent of the elements in $[A(t_i); B(t_i)]$ must be zero and, moreover, as shown

TABLE 5.7.

Comparison in efficiency of classical TLS, inverse iteration, ordinary Chebyshev iteration, and inverse Chebyshev iteration for solving Example 5.3. $[A(t_i); B(t_i)]$ *is a* 6×5 *sparse matrix with singular values* σ_i, $i = 1, \cdots, 5$ *and* $d = 1$. *The bounds* y *and* z *are given by* 0.95^2 *and* 1.25^2, *respectively.* ε *is set to* 10^{-4}. *The* $5 \times d$ *elements* $X_k^0(t_i)$ *are given by* $0.1 \sin(0.0001(k+9)i)$ *if* k *is even and* $0.1 \cos(0.0001(k+9)i)$ *if* k *is odd.*

problem specific.			cl.TLS	inv.it.		ord.Cheb.it.		inv.Cheb.it.	
σ_1	σ_4	σ_5	flops	flops	#it.	flops	#it.	flops	#it.
1.25	0.99	0.26	3926	864	5	746	6	731	3
1.20	1.02	0.23	3732	754	4	746	6	731	3
1.22	1.01	0.07	3558	644	3	746	6	731	3
1.18	0.97	0.22	3558	754	4	627	5	731	3
1.20	0.95	0.43	3916	864	5	746	6	861	4
1.24	0.97	0.03	3359	644	3	627	5	601	2
1.21	0.95	0.05	3844	644	3	746	6	601	2
1.20	1.02	0.41	3160	974	6	746	6	861	4
1.24	1.02	0.12	3543	644	3	627	5	731	3
1.20	0.97	0.27	3543	864	5	627	5	731	3
1.19	0.97	0.01	3640	534	2	746	6	601	2

in Table 5.7, the ordinary Chebyshev iteration Algorithm 5.3 must converge fast enough such that the work involved in performing the extra iteration steps do not overwhelm the initial gain in efficiency because of the omission of the QR factorization and the efficient computation of $C^T(CQ_k)$. Therefore, the spectrum of undesired singular values must be dense enough and good bounds must be available. Additionally, the larger the number of zeros in $C = [A(t_i); B(t_i)]$ (at least 50 percent), the fewer flops are needed per iteration step for computing $C^T(CQ_k)$ compared to the other iterative algorithms.

In general, we can conclude that despite its larger number of required iteration steps, the ordinary Chebyshev iteration algorithm will be more efficient than inverse (Chebyshev) iteration provided the data matrix $[A(t_i); B(t_i)]$ is sufficiently sparse (at least 50 percent zeros) and its convergence rate is only slightly slower. This difference in efficiency is more pronounced with increasing dimension $n + d$ of the matrix and decreasing number of required iteration steps, e.g., when the desired convergence accuracy ε is low.

5.7.2. Estimation of frequency response functions.

Problem description. In this section we compare the efficiency of the direct computation methods and the inverse iteration method in a practical problem concerning the *estimation of frequency response functions* (FRF). Until now, the FRF remains the most important measurement for experimental structural dynamical analysis. In particular, the application of multiple input measurement of FRFs [97], [129], coupled with broadband excitation signals, has been very successful.

Consider a structure excited simultaneously at n locations, called inputs.

Its response is measured at d other locations, called outputs. The idea is to estimate the FRFs $h_1, \cdots, h_n$ of a linear system between the n inputs and each output. Let, at a specific frequency f_i, the spectra of the inputs to the system be represented by $x_1, \cdots, x_n$ and of the response by y. Assuming a linear system, the following model is valid:

$$x_1 h_1(f_i) + \cdots + x_n h_n(f_i) = y. \tag{5.50}$$

The spectra that are actually measured will not represent the true inputs and outputs, due to measurement noise and other inaccuracies such as leakage [97]. Using m measurement samples at the specific frequency $f_i (m \geq n)$, a slightly incompatible set of m linear equations can be generated:

$$XH \approx Y. \tag{5.51}$$

$H = [h_1(f_i), \cdots, h_n(f_i)]^T$ contains the unknowns. The rows of X and Y contain the measured spectra of the n inputs and one output, respectively. Estimates of the FRFs $h_1, \cdots, h_n$ are then obtained by solving this *overdetermined* set of equations (5.51) at *each frequency point* f_i. The TLS technique is appropriate and is shown to be more accurate than the ordinary LS technique [129].

Method description. The TLS estimates of the true (unknown) values $h_j(f_i)$ of the FRFs at a specific output are always computed [97] from the coefficients of the eigenvector v_{n+1} (using (2.26)) associated with the smallest eigenvalue λ_{n+1} of the *cross-product* matrix $[X; Y]^H [X; Y]$, derived from (5.51), i.e., by solving

$$[X; Y]^H [X; Y] \begin{bmatrix} H \\ -1 \end{bmatrix} = \begin{bmatrix} X^H X & X^H Y \\ Y^H X & Y^H Y \end{bmatrix} \begin{bmatrix} H \\ -1 \end{bmatrix} \approx 0. \tag{5.52}$$

The superscript H denotes the Hermitian conjugate, i.e., the complex conjugate and transpose of the matrix. The elements $(X^H X)_{ij}$ of $X^H X$ are the crosspowers (or autopowers if i = j) between the measured input spectra x_i and x_j. The elements $(Y^H X)_i$ of $Y^H X$ are the crosspowers between the measured input spectrum x_i and measured output spectrum y. $Y^H Y$ is the autopower of the output spectrum y. Those *auto- and crosspowers* in (5.52) are immediately computed during data collection and are the *only available data*, so we have to compute the FRFs from the eigenvector equations. We thus use the relationship between the SVD of a matrix [X;Y] and the eigenvalue decomposition of its associated Hermitian matrix in (5.52). Hence, the use of Algorithm 5.1 is justified here. Using the interlacing property for eigenvalues [69, Cor. 8.1.4], it follows by contraposition that $[X; Y]^H [X; Y]$ has a nonrepeated smallest eigenvalue and has, therefore, a unique associated eigenvector if the submatrix $X^H X$ is quasi-nonsingular. This is so when the inputs are not correlated

[129]. In this case, the TLS problem is generic, hence iterative methods can be safely used. Moreover, this smallest eigenvalue is usually about two orders of magnitude smaller than the others, so inverse iteration is the best choice here.

Note that we deal with *complex* matrices. This does not create a problem since each complex SVD or complex set of equations can be transformed into the *real* domain. Splitting the complex set (5.51) into its real and imaginary part $(X_r+jX_c)(H_r+jH_c) \approx Y_r+jY_c$ it is immediately proved that the solution H is also obtained by solving the following real problem

$$(5.53)\; X_R \begin{bmatrix} H_r \\ H_c \end{bmatrix} \approx Y_R \quad \text{with } X_R = \begin{bmatrix} X_r & -X_c \\ X_c & X_r \end{bmatrix} \text{ and } Y_R = \begin{bmatrix} Y_r \\ Y_c \end{bmatrix}.$$

The $2(n+1) \times 2(n+1)$ real symmetric matrix $[X_R; Y_R]^T[X_R; Y_R]$ has the same eigenvalues as the $(n+1) \times (n+1)$ complex matrix in (5.52) but they are of multiplicity 2. However, all basis vectors of the eigensubspace, associated with one eigenvalue, correspond with the same complex eigenvector [83]. It is thus sufficient to calculate only one eigenvector of $[X_R; Y_R]^T[X_R; Y_R]$ associated with its smallest eigenvalue and then derive the corresponding complex eigenvector of $[X; Y]^H[X; Y]$. In fact, the upper and lower half of the eigenvector of $[X_R; Y_R]^T[X_R; Y_R]$ correspond with the real and imaginary part of the eigenvector of $[X; Y]^H[X; Y]$.

Computational results. To compare the efficiency of the direct computation methods classical and partial TLS and the inverse iteration method in the estimation of FRFs, we make use of the cross- and autopowers of tested structures obtained from Leuven Measurement Systems International(LMS). The calculations are performed in double precision on the IBM 4381 P03 of the Computer Center of the Katholieke Universiteit Leuven.

Application examples.

(a) As a first test structure, a *fuel tank* is considered. The structure is excited simultaneously at two locations, called inputs ($n = 2$), using broad band excitation signals and the frequency response is measured at four other locations, called outputs. The broad band measurements are taken at 512 frequency points f_i, spaced equally over the base band from 0 Hz to 100 Hz. Hence, the frequency resolution Δf is 100/512 Hz. The estimates of the FRFs h_1 and h_2 are then obtained by solving 512 sets of equations (5.52), corresponding to the 512 frequency points $f_i = i\Delta f$ for $i = 0, \cdots, 511$. By plotting the amplitude of these solutions $h_1(f_i)$ and $h_2(f_i)$ for $i = 0, \cdots, 511$ at each frequency point f_i, the amplitude curves of the FRFs are obtained. A typical curve is shown in Fig. 5.7.

(b) As a second test structure, a *motorbike frame* excited simultaneously at two inputs using broad band excitation signals is used. The response is also measured at four outputs. The broad band measurements are taken at 512 frequency points f_i, spaced equally over the base band from 0 Hz to 200 Hz. Hence, the frequency resolution Δf is now 200/512 Hz. Typical amplitude curves of the FRFs are given in [97].

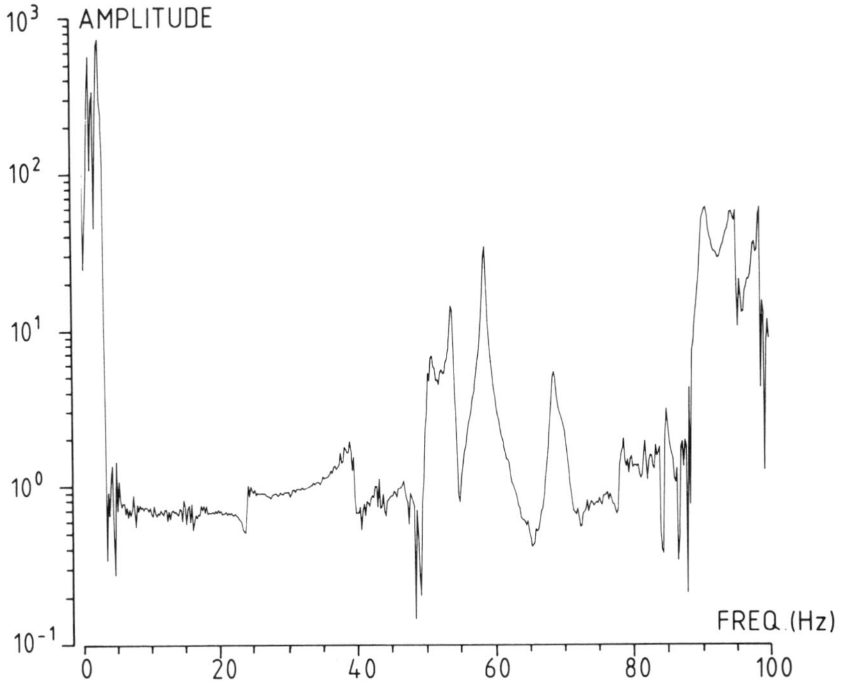

FIG. 5.7. *Typical curve showing the estimated amplitude of the frequency response function h_1 measured at one location (output 2 here) of the considered test structure (a fuel tank), due to excitation at another location (input 1 here) of the same structure. The curve is obtained by first computing estimates of h_1 at 512 equally spaced frequency points $f_i = i\Delta f$ for $i = 0, \cdots, 511$ and $\Delta f = 100/512$ Hz. The estimated amplitudes of $h_1(f_i)$ are plotted against f_i and linked with their neighbours through straight lines.*

To scale the data appropriately, LMS uses the following factor for all sets of equations:

$$\frac{1}{n} \frac{\sum_{i=1}^{n} \sqrt{\sum_{k=1}^{512} ((X^H X)_{ii})_k^2}}{\sqrt{\sum_{k=1}^{512} (Y^H Y)_k^2}}. \tag{5.54}$$

$((X^H X)_{ii})_k$ and $(Y^H Y)_k$ are the autopowers of the spectra of the ith input and the output, respectively (see (5.52)), corresponding to the kth frequency point. By multiplying the autopower $Y^H Y$ with (5.54) and all crosspowers

$(X^HY)_i = (Y^HX)_i$ with the square root of (5.54), the entries, as well as the errors in the matrix of (5.51), have approximately the same order of magnitude.

To compare the efficiency of the three methods in the aforementioned examples, we register the CPU times required for computing the FRFs at each output location. The desired convergence accuracy ε is 10^{-7}. All complex TLS problems are transformed into the real domain.

The FRFs at *each output* location are computed by solving 512 one-dimensional TLS problems (5.52) with two inputs ($n = 2$) and the considered output ($d = 1$). Tables 5.8(a) and 5.8(b) show the CPU times obtained by computing the response at each output location of the considered test structure, with the classical TLS Algorithm 3.1, the partial TLS Algorithm 4.5, and the inverse iteration Algorithm 5.1. For the inverse iteration Algorithm 5.1 we take as start vector the eigenvector associated with the smallest eigenvalue λ_n of the cross-product matrix $[X;Y]^H[X;Y]$ in (5.52), corresponding with the previous frequency point. For the first set only, we must start from an arbitrary vector. Looking at Fig. 5.7, we indeed observe that the solution of (5.52) obtained at a frequency point f_i is only slightly different from the solution of (5.52) at f_{i+1}, except around the resonance frequencies and the end points of the base band.

From Table 5.8, we observe that inverse iteration is more efficient than the direct computation methods classical TLS and PTLS . These differences are more pronounced if the desired convergence *accuracy* ε is *lowered.* Indeed, since the inverse iteration method converges linearly with constant factor $\sigma_n^2/\sigma_{n+1}^2$ (see Table 5.2), this method requires relatively less iteration steps K_i with decreasing convergence accuracy ε than the direct computation methods which converge ultimately cubically to the desired basis vectors in s QR or QL iteration steps. In other words, the ratio K_i/s decreases with decreasing ε and thus the efficiency of the inverse iteration method is relatively more pronounced.

According to our analysis given in Tables 5.2, 5.3, and 5.4, these differences in efficiency would be even more pronounced if the structure was excited at more than two locations (i.e., with increasing n). However, we do not have such data on hand.

Observe also that PTLS is about two times faster than the classical TLS algorithm. This result agrees with our analysis given in §4.5.

5.8. Other iterative algorithms

5.8.1. Rayleigh quotient iteration. If the gap between the desired ($\leq \sigma_{n-p+1}$) and undesired ($\geq \sigma_{n-p}$) singular values of a matrix C is not large and no adequate shift λ_0 can be computed such that the ratio $|\sigma_{n-p}^2 - \lambda_0|/|\sigma_{n-p+1}^2 - \lambda_0|$ is high enough, the convergence rate of the inverse iteration process is slowed down. In these cases, Rayleigh Quotient (RQ) iteration

TABLE 5.8.

Comparison of CPU *times (expressed in milliseconds) required for computing the two* FRF's h_1 *and* h_2 *at four output locations of* (a) *a fuel tank and* (b) *a motorbike frame, with the direct computation methods classical* TLS *and partial* TLS, *and the inverse iteration Algorithm* 5.1. The FRF's *are computed by solving* 512 TLS *problems* (5.52) *with two inputs* ($n = 2$) *and each of the four considered outputs* ($d = 1$). *For the inverse iteration Algorithm* 5.1, *the eigenvector, associated with the smallest eigenvalue of the matrix in* (5.52) *computed in frequency point* f_{i-1}, *is used as starting vector for computing the solution of* (5.52) *at* f_i *(at* f_0 *an arbitrary start vector is taken). The desired convergence accuracy* ε *is* 10^{-7}. *For this choice, the average number of iteration steps for solving* (5.52) *is approximately four.*

CPU time	OUTPUT 1	OUTPUT 2	OUTPUT 3	OUTPUT 4
classical TLS	5990	6042	6246	6093
partial TLS	3015	3026	3090	3069
inverse iteration	2457	2520	2560	2495

(a)

CPU time	OUTPUT 1	OUTPUT 2	OUTPUT 3	OUTPUT 4
classical TLS	6166	6331	6334	6332
partial TLS	3008	3069	3118	3147
inverse iteration	2448	2435	2670	2970

(b)

Table 5.8.

could be used for accelerating the convergence rate in cases where convergence to one singular vector of the desired singular subspace of C is required. RQ iteration [69, §8.4.2] is a variant of inverse iteration obtained by applying onto the matrix $S = C^T C$ the inverse power functions $f_k(\lambda) = (\lambda - \lambda_0(k))^{-1}$ with variable shift $\lambda_0(k)$, causing a well-defined motion of the vertical asymptote in each iteration step. This shift $\lambda_0(k)$ is the Rayleigh quotient $r(q_k)$ of the iteration vector q_k, defined by

$$r(q_k) = \frac{q_k^T S q_k}{q_k^T q_k} \quad \text{minimizing} \quad f(\lambda) = \|(S - \lambda I)q_k\|_2. \tag{5.55}$$

Clearly, because q_k is an approximate eigenvector of $S, \lambda_0(k) = r(q_k)$ is a reasonable choice for the corresponding eigenvalue. On the other hand, if $\lambda_0(k)$ is an approximate eigenvalue then inverse iteration tells us that the solution to

$$(S - \lambda_0(k)I)q_k = q_{k-1} \tag{5.56}$$

is almost always a good approximate eigenvector. Combining these two ideas in the natural way gives rise to the RQ iteration.

The RQ iteration method has several advantages. The shift λ_0 need *not* be estimated but is automatically generated by the algorithm. Because RQ iteration chooses the *best shift* in each step, it converges faster than inverse iteration. Parlett [118] has shown that the convergence of the RQ iteration method is ultimately *cubic*. However, on the other hand, RQ iteration requires *more computations* per iteration step since the matrix $S - \lambda_0(k)I$ in (5.56) changes in every step. Hence, its factorization must be updated (contrary to inverse iteration with fixed shift). Several implementations are discussed in [177, §10.5].

Furthermore, RQ iteration only applies to square symmetric matrices S, and there is no way to completely circumvent the explicit formation of $S = C^T C$ when we want to use RQ iteration and the data C are not available in the cross-product form S. As discussed in §5.3.1, this may affect the numerical accuracy of the computed basis vectors. Moreover, we must take care that the start vector q_0 is always correctly chosen to converge to the desired singular vector of C and to ensure a correct solution. For example, most time varying problems are characterized by abrupt changes at certain time instants that seriously affect the quality of the start vector and cause RQ iteration to converge to an undesired singular triplet. Hence, the TLS solution may be *incorrectly* computed. On the other hand, if we use inverse iteration with zero shift, convergence to the smallest singular triplet is always ensured. Of course, we can combine both methods to ensure a correct solution at an *optimal* convergence rate. No general rules can be set up but an optimal strategy must be worked out for each particular application.

If convergence to several basis vectors of the desired singular subspace is required, extensions of RQ iteration to subspaces must be used, e.g., inverse subspace iteration with Ritz acceleration [32] applied to R^{-1}, where R is obtained from a preceding QR factorization of C.

5.8.2. Lanczos methods. Lanczos methods are iterative methods that bidiagonalize an arbitrary matrix C or tridiagonalize a symmetric matrix S directly without any orthogonal updates of the matrix as in the Householder approach (see §4.2). The main advantages of the Lanczos methods are that the original matrix is not overwritten and that little storage is required since only matrix-vector products are computed. This makes the Lanczos methods interesting for large matrices especially if they are sparse and if there exist fast routines for computing matrix-vector products, such as for Toeplitz and Hankel matrices.

The Lanczos bidiagonalization (respectively, tridiagonalization) method generates a sequence of bidiagonal (respectively, tridiagonal) matrices with the property that its extremal singular values are progressively better estimates of the extremal singular values of the given matrix. Therefore, the Lanczos methods may be computationally more attractive than Householder bidiagonalization in the computation of the SVD of a matrix. Indeed, long before the bidiagonalization or tridiagonalization is complete, the extremal singular values of the given matrix can already be accurately obtained. From the associated singular vectors and the generated Lanczos vectors, good approximations to the singular vectors of the given matrix are derived. This is why it is worthwhile to compare the Lanczos methods with the classical Householder approach in bidiagonalizing a matrix C or tridiagonalizing the corresponding matrix C^TC, and evaluate their efficiency in solving TLS problems. In TLS applications, we seek the smallest singular values and they are often well separated from the others. In this case they may emerge before n iterations are performed. We wish to make use of this. Since we want to compute the singular triplets of the data matrix C itself rather than the eigenpairs of the cross-product matrix $S = C^TC$, we carry out the Lanczos *bidiagonalization* instead of a tridiagonalization.

The bidiagonal matrix $J_n = U^TCV$ is computed directly with $U = [u_1, \cdots, u_m], V = [v_1, \cdots, v_n]$ orthonormal and J_j given by

$$J_j = \begin{bmatrix} q_1 & e_2 & 0 & \cdots & 0 \\ 0 & q_2 & e_3 & & \vdots \\ \vdots & & \ddots & \ddots & 0 \\ & & & \ddots & e_j \\ 0 & & \cdots & 0 & q_j \end{bmatrix} \tag{5.57}$$

by equating columns in $CV = UJ_n$ and $C^T U = VJ_n^T$:

$$(5.58) \qquad Cv_j = q_j u_j + e_j u_{j-1}, \qquad e_1 u_0 = 0,$$

$$(5.59) \qquad C^T u_j = q_j v_j + e_{j+1} v_{j+1}, \qquad e_{n+1} v_{n+1} = 0,$$

for $j = 1, \cdots, n$. Defining $r_j = Cv_j - e_j u_{j-1}$ and $p_{j+1} = C^T u_j - q_j v_j$, we may conclude from the orthonormality that $q_j = \pm\|r_j\|_2, u_j = r_j/q_j, e_{j+1} = \pm\|p_{j+1}\|_2$ and $v_{j+1} = p_{j+1}/\|e_{j+1}\|_2$. Properly sequenced, these equations define the Lanczos method for bidiagonalizing a rectangular matrix.

ALGORITHM 5.5 (Lanczos bidiagonalization).

Given : an $m \times n$ matrix C,
a start vector $v_1 \in \mathcal{R}^n, \|v_1\|_2 = 1$

$r_1 = Cv_1$
$q_1 = \|r_1\|_2$
$u_1 = r_1/q_1$
For $j = 1, 2, \cdots$
  $p_{j+1} = C^T u_j - q_j v_j$
  $e_{j+1} = \|r_{j+1}\|_2$
  if $e_{j+1} = 0$ then stop
  $v_{j+1} = p_{j+1}/e_{j+1}$
  $r_{j+1} = Cv_{j+1} - e_{j+1} u_j$
  $q_{j+1} = \|r_{j+1}\|_2$
  if $q_{j+1} = 0$ then stop
  $u_{j+1} = r_{j+1}/q_{j+1}$
end for
END

This algorithm requires $K_{lb}(2mn + 3(m + n)) - m - n$ multiplications (and divisions) where K_{lb} is the number of iteration steps.

In exact arithmetic, the singular values of J_n coincide with those of C if at most n iterations are performed. In other words, the algorithm terminates as soon as $r(\leq n)$ orthogonal Lanczos vectors u_i and v_i have been generated. If the singular vectors s_j of J_r associated with the desired singular values of J_r (and C) are computed, e.g., by inverse iteration, then the desired left (respectively, right) singular vectors z_j of C (or good approximations) are obtained from $z_j = Q_r s_j$, where Q_r contains the r generated orthogonal Lanczos vectors u_i (respectively, v_i). The smaller r is compared to n, the more efficient the Lanczos method. However, because of roundoff errors, a *loss of orthogonality* among the computed Lanczos vectors occurs when some singular triplets have already converged and several undesirable phenomena may be observed [119], [69, §9.2]. This is usually coped with by introducing a complete or selective *reorthogonalization* procedure, which, of course, increases the complexity. If this is not done, then spurious singular values may appear and it is necessary to detect these. For more details, see [32], [114], [115], [33], [99], [143], and [60, §9.2].

As proved in [66] and experimentally verified, the bidiagonal J_j in (5.57), obtained after j steps of the Lanczos bidiagonalization Algorithm 5.5 on C, is related to the tridiagonal T_j, obtained after j steps of the standard Lanczos tridiagonalization method [69, §9.1] on $S = C^T C$, by

$$T_j = J_j^T J_j. \tag{5.60}$$

Hence, the eigenvalues and eigenvectors of T_j equal the squared singular values and right singular vectors of J_j. The only differences are that Lanczos tridiagonalization is computationally more efficient but the singular values of C may be computed less accurately from T_j because of the squaring. This also implies that Lanczos bidiagonalization of C and Lanczos tridiagonalization of $C^T C$ will have the same convergence properties with respect to the singular values of C. The well-known *Kaniel–Paige theory* reveals how fast the Lanczos process converges towards the desired singular values. Let us consider the case where the *smallest* singular value of C is desired, as required in TLS problems.

THEOREM 5.2 (Convergence of the Lanczos bidiagonalization method). *Let C be an $m \times n$ matrix, given by (4.1). If $\theta_1 \geq \cdots \geq \theta_j$ are the singular values of the bidiagonal matrix J_j obtained after j steps of the Lanczos iteration, then*

$$\sigma_n^2 \leq \theta_j^2 \leq \sigma_n^2 + \frac{(\sigma_1^2 - \sigma_n^2)\tan^2\varphi}{T_{j-1}^2(d_n)} \tag{5.61}$$

with $d_n = 1 + 2\rho_n, \rho_n = (\sigma_{n-1}^2 - \sigma_n^2)/(\sigma_1^2 - \sigma_{n-1}^2)$ and $\cos\varphi = |v_n^T q_0|$, q_0 start vector, $\|q_0\|_2 = 1$ and $T_{j-1}(d_n)$ the value in d_n of the $(j-1)$th Chebyshev polynomial.

Proof. For the proof see [69, Cor. 9.1.4]. □

The bounds are quite satisfactory in the case of convergence to an extremal singular value. If we wish to obtain the singular value σ_n of C, up to a precision ε, it follows from (5.61) that we must require

$$\frac{(\sigma_1^2 - \sigma_n^2)\tan^2\varphi}{T_{j-1}^2(d_n)} \leq \varepsilon^2. \tag{5.62}$$

Approximating $T_{j-1}(d_n)$ in (5.62) by (5.27) with $x = d_n$ and taking its logarithm, we can derive an estimate for the number K of required iteration steps or equivalently the dimension K_{lb} of the bidiagonal $J_{K_{lb}}$ for *convergence* of the Lanczos Algorithm 5.5 to the *smallest* singular value σ_n of C up to a precision ε:

$$K_{lb} \approx \frac{\frac{1}{2}\log(\sigma_1^2 - \sigma_n^2) + \log|\tan\varphi| + \log(2\varepsilon^{-1})}{\log(d_n + \sqrt{d_n^2 - 1})} + 1 \tag{5.63}$$

with d_n, φ, and σ_i defined in (5.61).

Comparing θ_j with the corresponding estimates of σ_n by ordinary Chebyshev iteration and inverse iteration (computed straightforwardly from the corresponding singular vector v_n as $\|Cv_n\|_2$) reveals interesting results about the convergence rate of the corresponding iterative methods. Indeed, using the proof and notation of Theorem 5.2, we prove that the ordinary Chebyshev iteration method has converged to σ_n, up to a precision ε, if

$$(5.64) \qquad (\sigma_1^2 - \sigma_n^2)\tan^2\varphi \frac{\max_{i=1,\cdots,n-1} T_{j-1}^2(d_i)}{T_{j-1}^2(d_n)} \leq \varepsilon^2$$

with $d_i = f(\sigma_i^2) = 1 + 2\rho_i$ and $\rho_i = \frac{\sigma_i^2 - y}{y-z}$ from (5.28). For inverse iteration, convergence to σ_n, up to a precision ε, is obtained if

$$(5.65) \qquad (\sigma_1^2 - \sigma_n^2)\tan^2\varphi \left(\frac{\sigma_n^2 - \lambda_0}{\sigma_{n-1}^2 - \lambda_0}\right)^{2j-2} \leq \varepsilon^2.$$

Taking optimal bounds (5.32) for ordinary Chebyshev iteration and since $\max_{i=1,\cdots,n-1} T_{j-1}^2(d_i) \leq 1$, (5.64) implies (5.62). Thus, the number K of iteration steps required by either method is equally bounded by (5.63). Hence, we can conclude that the *Chebyshev* iteration process with *optimal* bounds converges *as fast* as the *Lanczos* process with respect to an *extremal* singular value of a matrix (also for the largest singular value, we obtain the same result). Observe, however, that this result is derived from a comparison in upper bounds of the approximation error on the squared desired singular value. From the properties of the Lanczos process, we know that θ_j^2 in (5.61) is, in fact, the minimum of the *RQ* $r(x) = x^T Sx/x^T x$ with $S = C^T C$ over all vectors x of $R([q_0, \cdots, Sq_0, \cdots, S^{j-1}q_0])$. The Chebyshev estimate of σ_n corresponds with a particular vector x in the same span, namely, $x = T_{j-1}^{yz}(S)q_0$ with the bounds y, z given by (5.32). This proves that convergence of the Lanczos process will never be slower than the Chebyshev iteration convergence but only minimally faster, as can be deduced from the properties of the Chebyshev polynomials.

The *same* parameters influencing the convergence of the ordinary Chebyshev iteration Algorithm 5.3 to the smallest singular triplet of a matrix, also determine the *convergence* of the Lanczos method. Hence, the number of required iteration steps decreases if the desired accuracy ε is low, the quality of the start vector is good and the convergence rate $\log(d_n + \sqrt{d_n^2 - 1})$ is high. However, one important advantage of the Lanczos method compared to Chebyshev iteration is that no bounds on the undesired singular value spectrum are needed, and, the convergence rate is at least as fast as that of Chebyshev iteration with optimal bounds. Like ordinary Chebyshev iteration, the convergence rate of the Lanczos method depends not only on the *gap* between σ_{n-1}^2 and σ_n^2, but also on the *relative spread* of the undesired squared singular value spectrum $(\sigma_1^2 - \sigma_{n-1}^2)/\sigma_{n-1}^2$. This last parameter explains why the Lanczos bidiagonalization Algorithm 5.5 can only be efficient (and preferred to the classical Householder bidiagonalization in TLS problems) if the data matrix C has a dense singular value spectrum. This is usually not the case in TLS problems. Moreover, since many additional tests and/or orthogonalizations of Lanczos vectors are required for reasons of stability and convergence, we can conclude that the Lanczos methods are only worthwhile to be studied in large or sparse TLS problems.

Analogously to the inverse Chebyshev iteration method, we can accelerate the convergence of the Lanczos process to the smallest singular triplets of a matrix C by applying Algorithm 5.5 to R_C^{-1}, where R_C is the upper triangular factor obtained from a QR factorization of C. This so-called inverse Lanczos method requires a preceding QR factorization of C and a solution of two sets of linear equations of the form $R_C^T h = u_j$ and $R_C g = v_{j+1}$ at each step j, instead of two matrix-vector products. This algorithm first converges to the largest singular value of R_C^{-1}, which corresponds with the smallest singular value of C. Again, the Kaniel–Paige theory [69, Thm. 9.1.3] reveals how fast the inverse Lanczos method converges towards the desired singular value. If the SVD of C is given by (4.1) and $\theta_1 \geq \cdots \geq \theta_j$ are the singular values of the bidiagonal matrix J_j obtained after j steps of this inverse Lanczos method, then

$$\text{(5.66)} \qquad \frac{1}{\sigma_n^2} \geq \theta_1 \geq \frac{1}{\sigma_1^2} - \frac{(\frac{1}{\sigma_n^2} - \frac{1}{\sigma_1^2}) \tan^2 \varphi}{T_{j-1}^2(\tilde{d}_n)},$$

where $\tilde{d}_n = 1 + 2\tilde{\varrho}_n$, $\tilde{\varrho}_n = (\sigma_n^{-2} - \sigma_{n-1}^{-2})/(\sigma_{n-1}^{-2} - \sigma_1^{-2})$ and φ, q_0, T_{j-1} defined in (5.61). On the other hand, following the same proof, it is again proven that the inverse Chebyshev iteration algorithm has converged to σ_n, up to a precision ε, if

$$\text{(5.67)} \left(\frac{1}{\sigma_1^2} - \frac{1}{\sigma_n^2}\right) \tan^2 \varphi \frac{\max_{i=1,\cdots,n-1} T_{j-1}^2(\tilde{d}_i)}{T_{j-1}^2(\tilde{d}_n)} \leq \varepsilon^2 \text{ with } \tilde{d}_i = 1 + 2\frac{\sigma_i^{-2} - \tilde{y}}{\tilde{y} - \tilde{z}}.$$

Taking optimal bounds (5.34) and since $\max_{i=1,\cdots,n-1} T_{j-1}^2(\tilde{d}_i) \leq 1$, (5.67) coincides with the error bound $((1/\sigma_n^2) - (1/\sigma_1^2)) \tan^2 \varphi / T_{j-1}^2(\tilde{d}_n) \leq \varepsilon^2$ of the inverse Lanczos method, as can be deduced from (5.66). This proves that the inverse Chebyshev iteration method with optimal bounds converges approximately as fast as this inverse Lanczos method to an extremal singular value. Hence, the same conclusions hold with respect to solving TLS problems. The possible (dis)advantages of this inverse Lanczos method in solving slowly varying TLS problems need further investigation.

If we wish to compute several singular triplets, e.g., in multidimensional TLS problems, and if we have good initial guesses for more than one set, we can use a *block* recursion. Contrary to the single-vector recursion, the block recursion can deal with multiple singular values. The block recursions are defined in a very similar manner by *block Lanczos* algorithms.

The block Lanczos bidiagonalization method is discussed in [32] and [66], where extensions of the Kaniel–Paige theory are also presented, describing the convergence behavior of this block method. Loss of orthogonality also plagues the block Lanczos algorithms. However, all of the orthogonality enforcement schemes, discussed in [143], [99], and [69, §9.2] for the vector Lanczos methods, can also be applied in the block setting.

5.9. Concluding remarks

If a priori information about the solution of the problem to be solved is available, e.g., when solving slowly varying TLS problems, iterative methods are recommended. In this chapter, various iterative methods are investigated. In particular, inverse iteration, ordinary Chebyshev iteration and inverse Chebyshev iteration are discussed in detail: several algorithms are presented and the convergence properties are analyzed. Based on the theory of matrix functions (§5.2), estimates of the number of required iteration steps are deduced for each method. These estimates enable us to evaluate and compare the practical use and efficiency of each method in solving TLS problems (§§5.6 and 5.7).

If the ratio between the singular values corresponding to the desired and undesired singular subspace of the matrix $C = [A; B]$ is *high*, *inverse iteration* (§5.3) is shown to be the *best* choice. Indeed, *fast convergence* is ensured, efficient computations per iteration step, as well as a good numerical accuracy, and, moreover, no bounds whatsoever on the singular value spectrum are needed.

However, convergence is *slowed down* if the *gap* between the singular values associated with the desired ($\leq \sigma_{r+1}$) and undesired ($\geq \sigma_r$) singular subspace of the matrix C is *not* sufficiently *large*. Typically, this occurs for ratios $\sigma_r/\sigma_{r+1} < 10$. In these cases, convergence can be *accelerated* by applying the Chebyshev polynomials instead of the inverse power functions to the cross-product matrix $C^T C$ (respectively, $(C^T C)^{-1}$). This method is called *ordinary* (respectively, *inverse*) Chebyshev iteration (§5.4). Usually, inverse Chebyshev iteration is recommended. Indeed, this method is proven to always converge *faster* to the desired basis than ordinary Chebyshev iteration. Mostly, the gain in speed is very significant. In contrast to ordinary Chebyshev iteration and provided that the undesired singular value spectrum is not too small, say $\sigma_1/\sigma_r \geq 2$, the convergence rate of inverse Chebyshev iteration is hardly influenced by the spread of this spectrum, as well as by the quality of its upper bound $z \geq \sigma_1^2$. Moreover, this method is proved to always converge *faster* than inverse iteration, provided a lower bound y sufficiently close to the smallest undesired squared singular value σ_r^2 is known. The smaller the gap, the larger the gain in speed. Ordinary Chebyshev iteration is shown to only be efficient in problems characterized by a very dense singular value spectrum. For most TLS problems, this is not the case. However, if the matrix C is sparse enough such that the matrix products per iteration step can be computed very efficiently, this method can be the most efficient one despite its larger number of required iteration steps (§§5.6.2 and 5.7.1). Note, however, that the Chebyshev iteration algorithms require slightly more operations per iteration step than the inverse iteration algorithm. Moreover, these algorithms must be used with special care, especially for convergence to multidimensional singular subspaces, while the inverse iteration algorithm is nearly fullproof (§5.5).

Based on the convergence rate and number of operations in each iteration step, the efficiency of these iterative methods can be compared with that of the direct computation methods: classical and partial SVD (§5.6.1). It is concluded that iterative methods are *most efficient* when *only one* singular vector is to be computed, their convergence rate is sufficiently *high*, the desired *accuracy* is rather *low* and a *good start* vector is available. Moreover, we need to know the dimension of the desired subspace or equivalently, the number of desired vectors. Iterative methods are particularly *attractive* in solving *slowly varying* TLS problems (§5.7) since the solution of a previous set is then mostly a good initial guess for the solution of the next set. A practical problem concerning the estimation of frequency response functions illustrates the efficiency of the inverse iteration method.

Finally, Rayleigh quotient iteration and Lanczos methods are discussed (§5.8). These methods may be of interest for solving TLS problems where the gap between the desired ($\leq \sigma_{r+1}$) and undesired ($\geq \sigma_r$) singular triplets is not sufficiently large ($\sigma_r/\sigma_{r+1} < 10$). Rayleigh quotient iteration (§5.6.2) may improve the convergence rate of the iteration process if convergence to one singular vector is required, e.g., in one-dimensional TLS problems, and a good start vector is available. The convergence rate of the *Lanczos* methods (§5.8.2) is very similar to that of ordinary Chebyshev iteration with optimal bounds $y = \sigma_r^2$ and $z = \sigma_1^2$. Hence, the same conclusions hold. Similarly to inverse Chebyshev iteration, Lanczos methods can be applied implicitly to the inverse cross-product matrix $(C^TC)^{-1}$ to achieve a considerable improvement in convergence speed. These so-called "inverse" Lanczos methods have the same convergence behavior as inverse Chebyshev iteration in the presence of optimal bounds $\tilde{y} = 1/\sigma_r^2$ and $\tilde{z} = 1/\sigma_1^2$. Therefore, similar conclusions hold. Moreover, no bounds on the singular value spectrum are needed. Note, however that roundoff errors make the Lanczos methods difficult to use in practice. To maintain their stability and reliability, many additional computations are required, which reduce further their efficiency.

In this chapter, we mainly concentrate on slowly varying TLS problems $A(t_i)X(t_i) \approx B(t_i)$ in which the modifications in the data matrix at each time step are of small norm but still of full rank. This implies that at each step the refactorization of the new data matrix must be started from scratch. In other situations, where the modifications at each time step are of rank one (or two), e.g., when a new row is appended successively, efficient rank-one updating algorithms for the QR factorization [69, §12.6] can be applied for computing the new factorization from the previous one and can speed up the computation time considerably. In solving these problems, the same iterative algorithms as presented here can be used provided the QR factorization of the data matrix $[A(t_i); B(t_i)]$ is replaced by a QR updating algorithm. Hence, similar conclusions as those given here hold for these problems. The complexity per singular triplet is generally a quadratic function of the problem size but there exist faster algorithms whose complexity is linear. The main computational saving in the latter "fully adaptive" algorithms stems from the fact that at every time instant an *approximate* factorization is computed as opposed to the above updating techniques that require *exact* factorizations. Moreover, the possible structure or sparsity pattern is not taken into account. In this way, an acceptable approximation is computed at any stage at a low computational cost. Algorithms carrying out this task are surveyed in [32].

Recently, a fully adaptive algorithm was presented [107], [108] for updating the SVD of a matrix. At any time step a new row is appended to an $m \times n$ matrix and an approximate decomposition is computed from a previous one in $\mathcal{O}(n^2)$ operations. The updating consists in an interlacing of QR updatings and a Jacobi-type SVD algorithm applied to the triangular factor. A few SVD steps after each QR update suffice to restore an acceptable approximation of the SVD. When combined with exponential weighting, this algorithm is shown to be very useful in tracking slowly varying TLS problems [109]. Since only Givens or Jacobi rotations are used, efficient parallel implementations are possible. This algorithm can be elegantly mapped onto a *systolic* array with $\mathcal{O}(n^2)$ parallelism resulting in an $\mathcal{O}(n^2)$ throughput. Use is made of slight modifications of well-known parallel implementations for the matrix-vector product, the QR updating, and the SVD [109].

Chapter 6

Algebraic Connections Between Total Least Squares and Least Squares Problems

6.1. Introduction

In this chapter we compare the total least squares (TLS) and the ordinary least squares (LS) problem algebraically. Interesting connections between their solutions, their residuals, their corrections applied to data fitting, and their approximate subspaces are proved.

In §6.2, relationships between both *solutions* are presented, showing the influencing parameters that describe the main differences between both problems. These parameters also describe the connections between the *residuals*, as shown in §6.3. LS minimizes a sum of squared residuals, while TLS, in fact, minimizes a sum of weighted squared residuals. The *corrections* applied by TLS and LS to the data to obtain a compatible set of equations are compared analytically in §6.4. Expressions are derived relating the solution to the relative lengths of the corrections applied to each column of A and B. From the correction matrix of each technique, assumptions about the underlying perturbation model are deduced. Finally, in §6.5, the *approximate subspaces* into which the data A and B are projected to make the set of equations compatible are compared. Again, the closeness of both subspaces is shown to be determined by the same parameters that describe the connections between the solutions, the residuals, and the corrections applied by TLS and LS.

6.2. Connections between the solutions

Various relationships between the TLS and LS solutions, as well as upper bounds, have been presented for the basic TLS problem $Ax \approx b$ defined by (2.19)–(2.21) (see, e.g., [68]). These results can be straightforwardly extended to the multidimensional TLS problem $AX \approx B$ when $\sigma_{n+1} = \cdots = \sigma_{n+d}$ in (1.22). In addition, alternative relationships describing the connections between both solutions are proved in this section. The closed-form expressions of the TLS solution, derived in Theorems 2.7, 3.10, 3.13, and Corollary 3.1, are very useful. Indeed, recall that the TLS solution of $AX \approx B$ is given by

$$\widehat{X} = (A^T A - \sigma_{n+1}^2 I_n)^{-1} A^T B \tag{6.1}$$

provided $\sigma'_n > \sigma_{n+1} = \cdots = \sigma_{n+d}$ (let (1.21) and (1.22) be the SVD of A and $[A; B]$, respectively). Comparing (6.1) with the LS solution

$$X' = (A^T A)^{-1} A^T B \tag{6.2}$$

shows that σ_{n+1} completely determines the differences between both solutions which is exactly expressed below.

COROLLARY 6.1. *Let* (1.21) (*respectively,* (1.22)) *be the* SVD *of* A (*respectively,* $[A; B]$). *If* $\sigma'_n > \sigma_{n+1} = \cdots = \sigma_{n+d}$, *then*

$$\widehat{X} = [I - \sigma_{n+1}^2 (A^T A)^{-1}]^{-1} X' = [I + \sigma_{n+1}^2 (A^T A - \sigma_{n+1}^2 I)^{-1}] X'. \tag{6.3}$$

Proof. Substituting $A^T B = A^T A X'$ in (6.1) and using [138, Lemma 30] yields (6.3). □

Assume that A is of full rank ($\sigma'_n > 0$). $\sigma_{n+1} = 0$ means that the set $AX \approx B$ is compatible since $\text{rank}([A; B]) = \text{rank}(A)$ and thus $R(B) \subseteq R(A)$. In this case, both solutions coincide. As σ_{n+1} deviates from zero, the set $AX \approx B$ becomes more and more incompatible and the differences between the LS and TLS solutions are more and more pronounced. In a sense, we could say that σ_{n+1} measures the difference between both solutions, as well as the degree of *incompatibility* of the set $AX \approx B$ and thus indicates how *closely* the data A, B *fit* the so-called classical errors-in-variables model (see also chap. 8):

$$A_0 X = B_0, \qquad A = A_0 + \Delta A \quad \text{and} \quad B = B_0 + \Delta B. \tag{6.4}$$

This model assumes that the error-free data satisfy a compatible set of equations $A_0 X = B_0$ (i.e., the $(n+1)$th singular value σ_{n+1}^0 of $[A_0; B_0]$ is zero) but A_0, B_0 can only be observed with errors ΔA, ΔB and therefore $\sigma_{n+1} \neq 0$.

Using (6.3), the maximal difference between both solutions is bounded as follows.

COROLLARY 6.2. *Let* (1.21) (*respectively,* (1.22)) *be the* SVD *of* A (*respectively,* $[A; B]$) *and* B' *be the orthogonal projection of* B *onto* $R(A)$. *If* $\sigma'_n > \sigma_{n+1} = \cdots = \sigma_{n+d}$, *then*

$$\begin{aligned}
\|\widehat{X} - X'\|_F &= \sigma_{n+1}^2 \|(A^T A - \sigma_{n+1}^2 I)^{-1} X'\|_F && (6.5)\\
&\leq \sigma_{n+1}^2 \|B'\|_F \, \sigma_n'^{-1} (\sigma_n'^2 - \sigma_{n+1}^2)^{-1} && (6.6)\\
&\leq \sigma_{n+1}^2 \|X'\|_F \, (\sigma_n'^2 - \sigma_{n+1}^2)^{-1}, && (6.7)\\
\|\widehat{X}\|_F &\geq \|X'\|_F. && (6.8)
\end{aligned}$$

Proof. Equation (6.5) follows directly from (6.3) by taking norms. To prove (6.6), substitute (6.2) in (6.5). Using the dyadic decomposition $A = \sum_{i=1}^n \sigma'_i u'_i v_i'^T$ yields

$$\widehat{X} - X' = \sigma_{n+1}^2 \sum_{i=1}^{n} v'_i \sigma_i'^{-1} (\sigma_i'^2 - \sigma_{n+1}^2)^{-1} u_i'^T B. \tag{6.9}$$

Since $\sigma'_1 \geq \cdots \geq \sigma'_n > \sigma_{n+1}$, the Frobenius norm of (6.9) is maximized for $B'//u'_n$. In this case, $u_i'^T B = 0$ for all $i \neq n$ and $\|u_n'^T B\|_F = \|B'\|_F$. Also, $X' = v'_n \sigma_n'^{-1} u_n'^T B$ and thus $\|X'\|_F = \|B'\|_F \sigma_n'^{-1}$. Substituting these expressions into (6.9) and taking norms proves (6.6) and (6.7). Finally, substituting the dyadic decomposition $A = \sum_{i=1}^n \sigma'_i u'_i v_i'^T$ into (6.1) and (6.2) yields

$$\|\widehat{X}\|_F^2 = \sum_{i=1}^n \|u_i'^T B\|_2^2 \sigma_i'^2 (\sigma_i'^2 - \sigma_{n+1}^2)^{-2} \quad \text{and} \quad \|X'\|_F^2 = \sum_{i=1}^n \|u_i'^T B\|_2^2 \sigma_i'^{-2}.$$

Since each of the terms in the first sum is larger than the corresponding term in the second sum, we obtain (6.8). □

Compared to (6.6), the upper bound

$$\|\widehat{X} - X'\|_F \leq \|R'\|_F \|B\|_F (\sigma_n'^2 - \sigma_{n+1}^2)^{-1}$$

derived from [68, Cor. 4.2] under the same assumptions of Corollary 6.2, is usually too crude. Indeed, mostly $\|B\|_F >> \sigma_{n+1}$, and equality here is only attained when B is orthogonal to $R(A)$, which is very unlikely to occur. In contrast, the upper bound (6.6) is *met* whenever $B'//u'_n$ (or, equivalently, $X'//v'_n$), i.e., the orthogonal projection B' of B onto $R(A)$ is parallel with the singular vector of A associated with its smallest singular value. In this case, the LS solution is *most sensitive* to noise, and its accuracy can be seriously affected [204]. Moreover, the upper bound (6.6) is tight, not only if B is "close" to the lowest singular vector u'_n of A, but also if the condition number σ'_1/σ'_n of A is large and the projection of B onto u'_n is not too small (since X is then mainly oriented in the direction of v'_n). This implies that the upper bound (6.6) is indeed realistic.

The main contribution of Corollary 6.2 lies in the fact that it points out the *parameters* σ'_n, σ_{n+1}, $\|B\|_F$ and the orientation of B with respect to u'_n, which mainly *influence the expected differences in accuracy* between the TLS and LS solutions of the set $AX \approx B$ in the presence of *perturbations* in the measurements A and B.

First consider σ_{n+1}. σ_{n+1} not only influences the differences between both solutions but also the differences in condition between the full rank LS problem and the generic TLS problem and thus the difference in numerical accuracy between their respective solutions in the presence of worst-case perturbations, as explained in §7.2. Indeed, the bounds in Corollary 6.2 are large whenever σ_{n+1} is close to σ'_n. As pointed out in Chapters 2 and 3, this means that the TLS problem $AX \approx B$ becomes *close-to-nongeneric*. Nongeneric TLS problems occur whenever A is (nearly) rank-deficient ($\sigma'_n \approx 0$) or when the system $AX \approx B$ is highly incompatible ($\sigma'_n \approx \sigma_{n+1}$ large). Especially in the latter case, the generic TLS problem becomes more ill-conditioned than the full rank LS problem, which implies that its solution is more sensitive to data errors, as discussed in [68]. In a sense, we could say that $\sigma_n'^2 - \sigma_{n+1}^2$ is a measure of how *close* $AX \approx B$ is to the class of *nongeneric* TLS problems.

The influence of the remaining parameters σ_n', $\|B\|_F$ and the orientation of B with respect to u_n' on the relative accuracy of TLS with respect to LS is proved in §7.3. Assume that $A_0 X \approx B_0$ is the corresponding unperturbed set, A_0 is of rank n, and the perturbations in A and B have approximately the same size. Then, the TLS solution of $AX \approx B$ is expected to be *more accurate* than the LS solution provided the ratio $(\sigma_n - \sigma_{n+1}^0)/\sigma_n' > 1$ where σ_{n+1}^0 is the $(n+1)$th singular value of $[A_0; B_0]$. This implies that $A_0 X \approx B_0$ is sufficiently compatible. The advantages of TLS with respect to LS are more pronounced with increasing ratio. This is the case when A_0 tends to be rank-deficient ($\sigma_n' \approx 0$), when the number of observation vectors B_0 and their length is increasing ($\|B_0\|_F$ large), and when B_0 (or, equivalently, the solution X_0) becomes "close" to the singular vector u_n' (or v_n') of A_0 associated with its smallest singular value. These theoretical observations have been confirmed by experimental tests described in Figs. 7.4, 7.5, and 7.6 (see §7.4).

Finally, we compare the minimum norm TLS solution with the LS solution. Based on the expressions, proved in Theorem 3.10, the following relationships have been established [207].

THEOREM 6.1. *Consider the* TLS *problem* (3.1)–(3.2). *Let the* SVD *of* A *and* $[A; B]$ *be given by* (1.21) *and* (1.22), *respectively, but assume that* V *is partitioned differently as in* (3.55) *in which* V_{22} *is of full rank. Furthermore, assume that*

$$(6.10)\quad \sigma_p > \sigma_{p+1} = \cdots = \sigma_{n+e} > \sigma_{n+e+1} \geq \cdots \geq \sigma_{n+d} \quad \text{for } p \geq n \text{ and } e \geq 1.$$

If we compute $\widehat{X} = -V_{12}V_{22}^{\dagger}$ *or* (3.44) *as the minimum norm* TLS *solution and* $X' = A^{\dagger}B$ *is the* LS *solution, then*

$$\begin{aligned}
\|\widehat{X} - X'\|_2 &\leq \frac{\sigma_{n+1}^2}{\sigma_n'^2}\|\widehat{X}\|_2 \qquad \text{if } p = n \\
&\leq \frac{\sigma_{n+1}^2}{\sigma_r'^2}\|V_{12}^T\widehat{X} - V_{22}^T\|_2 + \|\widehat{X}\|_2 \\
&\qquad \text{if } p < n \text{ and } \operatorname{rank}(A) = r > p \\
&\leq \frac{\sigma_{n+1}^2}{\sigma_p'^2}\|V_{12}^T\widehat{X} - V_{22}^T\|_2 + \frac{\sigma_{n+1}^2}{\sigma_p'^2 - \sigma_{n+1}^2}\|\widehat{X}\|_2 \\
&\qquad \text{if } p < n \text{ and } \operatorname{rank}(A) = p.
\end{aligned}$$

Let $A_p = \sum_{i=1}^{p} \sigma_i' u_i' v_i'^T$, $p < n$ *and* $\operatorname{rank}(A) = r > p$. *If* $X_p' = A_p^{\dagger}B$, *then*

$$\|\widehat{X} - X_p'\|_2 \leq \frac{\sigma_{n+1}^2}{\sigma_p'^2}\|V_{12}^T\widehat{X} - V_{22}^T\|_2 + 2\frac{\sigma_{n+1}^2}{\sigma_p'^2 - \sigma_{n+1}^2}\|\widehat{X}\|_2.$$

Proof. For the proof see [207, Thm. 3.1]. □

Observe that $\widehat{X} = -V_{12}V_{22}^{\dagger}$ is the minimum norm TLS solution of a "nearby" TLS problem $\widetilde{A}X \approx \widetilde{B}$ in which $[\widetilde{A};\widetilde{B}] = \sum_{i=1}^{p} \sigma_i u_i v_i^T + \sigma_{n+1}\sum_{i=p+1}^{n+t} u_i v_i^T$, i.e., the smallest singular values $\sigma_{p+1} = \cdots = \sigma_{n+d}$ coincide. If these assumptions hold, we have the following theorem.

THEOREM 6.2. *Consider the same notation and assumptions of Theorem* 6.1 *but additionally assume that*

$$\sigma_p > \sigma_{p+1} = \cdots = \sigma_{n+d} \quad for\ p \le n. \tag{6.11}$$

In this case $X' = X'_p$ *and we have*

$$\begin{aligned} &\textit{if } \sigma_{n+1} = 0 \quad \textit{then} \quad \widehat{X} = X' \\ &\textit{if } \sigma_{n+1} > 0 \quad \textit{then} \quad \|\widehat{X} - X'\|_2 \le \frac{\sigma_{n+1}^2}{\sigma_p'^2}\|\widehat{X}\|_2 \\ &\qquad\qquad\qquad\qquad\ \ \|\widehat{X} - X'\|_2 \le \frac{\sigma_{n+1}^2}{\sigma_p'^2 - \sigma_{n+1}^2}\|X'\|_2. \end{aligned}$$

Proof. For the proof see [207, Thm. 3.2]. □

In practical applications, especially in rank-deficient problems $AX \approx B$, where the rank of $A - U_2\Sigma_2V_{12}^T$ equals p and $X' = X'_p$, the difference $\|\widehat{X} - X'\|_2$ is small provided $\sigma'_p >> \sigma_{n+1}$. If, however, a LS solution $X' \neq X'_p$ is considered, we expect that $\widehat{X}$ is not close to X' since $\|\widehat{X} - X'_p\|_2$ is much smaller than $\|\widehat{X} - X'\|_2$ when $\sigma_{n+1} << \sigma_p$. In these cases, the minimum norm LS solution X'_p rather than X' should be preferred as solution of the nearly rank-deficient LS problem. See [206] for a detailed discussion of rank-deficient LS problems.

6.3. Connections between the residuals

Consider the set $Ax \approx b$. While LS minimizes the sum of squared residuals $\|b - Ax\|_2^2$, TLS in fact minimizes a sum of *weighted* squared residuals. The results, derived from [59], are generalized for the multidimensional problem $AX \approx B$. Furthermore, various relationships between LS and TLS residuals are presented and upper bounds are given.

Consider the model (6.4) and assume that the rows of $[\Delta A; \Delta B]$ are independently and identically distributed with common zero mean and covariance matrix $\sigma_\nu^2 I_{n+d}$. The estimation of the parameters X, A_0, and σ_ν^2 has received much attention in the literature and various approaches have been suggested. Let us consider the *generalized* LS *estimation* (GLSE) approach. This approach makes no assumptions about the distribution of $[\Delta A; \Delta B]$ beyond those made above. Since under this model the rows of the residual matrix $B - AX$ have zero mean and common covariance matrix $\sigma_\nu^2(I_d + X^TX)$ [59], the GLS estimates of X are chosen to minimize some scalar function f of the *normalized* "error" or "*residual*" *matrix*, defined as

$$N(X) = (B - AX)(I + X^T X)^{-1/2}, \tag{6.12}$$

where $(I + X^T X)^{1/2}$ is any square root of $I + X^T X$.

If $d = 1$, (6.12) is an m-dimensional vector and minimization of the length of (6.12) yields the generalized LS estimator of Sprent [154]. This estimator is identical to the TLS solution of $Ax \approx b$. Indeed, the scalar function f is defined as

$$f(N(x)) = \|[A; b][x^T; -1]^T\|_2 / \|[x^T; -1]^T\|_2. \tag{6.13}$$

If the SVD of $[A; b]$ is given by (1.22), it is easy to verify that $f(N(x)) \geq \sigma_{n+1}$ since $\sigma_{n+1} = \min_{v \neq 0} \|[A; b]v\|_2 / \|v\|_2$ is the minimal singular value of $[A; b]$. Equality holds if and only if $v = [x^T; -1]^T$ is in the right singular subspace associated with σ_{n+1}. The fact that the TLS solution $\widehat{x}$, given by (2.23) or (3.33), is precisely defined in this way proves that the TLS problem amounts to finding an $\widehat{x} \in \mathcal{R}^n$ such that (6.13) is minimized and thus equals σ_{n+1}.

When $d > 1$, many scalar functions f of (6.12) could be used. Gleser and Watson [60] minimized the Frobenius norm of (6.12). Their estimator is identical to the TLS solution of $AX \approx B$. We can thus say that the generic TLS solution $\widehat{X}$ amounts to *minimize* the Frobenius norm of the *normalized residual matrix* (6.12). This result is generalized below.

THEOREM 6.3 (Normalized residual matrix norm minimization). *Let* (1.22) *be the* SVD *of* $[A; B]$ *and* V_{22} *nonsingular. Then, the* TLS *solution* $\widehat{X}$ *of* $AX \approx B$, *given by* (3.13), *minimizes any orthogonally invariant norm of the "normalized error" or "residual" matrix*

$$N(X) = (B - AX)(I + X^T X)^{-1/2},$$

where $(I + X^T X)^{1/2}$ *is any square root of* $I + X^T X$.

Proof. For the proof see [59, Thm. 2.1]. □

In addition to this GLSE approach, an approach analogous to ordinary LS could be used, based on minimizing some scalar function of a "*total error*" *matrix*, defined as

$$F(X, A_0) = [A - A_0; B - A_0 X] \tag{6.14}$$

over X, A_0. When based on minimizing the *Frobenius norm* of (6.14), we obtain *the* TLS *minimization function* (3.1). This result is generalized in the next theorem.

THEOREM 6.4 (Total error matrix norm minimization). *Consider model* (6.4). *Let* (1.22) *be the* SVD *of* $[A; B]$ *and* V_{22} *nonsingular. Then, the* TLS *solution* $\widehat{X}$ *of* $AX \approx B$ *and the* TLS *approximation* $\widehat{A}$, *given by* (3.13) *and* (3.11), *respectively, minimize any orthogonally invariant norm of the "total error" matrix*

$$F(X, A_0) = [A - A_0; B - A_0 X]$$

over X *and* A_0.

Proof. For the proof see [59, Thm. 2.1]. □

Another approach to estimating X comes from noting that $B - A\widehat{X}$ and $B\widehat{X}^T + A$ are *uncorrelated.* This suggests choosing X so that

$$(B - AX)^T(BX^T + A) = 0. \tag{6.15}$$

Thus, the TLS solution $\widehat{X}$ makes its residual $B - A\widehat{X}$ and $B\widehat{X}^T + A$ uncorrelated.

In the following, expressions for the TLS residual $\widehat{R}$, defined by $\widehat{R} \triangleq B - A\widehat{X}$, are derived and its connection with the LS residual $R' \triangleq B - AX'$ is described.

THEOREM 6.5. *Let* $\widehat{X}$ *be the solution of the* TLS *problem* (3.1)–(3.3) *and* $[\Delta\widehat{A}; \Delta\widehat{B}]$ *the corresponding* TLS *correction matrix, then*

$$\widehat{R} = \Delta\widehat{B}(\widehat{X}^T\widehat{X} + I) \tag{6.16}$$

$$= \Delta\widehat{B} - \Delta\widehat{A}\widehat{X}. \tag{6.17}$$

Proof. From the relation $(A - \Delta\widehat{A})\widehat{X} = B - \Delta\widehat{B}$, defined by (3.3) and (3.1), (6.17) follows immediately. From (3.40), we obtain

$$\Delta\widehat{B} = [A; B]\begin{bmatrix} Z \\ \Gamma \end{bmatrix}\Gamma^T. \tag{6.18}$$

Multiplying (6.18) with $(\widehat{X}^T\widehat{X} + I)$ from the right and substituting (3.43) and (3.38) into it gives $\widehat{R}$ in (6.16). □

Observe from (6.17) that $R(\widehat{R}) \subseteq R([\Delta\widehat{A}; \Delta\widehat{B}])$. This implies that the TLS residual is *orthogonal* to the TLS approximate subspace $R([\widehat{A}; \widehat{B}])$. The LS residual R' is orthogonal to the LS approximate subspace $R([A; B']) = R(A)$. Using (6.16) and (6.17), the *length* of the TLS residual follows immediately for the one-dimensional case:

$$\|\widehat{r}\|_2 = \|\widehat{x}\|_2\|\Delta\widehat{A}\|_F + \|\Delta\widehat{b}\|_2 = \sigma_{n+1}\sqrt{\|\widehat{x}\|_2^2 + 1}. \tag{6.19}$$

Equation (6.19) shows that $\|\widehat{r}\|_2$ is precisely composed of the minimized TLS ingredients $\|\widehat{x}\|_2, \|\Delta\widehat{A}\|_F$, and $\|\Delta\widehat{b}\|_2$.

Let us now compare the TLS residual $\widehat{R}$ and LS residual R' in the following theorem.

THEOREM 6.6. *Let* (1.21) (*respectively,* (1.22)) *be the* SVD *of* A (*respectively,* $[A; B]$) *and* B' *be the orthogonal projection of* B *onto* $R(A)$. *If* $\sigma'_n > \sigma_{n+1} = \cdots = \sigma_{n+d}$, *then*

$$\widehat{R} - R' = \sigma_{n+1}^2 A(A^T A - \sigma_{n+1}^2 I)^{-1} X', \tag{6.20}$$

$$\|\widehat{R} - R'\|_F \leq \sigma_{n+1}^2 \|B'\|_F (\sigma_n'^2 - \sigma_{n+1}^2)^{-1} \tag{6.21}$$

$$\leq \sigma_{n+1}^2 \|X'\|_F \sigma_n'^{-1} (\sigma_n'^2 - \sigma_{n+1}^2)^{-1}, \tag{6.22}$$

$$\|\widehat{R}\|_F \geq \|R'\|_F. \tag{6.23}$$

Proof. Since $\widehat{R} - R' = A(\widehat{X} - X')$, we obtain (6.20) by multiplying (6.3) with A from the left. Using the dyadic decomposition $A = \sum_{i=1}^{n} \sigma_i' u_i' v_i'^T$ and taking norms,

$$\|\widehat{R} - R'\|_F^2 = \sum_{i=1}^{n} \|u_i'^T B\|_2^2\, \sigma_{n+1}^4 (\sigma_i'^2 - \sigma_{n+1}^2)^{-2}. \tag{6.24}$$

Since $\sigma_1' \geq \cdots \geq \sigma_n' > \sigma_{n+1}$, (6.24) is maximized if $B' // u_n'$. In this case, $\|u_i'^T B\| = 0$ for all $i \neq n$ and $\|u_n'^T B\|_F = \|B'\|_F$. Substituting this in (6.24) gives (6.21). Furthermore, $B' // u_n'$ implies that $X' = v_n' \sigma_n'^{-1} u_n'^T B$ and thus, $\|B'\|_F = \|X'\|_F / \sigma_n'$. Substituting this in (6.21) gives (6.22). Finally, since LS always minimizes any orthogonally invariant norm of its residual, (6.23) follows immediately. □

Compared to (6.21), the upper bound

$$\|\widehat{R}\|_F - \|R'\|_F \leq \|\widehat{R} - R'\|_F \leq \|R'\|_F \|B\|_F (\sigma_n' - \sigma_{n+1})^{-1}$$

derived from [68, Cor 4.2] under the same assumptions of Theorem 6.6, is mostly too crude, e.g., when $\|B\|_F >> \sigma_{n+1}$, and is never attained (unless $B \in R(A)$). In contrast, (6.21) is met whenever $B' // u_n'$. Moreover, (6.21) allows us to point out the parameters that mainly influence the difference between both residuals. Comparing (6.6) with (6.21), it is clear that the same parameters that decrease the differences between both solutions (as illustrated in Figs. 7.4, 7.5, and 7.6), also decrease the differences between both residuals. Hence, figures for this relationship would give similar graphs. Indeed, the TLS and LS residuals *approach* each other if

1. σ_{n+1} is small, i.e., the set of equations $AX \approx B$ is only slightly incompatible.
2. $\|B\|_F$ is small, i.e., the TLS solution becomes close to the LS solution.
3. $\sigma_n' >> \sigma_{n+1}$, i.e., A may not be (nearly) rank-deficient.
4. B is "close" to the largest singular vectors of A.

Finally, we compare the residuals corresponding to the TLS and LS minimum norm solution of multidimensional problems $AX \approx B$ with more than one solution.

THEOREM 6.7. *Consider the same notation and assumptions of Theorem 6.1. Then*

$$\begin{aligned}
\|\widehat{R}\|_2 &\leq \|R'\|_2 + \frac{\sigma_{n+1}^2}{\sigma_n'} \|\widehat{X}\|_2 && \textit{if } p = n \\
&\leq \|R'\|_2 + \frac{\sigma_{n+1}^2}{\sigma_r'} \|V_{12}^T \widehat{X} - V_{22}^T\|_2 && \textit{if } p < n \textit{ and } \mathrm{rank}(A) = r > p \\
&\leq \|R'\|_2 + \frac{\sigma_{n+1}^2}{\sigma_p'} \|V_{12}^T \widehat{X} - V_{22}^T\|_2 && \textit{if } p < n \textit{ and } \mathrm{rank}(A) = p.
\end{aligned}$$

Let $A_p = \sum_{i=1}^{p} \sigma'_i u'_i v'^T_i$, $p < n$ *and* $\text{rank}(A) = r > p$. *If* $X'_p = A_p^\dagger B$ *and* $R'_p = B - AX'_p$, *then*

$$\|\widehat{R}\|_2 \leq \|R'_p\|_2 + \frac{\sigma_{n+1}^2}{\sigma'_p}\|V_{12}^T\widehat{X} - V_{22}^T\|_2 + \sigma_{n+1}\|\widehat{X}\|_2.$$

Proof. For the proof see [207, Thm. 3.1]. □

Here, $\widehat{R}$ is the residual corresponding to the minimum norm TLS solution of a "nearby" TLS problem $\widetilde{A}X \approx \widetilde{B}$ in which $[\widetilde{A}; \widetilde{B}]$ is derived from $[A; B]$ by making the smallest singular values $\sigma_{p+1} = \cdots = \sigma_{n+d}$ equal to σ_{n+d}. If these singular values indeed coincide, we have the following theorem.

THEOREM 6.8. *Consider the same notation and assumptions of Theorem* 6.2, *then*

$$\begin{aligned} &\textit{if } \sigma_{n+1} = 0 \quad \textit{then} \quad \widehat{R} = R' = 0; \\ &\textit{if } \sigma_{n+1} > 0 \quad \textit{then} \quad \|\widehat{R}\|_2 \leq \|R'\|_2 + \frac{\sigma_{n+1}^2}{\sigma'_p}\|\widehat{X}\|_2. \end{aligned}$$

Proof. For the proof see [207, Thm. 3.2]. □

6.4. Connections between the corrections applied to the data

LS minimizes the Frobenius norm of the residual matrix of $AX \approx B$, while TLS minimizes the Frobenius norm of the correction matrix $[\Delta\widehat{A}; \Delta\widehat{B}]$ applied to this set. In this section, we analyze the corrections applied by TLS and LS to the data in $AX \approx B$ to make the set compatible. We also deduce the assumptions about the underlying perturbation model. Furthermore, the relative TLS corrections applied to A with respect to B and with respect to LS are evaluated.

Let us first show the relation between the *solution* $\widehat{X}$ and the TLS *corrections* $\Delta\widehat{A}$ and $\Delta\widehat{B}$ applied to the data. From this, the underlying TLS perturbation model can be deduced.

COROLLARY 6.3. *Let* $\widehat{X}$ *be a solution of the* TLS *problem* (3.1)–(3.3) *with corresponding correction matrix* $[\Delta\widehat{A}; \Delta\widehat{B}]$, *then*

$$\Delta\widehat{B}\widehat{X}^T = -\Delta\widehat{A}. \tag{6.25}$$

Proof. Let (1.22) be the SVD of $[A; B]$. From (3.40), we obtain

$$-\Delta\widehat{A} = -[A; B]\begin{bmatrix} Z \\ \Gamma \end{bmatrix} Z^T, \tag{6.26}$$

$$\Delta\widehat{B} = [A; B]\begin{bmatrix} Z \\ \Gamma \end{bmatrix} \Gamma^T. \tag{6.27}$$

Multiplying (6.27) with $\widehat{X}^T$ and substituting (3.38) into it gives (6.26), which proves (6.25). □

Observe that for the *one-dimensional* case, (6.25) reduces to

$$(6.28) \qquad \widehat{x}_i \Delta \widehat{b} = -\Delta \widehat{a}_i \qquad i = 1, \cdots, n,$$

where $\Delta \widehat{a}_i$ is the ith column of $\Delta \widehat{A}$. Equation (6.28) implies that the ith component $\widehat{x}_i$ of the solution $\widehat{x}$

is given by the *ratio* $\|\Delta \widehat{a}_i\|_2 / \|\Delta \widehat{b}\|_2$ of the length of the correction vectors $\Delta \widehat{a}_i$ and $\Delta \widehat{b}$.

Corollary 6.3 allows us to analyze the perturbation model imposed by TLS to obtain a unique solution. In fact, the solution of a perturbed set of linear equations $AX \approx B$ is intrinsically *nonunique*. Indeed, any uncertainty in the data A, B implies that no particular solution can be identified with certainty. However, by imposing *additional* a priori *assumptions* on the data perturbations ΔA and ΔB, *uniqueness* of the solution is enforced. Let us now compare these assumptions imposed by TLS and LS.

Equation (6.25) implies that TLS assumes a *perturbation model* $[\Delta \widehat{A}; \Delta \widehat{B}]$ of rank $\leq d$ (a rank 1 model if $d = 1$ and $\sigma_{n+1} \neq 0$). From the statistical point of view, this is very unrealistic [43], [44], and [38, §5.3]. Indeed, mostly the noise vectors added to the data are independently and identically distributed, and thus correspond to a full rank perturbation model. However, it is easily proved that every perturbation model $[\Delta \widetilde{A}; \Delta \widetilde{B}]$, which makes the null space of $[A - \Delta \widetilde{A}; B - \Delta \widetilde{B}]$ equal to the null space of the TLS approximation $[\widehat{A}; \widehat{B}]$, makes the set $(A - \Delta \widetilde{A})X = B - \Delta \widetilde{B}$ compatible and gives rise to the same TLS solution. Indeed, the TLS solution $\widehat{X}$ in (3.38) is entirely determined by the null space of $[\widehat{A}; \widehat{B}]$ given by $R([Z^T; \Gamma^T]^T)$, and is invariant with respect to any basis transformation P in this space, i.e., $\widehat{X} = -(ZP)(\Gamma P)^{-1} = -Z\Gamma^{-1}$. Assume that (1.22) is the SVD of $[A; B]$ with $\sigma_n > \sigma_{n+1}$ and V_{22} nonsingular for simplicity. Then, this *class* of perturbation models producing the *same* TLS solution $\widehat{X}$ is given by

$$(6.29) \quad [\Delta \widetilde{A}; \Delta \widetilde{B}] = U_2 \Sigma_2 V_2^T + H^T \text{ with } V_2 = \begin{bmatrix} V_{21} \\ V_{22} \end{bmatrix} \text{ and } R(H) \perp R(V_2).$$

From (6.29) it is clear that TLS only makes *assumptions* about the *noise distribution* in the singular subspaces associated with the *smallest* singular values $\sigma_{n+1}, \cdots, \sigma_{n+d}$ of $[A; B] = [A_0; B_0] + [\Delta A; \Delta B]$. In particular, the TLS correction matrix $[\Delta \widehat{A}; \Delta \widehat{B}]$, given in (3.12), equals the perturbation model $[\Delta \widetilde{A}; \Delta \widetilde{B}]$ with H set to zero in (6.29). Alternatively, consider the statistically oriented white noise approach [10], [43]. If $Ax \approx b$, this approach proves that a perturbation model $[\Delta \widetilde{A}; \Delta \widetilde{b}]$, satisfying $\mathcal{E}([\Delta \widetilde{A}; \Delta \widetilde{b}]^T [\Delta \widetilde{A}; \Delta \widetilde{b}]) = \sigma_\nu^2 I$, is statistically the best choice whenever the errors are independently and identically distributed with zero mean and equal variance σ_ν^2, e.g., $[\Delta \widetilde{A}; \Delta \widetilde{b}] = \sigma_{n+1} U V^T$. Observe that this perturbation model also satisfies (6.29) and thus produces the same solution as the TLS approach. The only difference is that condition (3.1) is no longer satisfied. For the white noise approach, $\|[\Delta\widetilde{A};\Delta\widetilde{b}]\|_F = \sqrt{n+1}\sigma_{n+1} > \sigma_{n+1} = \|[\Delta\widehat{A};\Delta\widehat{b}]\|_F$. Only the TLS correction matrix minimizes (3.1). We can thus conclude that depending on the initial approach, either algebraic or statistical, a different perturbation model $[\Delta\widetilde{A};\Delta\widetilde{B}]$ can be applied to the data matrix $[A;B]$ but these are led to the same TLS solution if they are contained in the model set (6.29). If we single out the perturbation matrix with *minimal* norm, then we obtain the TLS *correction matrix*. Thus, this matrix need not be considered as the perturbation model, expected in practice for the data $[A;B]$.

The TLS solution $\widehat{X}$ equals the exact solution X_0 of the corresponding unperturbed problem $A_0X = B_0$ if the error matrix $[\Delta A;\Delta B]$ added to the data $[A_0;B_0]$, i.e., $[A;B] = [A_0;B_0] + [\Delta A;\Delta B]$, satisfies one of the perturbation models $[\Delta\widetilde{A};\Delta\widetilde{B}]$ given by (6.29). In other words, noise is completely rejected if the null space of $[A-\Delta A; B-\Delta B] = [A_0;B_0]$ equals the null space of the TLS approximation $[\widehat{A};\widehat{B}]$. These assumptions are more and more satisfied with increasing number of equations provided the data A, B satisfy the model (6.4) and $[A;B]$ is row-wise perturbed by independently and identically distributed errors with zero mean and common covariance matrix $\sigma_\nu^2 I_{n+d}$. Indeed, as proved in §8.3, the TLS solution of the set $AX \approx B$ of m linear equations then converges with probability one to the exact solution X_0 as $m \to \infty$.

Analogously, it can be seen that LS assumes a *perturbation model* of the form $[\Delta\widetilde{A};\Delta\widetilde{B}] = [0;\Delta B']$. For LS, however, only perturbations $[0;\Delta B]$ of $[A_0;B_0]$ satisfying $\Delta B \perp R(A)$ correspond with the LS solution X' of $AX \approx B$. X' equals the exact solution X_0 of $A_0X = B_0$ if and only if A is error-free ($A = A_0$), and the projection $A_0A_0^TB$ of B onto $R(A_0)$ equals B_0.

Another statistical approach to look at these perturbations is given in [162]. Stewart recognizes that problems $AX \approx B$ with errors $\Delta A, \Delta B$ in all data can in many aspects be regarded as a model $B = AX + E$ with an altered error $E = \Delta B - \Delta AX$ in the response. Contrary to our data-oriented approach, his underlying perturbation model considers the equation error E.

In the general case, where the number of a priori assumptions is minimal, the solution is intrinsically nonunique. Kalman [86] and Patefield [120] obtained interesting results for the case that maximally one linear relation is present in the data. Using the classical Perron–Frobenius theorem [56] for positive matrices, they both proved that the solution set is a convex simplex generated by all the least squares solutions to the linear problem. A geometrical approach to this problem, which leads to a general solution for any number of linear relations, is given in [38].

We now compare the corrections $[\Delta\widehat{A};\Delta\widehat{B}]$ and $[0\,;\Delta B']$ applied to the data by TLS and LS, respectively. As proved in Theorem 2.8, the corrections applied by TLS to make the set of equations $AX \approx B$ compatible are *smaller* than for LS. It is possible now to express *exactly* this *advantage* of the TLS approximation $[\widehat{A};\widehat{B}]$ over the LS approximation $[A;B']$ by the following theorem.

THEOREM 6.9. *Let (1.21) (respectively, (1.22)) be the SVD of A (respectively, $[A;B]$). Denote the singular values of $[\Delta\widehat{A};\Delta\widehat{B}]$ by σ_i''. If $\sigma_n > \sigma_{n+1}$ and V_{22} is nonsingular, then*

$$\prod_{i=1}^{d}\sigma_i''^2 = \prod_{i=1}^{n}(\sigma_i'/\sigma_i)^2\det(\Delta B'^T\Delta B'). \tag{6.30}$$

Proof. If σ_i (respectively, σ_i') are the singular values of $[A;B]$ (respectively, A), then σ_i^2 (respectively, $\sigma_i'^2$) are the eigenvalues of $[A;B]^T[A;B]$ (respectively, A^TA). Hence,

$$\det(A^TA) = \prod_{i=1}^{n}\sigma_i'^2 \quad\text{and}\quad \det([A;B]^T[A;B]) = \prod_{i=1}^{n+d}\sigma_i^2. \tag{6.31}$$

We also have [56]

$$\begin{aligned}\det([A;B]^T[A;B]) &= \det(A^TA)\det(B^TB - B^TA(A^TA)^{-1}A^TB)\\ &= \det(A^TA)\det(B^TB - B^TAX')\\ &= \det(A^TA)\det(B^T\Delta B')\\ &= \det(A^TA)\det(\Delta B'^T\Delta B'). \qquad (6.32)\end{aligned}$$

Substituting the expressions of (6.31) into (6.32) produces

$$\prod_{i=1}^{n+d}\sigma_i^2 = \prod_{i=1}^{n}\sigma_i'^2\det(\Delta B'^T\Delta B') \quad\text{and}\quad \prod_{i=1}^{d}\sigma_{n+i}^2 = \prod_{i=1}^{n}(\sigma_i'/\sigma_i)^2\det(\Delta B'^T\Delta B').$$

Since $\sigma_n > \sigma_{n+1}$ and V_{22} nonsingular, $[\Delta\widehat{A};\Delta\widehat{B}]$ is given by $U_2\Sigma_2[V_{12}^T;V_{22}^T]$ and thus $\sigma_i'' = \sigma_{n+i}$ for $i = 1,\cdots,d$ (if $m < n+d$, $\sigma_j'' = \sigma_{n+j} = 0$ for all $j = m-n+1,\cdots,d$). Hence, $\prod_{i=1}^{d}\sigma_i''^2 = \prod_{i=1}^{d}\sigma_{n+i}^2$, which proves (6.30). □

For the one-dimensional case, (6.30) reduces to

$$\|[\Delta\widehat{A};\Delta\widehat{b}]\|_F = \prod_{i=1}^{n}(\sigma_i'/\sigma_i)\|\Delta b'\|_2. \tag{6.33}$$

Theorem 6.9 implies that the *relative position* of B with respect to the singular vectors u_i' of A characterizes the *ratio* of the corrections applied by TLS and LS. If $R(B) \subseteq R(A)$, $AX \approx B$ has an exact solution $X_0 = \widehat{X} = X'$ and hence both corrections are zero. If $R(B) \not\subseteq R(A)$ the TLS correction is relatively smaller than the LS correction when B' lies more along the singular vectors associated with the smallest singular values of A. It is minimal if B' lies along u'_n. Especially, then, the TLS solution is expected to be more accurate than the LS solution. This is illustrated in Fig. 7.4. We can now derive the following *bounds* for the TLS correction.

THEOREM 6.10. *Let* (1.21) (*respectively,* (1.22)) *be the* SVD *of* A (*respectively,* $[A; B]$). *Assume that* A *and* $\sigma_{n+1}, \cdots, \sigma_{n+d}$ *are fixed and* $m \geq n + d$. *If* $\sigma'_n > \sigma_{n+1} = \cdots = \sigma_{n+d}$, *then*

$$\max_{\sigma_1,\cdots,\sigma_n} \frac{\|[\Delta\widehat{A}; \Delta\widehat{B}]\|_F^2}{\|\Delta B'\|_F^2} = 1, \tag{6.34}$$

$$\min_{\sigma_1,\cdots,\sigma_n} \frac{\|[\Delta\widehat{A}; \Delta\widehat{B}]\|_F^2}{\|\Delta B'\|_F^2} = \frac{\sigma_n'^2}{\sigma_n'^2 - \sigma_{n+1}^2 + \|B\|_F^2/d}. \tag{6.35}$$

Proof. The interlacing Theorem 2.4 for singular values implies that $\sigma_i \geq \sigma'_i$ for $i = 1, \cdots, n$. Equality is obtained by choosing $[A; B]$ so that $\sigma'_i = \sigma_i$. This implies that $B \perp R(A)$. Hence, $\|[\Delta\widehat{A}; \Delta\widehat{B}]\|_F^2 = \|B\|_F^2 = \|\Delta B'\|_F^2$, which proves (6.34).

To prove (6.35), we start from the following equality, proved in [68] and extended to multidimensional problems which satisfy the above assumptions:

$$\Delta B'^T \Delta B' = \sigma_{n+1}^2 [I_d + B^T U'_1 (\Sigma'^2 - \sigma_{n+1}^2 I_n)^{-1} U_1'^T B]. \tag{6.36}$$

Since $\|\Delta B'\|_F^2 = \mathrm{tr}(\Delta B'^T \Delta B')$, by taking the trace of both sides of (6.36) we obtain

$$\|\Delta B'\|_F^2 / \sigma_{n+1}^2 = \mathrm{tr}[I + B^T U'_1 (\Sigma'^2 - \sigma_{n+1}^2 I)^{-1} U_1'^T B]. \tag{6.37}$$

Since $\mathrm{tr}(A + B) = \mathrm{tr}(A) + \mathrm{tr}(B)$ and $\|[\Delta\widehat{A}; \Delta\widehat{B}]\|_F^2 = d\sigma_{n+1}^2$, this yields

$$\frac{\|\Delta B'\|_F^2}{\|[\Delta\widehat{A}; \Delta\widehat{B}]\|_F^2} = 1 + \sum_{i=1}^{n} \frac{\|u_i'^T B\|_2^2}{d(\sigma_i'^2 - \sigma_{n+1}^2)}. \tag{6.38}$$

To prove (6.35), the expression in (6.38) must be maximized. This is the case if the orthogonal projection B' of B onto $R(A)$ satisfies $B' // u'_n$. In this case, (6.38) equals

$$\max_{\sigma_1,\cdots,\sigma_n} \frac{\|\Delta B'\|_F^2}{\|[\Delta\widehat{A}; \Delta\widehat{B}]\|_F^2} = 1 + \frac{\|B'\|_F^2}{d(\sigma_n'^2 - \sigma_{n+1}^2)}. \tag{6.39}$$

Since $\|B'\|_F^2 = \|B\|_F^2 - \|\Delta B'\|_F^2$, (6.39) yields

$$\max_{\sigma_1,\cdots,\sigma_n} \frac{\|\Delta B'\|_F^2}{\|[\Delta\widehat{A}; \Delta\widehat{B}]\|_F^2} = \frac{\sigma_n'^2 - \sigma_{n+1}^2 + \|B\|_F^2/d - \|\Delta B'\|_F^2/d}{\sigma_n'^2 - \sigma_{n+1}^2}. \tag{6.40}$$

By moving the last term on the right of (6.40) to the left and since $\|[\Delta\widehat{A}; \Delta\widehat{B}]\|_F^2 = d\sigma_{n+1}^2$, we obtain

$$\max_{\sigma_1,\cdots,\sigma_n} \frac{\|\Delta B'\|_F^2}{\|[\Delta\widehat{A}; \Delta\widehat{B}]\|_F^2} \cdot \frac{\sigma_n'^2}{\sigma_n'^2 - \sigma_{n+1}^2} = \frac{\sigma_n'^2 - \sigma_{n+1}^2 + \|B\|_F^2/d}{\sigma_n'^2 - \sigma_{n+1}^2}. \tag{6.41}$$

By multiplying both sides of (6.41) with $(\sigma_n'^2 - \sigma_{n+1}^2)\sigma_n'^{-2}$, we obtain (6.35). □

Observe that the maximum (6.34) is obtained for $B \in R(U_2)$. Since $B \perp R(A)$, TLS as well as LS give the zero solution. The minimum (6.35) is obtained for B belonging to the space generated by u_n' and $U(\sigma_{n+1})$, i.e., $B'//u_n'$, implying that B is mainly directed along the singular direction of A that is most sensitive to noise. The better accuracy of TLS in those cases with respect to perturbations of the data is clearly illustrated in Fig. 7.4.

Observe again that (6.35) points out the same parameters that influence the differences between the solutions (see (6.6)) and the residuals (see (6.21)) of both problems, namely $\|B\|_F, \sigma_n', \sigma_{n+1}$ and the orientation of B with respect to u_n'. These parameters now determine the ratio of the corrections applied by both methods to the data in $AX \approx B$, and hence the closeness of both problems. Figures 7.4 and 7.5 illustrate the influence of the orientation of B with respect to u_n' and the influence of its length $\|B\|_F$ on the expected accuracy of both methods in the presence of perturbations. From the TLS definition, we know that TLS projects A and B onto the n-dimensional subspace $R([\widehat{A}; \widehat{B}])$ with minimal correction $[\Delta\widehat{A}; \Delta\widehat{B}]$. Depending on the location of $R([\widehat{A}; \widehat{B}])$, the angular correction of A or B will be larger. The following theorem shows how $\widehat{X}$ determines *the relative orientation* of the space $R([\widehat{A}; \widehat{B}])$ with respect to the columns of A and B.

THEOREM 6.11. *Let* (1.22) *be the* SVD *of* $[A; B]$. *If* $\sigma_p > \sigma_{p+1} = \cdots = \sigma_{n+d}$ *with* $p \le n$ *and* V_γ *in* (1.23) *is nonsingular, then a solution* $\widehat{X}$ *of the* TLS *problem* (3.1)–(3.3) *exists and*

$$\Delta\widehat{B}^T\Delta\widehat{B} = \sigma_{n+1}^2(\widehat{X}^T\widehat{X} + I)^{-1}, \tag{6.42}$$

$$\Delta\widehat{A}^T\Delta\widehat{A} = \sigma_{n+1}^2\widehat{X}(\widehat{X}^T\widehat{X} + I)^{-1}\widehat{X}^T. \tag{6.43}$$

Proof. From (3.40) we obtain

$$\Delta\widehat{B}^T\Delta\widehat{B} = \Gamma[Z^T; \Gamma^T][A; B]^T[A; B]\begin{bmatrix} Z \\ \Gamma \end{bmatrix}\Gamma^T, \tag{6.44}$$

$$\Delta\widehat{A}^T\Delta\widehat{A} = Z[Z^T; \Gamma^T][A; B]^T[A; B]\begin{bmatrix} Z \\ \Gamma \end{bmatrix}Z^T. \tag{6.45}$$

Using the condition $\sigma_{p+1} = \cdots = \sigma_{n+d}$ and the orthonormality of the matrix $[v_1, \cdots, v_{n-p}; V_\gamma Q]$ with $V_\gamma = [v_{p+1}, \cdots, v_{n+d}]$ and Q defined in (3.42), we can easily prove that

$$[Z^T; \Gamma^T][A; B]^T[A; B]\begin{bmatrix} Z \\ \Gamma \end{bmatrix} = \sigma_{n+1}^2 I_d. \tag{6.46}$$

Substituting (6.46) and (3.43) into (6.44) proves (6.42).

Using (3.43) and (3.38), we have

$$ZZ^T = (\widehat{X}\Gamma)(\widehat{X}\Gamma)^T = \widehat{X}(\widehat{X}^T\widehat{X} + I)^{-1}\widehat{X}^T. \tag{6.47}$$

Substituting (6.46) and (6.47) into (6.45), proves (6.43). □

In the one-dimensional case and using $\sigma_{n+1}^2 = \|[\Delta\widehat{A}; \Delta\widehat{b}]\|_F^2$, (6.42) and (6.43) reduce to

$$\|\Delta\widehat{b}\|_2^2 = \|[\Delta\widehat{A}; \Delta\widehat{b}]\|_F^2/(\|\widehat{x}\|_2^2 + 1), \tag{6.48}$$

$$\|\Delta\widehat{A}\|_F^2 = \|[\Delta\widehat{A}; \Delta\widehat{b}]\|_F^2\|\widehat{x}\|_2^2/(\|\widehat{x}\|_2^2 + 1). \tag{6.49}$$

Equations (6.48) and (6.49) have nice *geometric* properties. Depending on the norm of the solution $\widehat{x}$, TLS forces $\widehat{A}$ to move more toward b or conversely to move $\widehat{b}$ more toward A. If $\|\widehat{x}\|_2$ is large, $\|\Delta\widehat{b}\|_2$ is small: $\widehat{A}$ is bent toward b. If $\|\widehat{x}\|_2 = 1$, then $\|\Delta\widehat{b}\|_2 = \|\Delta\widehat{A}\|_F$: $\widehat{A}$ and $\widehat{b}$ move equally toward each other. If $\|\widehat{x}\|_2$ is small, $\|\Delta\widehat{A}\|_F$ is small and $\widehat{b}$ is bent toward A. Observe that in the last case TLS comes close to LS where b' is completely bent toward A.

6.5. Connections between the approximate subspaces

In this section we want to find parameters that determine *the closeness of the approximate subspaces*, $R([A; B']) = R(A)$ of LS and $R([\widehat{A}; \widehat{B}]) = R(\widehat{A})$ of TLS, into which the data A and B are projected to make the set of equations $AX \approx B$ compatible. A natural way of describing the closeness of any two subspaces S_1 and S_2 is to compute the *distance* between these subspaces which is defined by [69, §2.6.3]:

$$\operatorname{dist}(S_1, S_2) = \|P_1 - P_2\|_2, \tag{6.50}$$

where P_i is the orthogonal projection onto S_i. In the case of one-dimensional subspaces, $\operatorname{dist}(S_1, S_2)$ is geometrically given by the *sine* of the angle between two vectors [69, §2.6.3]. For higher dimensions, the distance between any two subspaces is just the sine of the *largest canonical* (or principal) *angle* θ, which is defined by [69, §12.4.3]:

$$\cos\theta = \max_{\substack{u \in S_1 \\ \|u\|_2 = 1}} \max_{\substack{v \in S_2 \\ \|v\|_2 = 1}} u^T v. \tag{6.51}$$

The distance between the subspaces $R(A)$ and $R(\widehat{A})$ can be computed as follows.

COROLLARY 6.4. *Let* (1.21) (*respectively,* (1.22)) *be the* SVD *of* A (*respectively,* $[A; B]$). *Assume* $m \leq 2n, \sigma_n > \sigma_{n+1}$ *and* V_{22} *nonsingular. Then* $R(A) = R(U_1')$ *and* $R(\widehat{A}) = R(U_1)$ *and the distance between both spaces is given by*

$$\operatorname{dist}(R(A), R(\widehat{A})) = \|U_1'^T U_2\|_2 = \|U_2'^T U_1\|_2. \tag{6.52}$$

Proof. By invoking the C-S decomposition [69, Thm. 2.6.1] and applying [69, Cor. 2.6.2] straightforwardly, (6.52) follows immediately. □

If $m > 2n$, we may work with the orthogonal complements of $R(A)$ and $R(\hat{A})$. Corollary 6.4 can be easily extended if $\text{rank}(A) = p < n$.

In the *one-dimensional case*, i.e., $d = 1$, the *distance* between $R(A)$ and $R(\hat{A})$ can be more nicely characterized.

COROLLARY 6.5. *Let* (1.21) (*respectively,* (1.22)) *be the* SVD *of* A (*respectively,* $[A; b]$). *Assume* $m \le 2n, \sigma_n > \sigma_{n+1}$, $d = 1$ *and* $v_{n+1,n+1} \neq 0$. *Then the distance between* $R(A)$ *and* $R(\hat{A})$ *is given by*

$$\text{(6.53)} \quad \text{dist}(R(A), R(\hat{A})) = \sigma_{n+1}\|(\Sigma'^2 - \sigma_{n+1}^2 I)^{-1} U'^T b\|_2 / \sqrt{\|\hat{x}\|_2^2 + 1}$$

$$\text{(6.54)} \quad = \|\hat{r} - r'\|_2 / \|\hat{r}\|_2.$$

Proof. Since $d = 1$, it follows that $R(U_1') \perp R([u_{n+2}, \cdots, u_m])$. Hence, (6.52) reduces to $\|U_1'^T u_{n+1}\|_2$. Using the SVD (1.21) and (1.22) of A and $[A; b]$, respectively, we have

$$\text{(6.55)} \quad U_1'^T u_{n+1} = \Sigma'^{-1} V'^T A^T u_{n+1} = \sigma_{n+1} \Sigma'^{-1} V'^T [v_{1,n+1}, \cdots, v_{n,n+1}]^T.$$

Since $\sigma_n > \sigma_{n+1}$, $d = 1$ and $v_{n+1,n+1} \neq 0$, (2.23) equals

$$\text{(6.56)} \qquad \begin{bmatrix} \hat{x} \\ -1 \end{bmatrix} = -\frac{1}{v_{n+1,n+1}} v_{n+1} = -\sqrt{\|\hat{x}\|_2^2 + 1}\, v_{n+1}.$$

Substituting (6.56) into (6.55) yields

$$\text{(6.57)} \qquad \begin{aligned} U_1'^T u_{n+1} &= -\sigma_{n+1} \Sigma'^{-1} V'^T \hat{x} v_{n+1,n+1} \\ &= -\sigma_{n+1} \Sigma'^{-1} V'^T \hat{x} / \sqrt{\|\hat{x}\|_2^2 + 1} \end{aligned}$$

Taking the SVD of (2.27) yields

$$\text{(6.58)} \qquad \hat{x} = V' \Sigma' (\Sigma'^2 - \sigma_{n+1}^2 I)^{-1} U'^T b.$$

Substituting (6.58) in (6.57) and taking the 2-norm proves (6.53).

To prove (6.54), take the 2-norm of (6.16) for $d = 1$ and substitute the 2-norm of (6.42) into it to obtain $\|\hat{r}\|_2 = \sigma_{n+1}\sqrt{\|\hat{x}\|_2^2 + 1}$. Then substituting this and (6.24) into (6.53) proves (6.54). □

Some other expressions of $\text{dist}(R(A), R(\hat{A}))$ are proven in [177, Cors. 2-19 and 2-20] for the case $d = 1$. Observe from (6.53) that the distance between $R(A)$ and $R(\hat{A})$ depends on the *same* parameters that determine the closeness of the solutions, the residuals and the applied corrections of both methods.

First consider $\sigma_{n+1} = \|[\Delta\hat{A}; \Delta\hat{b}]\|_F$. If $b \in R(A)$, then $\sigma_{n+1} = 0$ and $\text{dist}(R(A), R(\hat{A})) = 0$, i.e., the set $Ax = b$ is compatible. If σ_{n+1} is increasing, the set of equations becomes *less compatible* and the distance between $R(A)$ and $R(\widehat{A})$ *enlarges*. Furthermore, the distance enlarges not only if b' is close to the lowest singular vectors $u'_n, \cdots$ of A but also if the condition number σ'_1/σ'_n of A is large or A tends to be rank-deficient ($\sigma'_n \approx \sigma_{n+1} \approx 0$) since then, $\widehat{x}$ is mainly oriented in the direction of v'_n (if $u'^T_n b$ is not too small). Enlarging $\|b\|_2$ strengthens these effects. The distance is maximal if $b'//u'_n$ for a given σ_{n+1} and A. As shown in Figs. 7.4 and 7.5, the differences between the LS and TLS approach are then most pronounced.

6.6. Concluding remarks

In this chapter, the TLS technique is compared analytically with the classical LS method. *Algebraic* connections between their *solutions* (§6.2), their *residuals* (§6.3), their *corrections* applied to the fitting of the data (§6.4) and their *approximate subspaces* (§6.5) are proved. They all reveal that the *same* parameters mainly determine the correspondences and differences between both techniques. These parameters are, moreover, useful tools for getting more insight into the *sensitivity* of both techniques with respect to perturbations in all data, as discussed in Chapter 7.

In particular, various upper bound expressions show that the difference between the TLS and LS solution of the set $AX \approx B$, and their corresponding residuals, is maximal when the orthogonal projection B' of B onto $R(A)$ is parallel with the left singular vector of matrix A associated with its smallest singular value. In this case, the corrections applied by TLS are minimal with respect to the LS corrections while the distance between the TLS and LS approximate subspaces is maximal. As illustrated in §7.4, the better accuracy of the TLS solution in the presence of perturbations in all data is then most pronounced.

Several other properties clearly show how the differences between the TLS and LS approach increase when the set $AX \approx B$ becomes less compatible, when the norm of matrix B or the solution X is growing or when A tends to be rank-deficient. Moreover, these relationships point out the most influencing parameters that determine the relative differences in sensitivity between TLS and LS problems, namely, the minimal singular values of A and $[A; B]$ and the orientation of B within the range of A and the length of B or X.

Furthermore, it is shown how TLS can be considered as a *weighted* LS problem (§6.3). Also, assumptions about the underlying perturbation models of the two problems are deduced. As TLS makes only assumptions about the noise distribution in the singular subspaces of the data $[A; B]$ associated with the smallest singular values, many perturbation models correspond with the same TLS solution (§6.4).

Finally, observe that several properties given in this chapter require that the smallest singular values of the data matrix $[A; B]$ in a multidimensional

TLS problem are equal. This condition may appear too restrictive. Indeed, even if the data matrix $[A; B]$ represents m observations of an exact relation and the errors in the observations are independently and identically distributed with zero mean and equal variance, this condition is only satisfied for infinite m [59]. However, in these problems the singular values $\sigma_{n+1}, \cdots, \sigma_{n+d}$ approach equality with increasing probability for increasing sample size m (as proven in [38, §5.2], [59] and confirmed by simulations), so that for moderate sample sizes m the assumption of equal singular values is a fairly good approximation to reality.

Chapter 7

Sensitivity Analysis of Total Least Squares and Least Squares Problems in the Presence of Errors in all Data

7.1. Introduction

The connections between the TLS problem and the ordinary LS problem, discussed in the previous chapter, are useful tools for the sensitivity study of this chapter. Here, we compare the *relative deviations* of the TLS and LS approach in the presence of errors with respect to the corresponding unperturbed problem $A_0 x \approx b_0$.

First, the sensitivity of *both solutions* is compared in §7.2. This can be measured by computing the *condition number* of the problem $A_0 x \approx b_0$, quantifying the maximal error effect on its solution by solving the corresponding perturbed problem $Ax \approx b$. Based on the classical worst perturbation results for LS problems, we derive the *worst perturbation* effects on the TLS solution with respect to the solution of the unperturbed problem $A_0 x \approx b_0$. *Zero residual* (i.e., $b_0 \in R(A_0)$) as well as nonzero residual (i.e., $b_0 \notin R(A_0)$) problems are investigated and emphasize the impact of the residual on the sensitivity.

Second, the sensitivity properties of the SVD are discussed in §7.3. First, we recapitulate the main perturbation results for the singular values and their associated singular subspaces. They allow us to measure for the LS as well as for the TLS problem, the *maximal distance* between their perturbed singular subspaces and the corresponding ones of the unperturbed problem. These expressions are the key to understanding the relative differences in sensitivity between the TLS and the LS method. Additionally, we investigate the effects of some *special* subspace perturbations, which *maximally* affect the accuracy of the TLS problem, and also compute the effect of singular subspace perturbations on the TLS solution for the one-dimensional problem.

Finally in §7.4, the *expected differences* between the TLS and LS approach are illustrated by *simulations*. They not only confirm the main results of the sensitivity analysis but also show the most *influencing parameters* and their importance with respect to the applicability of the TLS technique. In particular, they clearly exhibit the relative gains in accuracy we can typically expect in numerical applications.

7.2. Sensitivity properties of the solution

7.2.1. Sensitivity to small perturbations. Throughout this section we consider the one-dimensional problem $Ax \approx b$, i.e., $d = 1$ in (1.17). Our aim is to examine how perturbations in A and b affect the solution x. In this analysis the condition number of the matrix A plays a significant role. The following definition generalizes the condition number of a square nonsingular matrix [69, §5.3.2].

DEFINITION 7.1 (Condition number of $A_{m \times n}$). *Let $A \in \mathcal{R}^{m \times n}$ have rank r and denote its SVD by* (1.21). *The condition number of A is*

$$\kappa(A) = \|A\|_2 \|A^\dagger\|_2 = \frac{\sigma_1'}{\sigma_r'}. \tag{7.1}$$

Matrices with small (respectively, large) condition numbers are said to be *well*-conditioned (respectively, *ill*-conditioned). Orthogonal matrices Q are perfectly conditioned in that $\kappa(Q) = 1$. Analogously to the square linear equation problem, the condition number (7.1) measures the sensitivity of the solution to errors in the matrix A and right-hand side b. It shows that small changes in the elements of A can induce large changes in the solution x of $Ax \approx b$ if A's columns are nearly dependent, i.e., $\sigma_n' \approx 0$.

When studying the sensitivity in *overdetermined* linear equation sets, it is important to distinguish between zero and nonzero residual problems. Figure 7.1 helps to clarify this relationship. Define the angle θ_0 between the unperturbed b_0 and $R(A_0)$ by

$$\sin \theta_0 = \frac{\|b_0 - A_0 x\|_2}{\|b_0\|_2}. \tag{7.2}$$

Then a problem is said to be *zero residual* if $b_0 \in R(A_0)$, i.e., θ_0 is zero. In other words, in the absence of any perturbation in the data, an *exact* but unobservable linear relationship $A_0 x_0 = b_0$ exists. The TLS problem typically applies to those cases. The sensitivity of the solution depends *linearly* on the condition number and it is easily shown that LS and TLS may be *equally* sensitive.

In *nonzero residual* problems, $b_0 \notin R(A_0)$ and thus, the angle θ_0 in (7.2) is no longer zero. This means that, even in the absence of any perturbation in the data, *no exact* linear relationship $A_0 x_0 = b_0$ can be found. As shown in Corollary 7.1, it is the *square* of the condition number that measures the sensitivity of nonzero residual LS problems and it is proved that the TLS solution may be *more* sensitive than the LS solution. Those problems occur, for example, in model fitting and prediction when we want to approximate nonlinear data by linear models or predict the systems response by a simplified

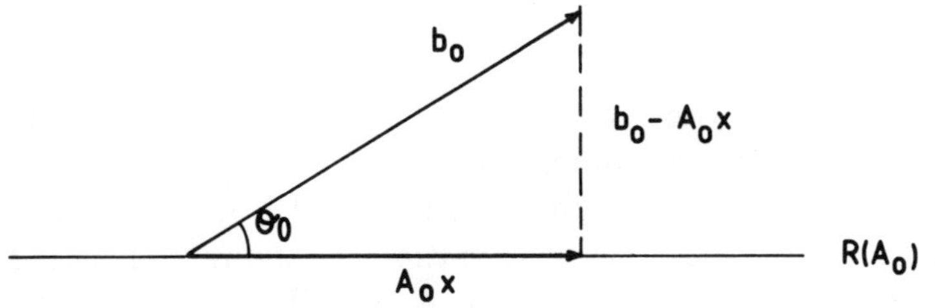

FIG. 7.1. *Geometric illustration of the angle θ_0 between b_0 and $R(A_0)$, given by* $\sin\theta_0 = \|b_0 - A_0x\|_2/\|b_0\|_2$.

model of the system. In those cases LS should be applied rather than TLS if minimal sensitivity is primordial.

In the remainder of this section, we only consider zero residual problems and investigate the accuracy of the LS and TLS solution in the presence of *small* perturbations $\Delta A, \Delta b$ in the data A, b.

Several authors have studied these problems, see, e.g., [175], [206], and [160]. Van der Sluis and Veltkamp [175] supposed that $Ax \approx b$ is only slightly incompatible, say $\mathcal{O}(\varepsilon)$ from a fixed zero residual problem $A_0x = b_0$ with rank $(A_0) = r$, and considered all compatible sets $\widetilde{A}x = \widetilde{b}$ with $A - \widetilde{A} = \mathcal{O}(\varepsilon)$ and rank $(\widetilde{A}) = r$, such that $[\widetilde{A}; \widetilde{b}]$ is the orthogonal projection of $[A; b]$ onto $R(\widetilde{A})$. Under suitable circumstances, the rank r approximation nearest to $[A; b]$ in Frobenius norm belongs to this class, as well as the TLS approximation $[\widehat{A}; \widehat{b}]$ and the LS approximation $[A; b']$ provided $r = n$. If $\|[A; b] - [A_0; b_0]\|_F = \varepsilon$, the minimum norm solutions of all sets $\widetilde{A}x = \widetilde{b}$ so related to $[A; b]$ are shown to differ mutually only by $\mathcal{O}(\varepsilon^2)$. Indeed the following theorem holds.

THEOREM 7.1. *Let* $A_0x = b_0$, $\mathrm{rank}(A_0) = r$, *and* $x_0 = A_0^\dagger b_0$, *so* $b_0 = A_0x_0$. *Assume that* $[A; b] = [A_0; b_0] + [\Delta A; \Delta b]$ *and consider a compatible set* $\widetilde{A}x = \widetilde{b}$ *with* $\mathrm{rank}(\widetilde{A}) = r$ *and* $[\widetilde{A}; \widetilde{b}]$ *the orthogonal projection of* $[A; b]$ *onto* $R(\widetilde{A})$. *Then* $\widetilde{x} = \widetilde{A}^\dagger \widetilde{b} = \widetilde{A}^\dagger b$. *Let* $\|[\Delta A; \Delta b]\|_F \le \varepsilon$, $\|A - \widetilde{A}\|_F \le \varepsilon$ *with* $\varepsilon < \frac{1}{2}\|A_0^\dagger\|_2^{-1}$. *Then,* $\widehat{A}x = \widehat{b}$ *is also compatible with* $[\widehat{A}; \widehat{b}]$, *the rank* r *approximation nearest to* $[A; b]$ *in Frobenius norm, and the minimum norm solution* $\widehat{x} = \widehat{A}^\dagger \widehat{b}$ *satisfies*

$$\|\widehat{x} - \widetilde{x}\|_2 \le 2\varepsilon^2\alpha^2\left(\sqrt{1 + \|x_0\|_2^2} + \|x_0\|_2\right) \text{ with } \alpha = \frac{\|A_0^\dagger\|_2}{1 - 2\varepsilon\|A_0^\dagger\|_2}. \tag{7.3}$$

Hence

$$\begin{aligned} \widehat{x} &= \widetilde{x} + \mathcal{O}(\varepsilon^2), \\ \widehat{r} &= \widetilde{r} + \mathcal{O}(\varepsilon^2) \text{ with } \widetilde{r} = b - A\widetilde{x}, \end{aligned}$$

$$\begin{aligned}\Delta\widehat{b} &= (1+\|\widetilde{x}\|_2^2)^{-1}\widetilde{r} + \mathcal{O}(\varepsilon^2)\\ \Delta\widehat{A} &= -(1+\|\widetilde{x}\|_2^2)^{-1}\widetilde{r}\widetilde{x}^T + \mathcal{O}(\varepsilon^2). \quad \textit{if } r = n.\end{aligned}$$

Proof. For the proof see [175, Thm. 5.9 and Cor. 5.12]. □

Observe that $\widehat{x}$ is the minimum norm TLS solution of $Ax \approx b$, as computed by Algorithm 3.1 with r the numerical rank of $[A; b]$. If $r = n$ then $[\widehat{A}; \widehat{b}]$ in Theorem 7.1 equals the TLS approximation satisfying (2.19)–(2.21). Thus by solving a compatible rank r set $\widetilde{A}x = \widetilde{b}$ near (but not nearest) to $Ax \approx b$, e.g., by solving the LS problem, we can obtain an $\mathcal{O}(\varepsilon^2)$ approximation to the TLS solution $\widehat{x}$ and hence also to $\widehat{r} = b - A\widehat{x}$. Observe, however, that these conclusions are only meaningful for sufficiently *small* ε. Indeed, assume that $\varepsilon \to \frac{1}{2}\|A_0^\dagger\|_2^{-1}$, then the bound (7.3) goes to infinity and, hence, the differences in accuracy between $\widehat{x}$ and $\widetilde{x}$ can be very significant!

First-order approximations for the best rank r approximation $[\widehat{A}; \widehat{b}]$ to $[A; b]$ in terms of $A_0, b_0, \Delta A$ and Δb are readily given by the following theorem.

THEOREM 7.2. *Let $A_0 x = b_0$ be compatible with solution $x_0 = A_0^\dagger b_0$, $\operatorname{rank}(A_0) = r$, and $[A; b] = [A_0; b_0] + [\Delta A; \Delta b]$ with $\|[\Delta A; \Delta b]\|_F \leq \varepsilon$. Then, the rank r approximation $[\widehat{A}; \widehat{b}]$ nearest to $[A; b]$ in Frobenius norm satisfies*

$$\begin{aligned}[\widehat{A}; \widehat{b}] &= [A; b] - (I - A_0 A_0^\dagger)[\Delta A; \Delta b]\\ &\quad \cdot \left\{ \begin{bmatrix} I - A_0 A_0^\dagger & 0 \\ 0 & 0 \end{bmatrix} + (1+\|x_0\|_2^2)^{-1} \begin{bmatrix} x_0 \\ -1 \end{bmatrix} [x_0^T; -1] \right\} + \mathcal{O}(\varepsilon^2).\end{aligned}$$

Proof. For the proof see [175, Thm. 5.14]. □

$I - A_0 A_0^\dagger = I - [A_0; b_0][A_0; b_0]^\dagger$ is the orthogonal projector on $N(A_0^T) = N([A_0; b_0]^T)$, while the factor between braces is equal to $I - [A_0; b_0]^\dagger [A_0; b_0]$, i.e., the orthogonal projector on $N([A_0; b_0])$.

First-order perturbation formulas for $\Delta\widehat{x}$ and $\widehat{r}$ are also given in [175]. Similar conclusions have been obtained by Stewart [160], [163] and Wei [206]. In particular, the latter proved that the LS solution x' of $Ax \approx b$ is close to a solution x_0 of the zero residual problem $A_0 x = b_0$, even when A_0 is not of full rank and $\operatorname{rank}(A) \neq \operatorname{rank}(A_0)$, provided $\varepsilon\kappa << 1$ with $\kappa = \kappa(A_0)$ and $\varepsilon \leq \min(\|\Delta A\|_2/\|A_0\|_2, \|\Delta b\|_2/\|b_0\|_2)$. Under these conditions, Wei showed that $x' = x_0 + \mathcal{O}(\varepsilon\kappa)$.

Summarizing the results of Theorems 7.1 and 7.2, we can conclude that, in the *absence of any further information* about the data A, b, such as the error distribution, the orientation of b with respect to $R(A)$, etc., it will usually *not* pay to compute the solution of the nearest set, i.e., the TLS solution of $Ax \approx b$, provided the unperturbed problem $A_0 x = b_0$ is *zero residual* and the *perturbations* in the data are sufficiently *small*, i.e., $\|[A; b] - [A_0; b_0]\|_F \leq \varepsilon$. Under these conditions, the solution $\widetilde{x}$ of $\widetilde{A}x = \widetilde{b}$, where $[\widetilde{A}; \widetilde{b}]$ is obtained by a suitable orthogonal projection of $[A; b]$, is already adequate, e.g., the LS solution.

7.2.2. Upper bounds of the LS solution perturbation. For any given LS minimization problem (2.1) we will investigate the *maximum (or worst) perturbation* caused in its solution. The importance of this notion of worst perturbation is that in the absence of further information about ΔA and Δb, the worst perturbation should be considered as an *attainable* or realistic perturbation of the minimization problem at hand.

Let the SVD of the unperturbed A_0 be given by (1.21). We assume $\sigma_n' \neq 0$ throughout this section and investigate the effect of perturbations for given $\varepsilon > 0$ satisfying

$$\|\Delta A\|_2 \leq \varepsilon \|A_0\|_2 = \varepsilon \sigma_1' \text{ and } \|\Delta b\|_2 \leq \varepsilon \|b_0\|_2. \tag{7.4}$$

It is reasonable to limit ourselves to $\varepsilon\kappa(A_0) < 1$ to ensure uniqueness of the solution of the perturbed LS problem. Let x_0' minimize $\|A_0 x - b_0\|_2$. For given ΔA and Δb we define $\Delta x'$ as the vector that minimizes

$$\|(A_0 + \Delta A)(x_0' + \Delta x') - (b_0 + \Delta b)\|_2. \tag{7.5}$$

Since (7.5) equals

$$\|(A_0 + \Delta A)\Delta x' - r_0' + \Delta A x_0' - \Delta b\|_2 \tag{7.6}$$

with r_0' given by $r_0' = b_0 - A_0 x_0'$, the minimizing $\Delta x'$ must be such that $(A_0 + \Delta A)\Delta x'$ is the projection of $r_0' - \Delta A x_0' + \Delta b$ on $R(A_0 + \Delta A)$. Writing $A = A_0 + \Delta A$, we have

$$\Delta x' = (A^T A)^{-1} A^T (r_0' - \Delta A x_0' + \Delta b). \tag{7.7}$$

In the case of finite perturbations, the following *upper bounds on the absolute error* $\|\Delta x'\|_2$ have been proved.

THEOREM 7.3 (Upper bound on the absolute error in the LS solution). *Consider $A_0 x \approx b_0$ with* LS *solution x_0' and* $\text{rank}(A_0) = n$. *Let* (1.21) *be the* SVD *of A_0 and $\mu = \varepsilon\kappa(A_0) = \varepsilon\sigma_1'/\sigma_n' < 1$. Then, for any perturbation $[\Delta A; \Delta b]$ satisfying* (7.4), *the absolute error $\Delta x' = x' - x_0'$ in the* LS *solution x' of the set $(A_0 + \Delta A)x \approx b_0 + \Delta b$ satisfies*

$$\|\Delta x'\|_2 \leq \frac{\varepsilon}{\sigma_n'}\left[\frac{\sigma_1'\|r_0'\|_2}{\sigma_n'(1-\mu^2)} + \frac{\sigma_1'\|x_0'\|_2}{1-\mu} + \frac{\|b_0\|_2}{1-\mu}\right]. \tag{7.8}$$

Proof. For the proof see [174, Thm. 4.3]. □

For zero residual problems, the first term vanishes since $r_0' = 0$. Observe that the bound (7.8) is tighter than the bound given in [205] if $\text{rank}(A_0) = n$. If $\text{rank}(A_0) = r < n$, replace σ_n' by σ_r' in (7.8) and add the term $\varepsilon\sigma_1'\|y_0\|_2$ with $y_0 = A_0^{\dagger T} x_0' = (A_0 A_0^T)^\dagger b_0$ [205].

Theorem 7.3 assesses the effect of the *residual* on $\Delta x'$ *exactly*, if the projection of r'_0 onto A has the direction of the image of the right singular vector v'_n, associated with its minimal singular value, i.e.,

$$AA^T r'_0 // Av'_n \text{ and } \|\Delta A\|_2 = \varepsilon\sigma'_1. \tag{7.9}$$

This happens if $A = A_0 + qv_n'^T$ with $q \perp R([u'_1, \cdots, u'_{n-1}])$. Then:

1. $Av'_i = \sigma'_i u'_i$, for all $i < n$; $Av'_n = \sigma'_n u'_n + q$,
2. $v'_1, \cdots, v'_n$ are singular vectors of A also, and
3. $R(A)$ may be obtained from $R(A_0)$ by a rotation around $R([u'_1, \cdots, u'_{n-1}])$.

This is illustrated in Fig. 7.2 for the case $n = 2$. Van der Sluis [174] shows that this mechanism in fact *maximizes* the effect of the *residual* on $\Delta x'$. More generally, we may say that this is due to the coincidence of two events: *rotation* of the projection plane around the image of a *dominant* singular vector produces a *large* projection of the residual and this projection has the direction of the image of the *lowest* singular vector.

The contribution of $\Delta b - \Delta A x'_0$ to $\Delta x'$, i.e., the last two terms of (7.8), is *majorized* if

$$\Delta A = -\varepsilon\sigma'_1 u'_n v_n'^T \in R(A_0), \quad \Delta b = \varepsilon\|b_0\|_2 u'_n \in R(A_0) \text{ and } x'_0 = \|x'_0\|_2 v'_n. \tag{7.10}$$

For *zero residual* problems, i.e., $\|r'_0\|_2 = 0$, the upper bound of $\|\Delta x'\|_2$ in (7.8) reduces to the following:

$$\|\Delta x'\|_2 \leq \frac{\varepsilon}{\sigma'_n} \frac{\sigma'_1\|x'_0\|_2 + \|b_0\|_2}{1-\mu} \tag{7.11}$$

and is indeed *attained* under the conditions given by (7.10). In this case, $\Delta x' = \|\Delta x'\|_2 v'_n$ with $\|\Delta x'\|_2$ given by (7.11).

Observe from (7.9) and (7.10) that the worst perturbations on $\Delta x'$ are obtained for $\Delta A = qv_n'^T$ with $q \perp R([u'_1, \cdots, u'_{n-1}])$. For those perturbation effects, it is possible to calculate analytically the LS solution of the perturbed problem $Ax \approx b$, as proven more generally in the following theorem.

THEOREM 7.4 (Perturbation formula for the LS solution). *Let* (1.21) *be the* SVD *of* A_0, rank$(A_0) = n$, *and* $b_0 = \sum_{i=1}^m \beta_i u'_i$. *Consider perturbations* $\Delta A = qv_k'^T, q = \sum_{j=1}^m q_j u'_j$, *and* $\Delta b = \sum_{i=1}^m \nu_i u'_i$ *on the (non)zero residual problem* $A_0 x \approx b_0$ *with LS solution* x'_0. *Then, the LS solution* x' *of the perturbed set* $(A_0 + \Delta A)x \approx b_0 + \Delta b$ *is given by*

$$x' = x'_0 + \Delta x' = \sum_{i=1}^n \lambda_i v'_i \text{ and } \|\Delta x'\|_2^2 = \sum_{i=1}^n (\lambda_i - \beta_i/\sigma'_i)^2 \tag{7.12}$$

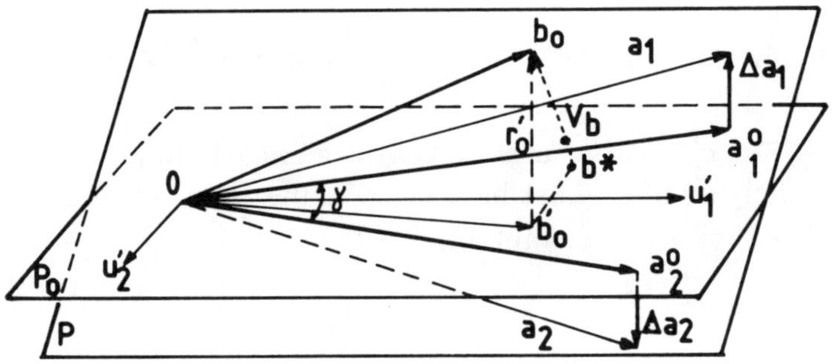

FIG. 7.2. *Geometric illustration of the maximal effect of the residual r_0' on the error $\Delta x'$ of the* LS *solution x_0' of $A_0 x \approx b_0$, due to perturbations Δa_1 and Δa_2 on the columns a_1^0 and a_2^0 of A_0. Take Δa_1, Δa_2 of size $\|\Delta a_1\|_2 = \|\Delta a_2\|_2 = \varepsilon$ orthogonal to plane $P_0 = R([a_1^0, a_2^0])$ and of opposite direction so that plane $P = ([a_1, a_2])$ is obtained by a rotation of P around the bisector $\frac{1}{2}(a_1 + a_2)$ which, according to Example 2.1, is an approximate left singular vector u_1' provided the angle γ between a_1^0 and a_2^0 is small. The angle of rotation $= \varepsilon\frac{\gamma}{2}$. b (respectively, b_0') is the projection of b_0 on plane P (respectively, P_0). Rotating P back to P_0, moves b to $b^\star$. Observe that $b^\star - b_0'$ is parallel with the lowest singular vector u_2' of A_0 [174].*

$$\text{with} \quad \begin{cases} \lambda_i = \frac{\beta_i + \nu_i - \lambda_k q_i}{\sigma_i'} & \forall i \neq k \\ \lambda_k = \dfrac{\beta_k + \nu_k + \sum_{i=n+1}^{m} q_i \frac{\beta_i + \nu_i}{\sigma_k' + q_k}}{\sigma_k' + q_k + \frac{\|q^\perp\|_2^2}{\sigma_k' + q_k}} & \text{and } \|q^\perp\|_2^2 = \sum_{i=n+1}^{m} q_i^2 \end{cases} .$$

Proof. Using the dyadic decomposition and the expressions above, we obtain

$$\begin{aligned} (A_0 + q v_k'^T)x' - (b_0 + \Delta b) &= \sum_{i=1}^{n} \lambda_i \sigma_i' u_i' + \lambda_k \sum_{i=1}^{m} q_i u_i' - \sum_{i=1}^{m} (\beta_i + \nu_i) u_i' \\ &= \sum_{i=1}^{n} (\lambda_i \sigma_i' + \lambda_k q_i - \beta_i - \nu_i) u_i' \qquad (7.13) \\ &\quad + \sum_{i=n+1}^{m} (\lambda_k q_i - \beta_i - \nu_i) u_i'. \end{aligned}$$

The LS solution x' is obtained by multiplying (7.13) with $A^T = (A_0 \mid \Delta A)^T$:

$$(7.14)\ \sum_{i=1}^{n} v_i'\sigma_i'(\lambda_i\sigma_i' + \lambda_k q_i - \beta_i - \nu_i) + v_k' q^T[(A_0 + qv_k'^T)x' - (b_0 + \Delta b)] = 0.$$

The vectors $v_1', \cdots, v_n'$ form a basis, so they are linearly independent. To satisfy (7.14), the coefficients of v_i' must be zero. Making the coefficients of v_i' zero for all $i \neq k$ yields ($\sigma_i' \neq 0$ since rank$(A_0) = n$)

$$(7.15) \qquad \lambda_i\sigma_i' + \lambda_k q_i - \beta_i - \nu_i = 0.$$

Hence, we obtain the expression of λ_i in (7.12) for $i \neq k$. Substituting (7.13) in (7.14) and making the coefficient of v_k' zero yields

$$\begin{aligned} \sigma_k'(\lambda_k\sigma_k' + \lambda_k q_k - \beta_k - \nu_k) + \sum_{i=1}^{n} q_i(\lambda_i\sigma_i' + \lambda_k q_i - \beta_i - \nu_i) & \\ + \sum_{i=n+1}^{m} q_i(\lambda_k q_i - \beta_i - \nu_i) &= 0. \end{aligned}$$

With (7.15) we obtain

$$\begin{aligned} \sigma_k'\lambda_k(\sigma_k' + q_k) - \sigma_k'(\beta_k + \nu_k) + \lambda_k q_k(\sigma_k' + q_k) - q_k(\beta_k + \nu_k) & \\ + \sum_{i=n+1}^{m} \lambda_k q_i^2 - \sum_{i=n+1}^{m} q_i(\nu_i + \beta_i) &= 0. \end{aligned}$$

By setting $\sum_{i=n+1}^{m} q_i^2 = \|q^\perp\|_2^2$ and arranging the terms we obtain (7.12). The projection b_0' of b_0 onto $R(A_0)$ is given by $\sum_{i=1}^{n} \beta_i u_i'$. Hence $x_0' = \sum_{i=1}^{n}(\beta_i/\sigma_i')v_i'$ and $x' = \sum_{i=1}^{n}(\lambda_i - \beta_i/\sigma_i')v_i'$. Taking the norm of $\Delta x'$, we obtain $\|\Delta x'\|_2^2$ given in (7.12). □

Observe from (7.12) that the error $\Delta x'$ will be larger if the residual $r_0' = \sum_{i=n+1}^{m} \beta_i u_i'$ is large and if the perturbations ΔA and Δb are more oriented along the *lowest* singular vectors, i.e., k is large (e.g., $k = n$) and the projections of q and Δb in the direction of u_n' and orthogonal to $R(A)$ become more important (e.g., q and $\Delta b \perp R([u_1', \cdots, u_{n-1}'])$).

For *nonzero residual* problems, (7.9) and (7.10) *cannot* be *simultaneously* satisfied. Hence, we can assess the effect of the residual exactly and the bounds of the other effects as nearly as possible, by majorizing the contribution of the projection of $\Delta b - \Delta A x_0'$ on A to $\Delta x'$. These upper bounds are quite close for $\mu < 0.5$. If r_0' is large, the first term in (7.8) would dominate $\|\Delta x'\|_2$, i.e., if the overdetermined set $A_0 x \approx b_0$ is very incompatible (see Example 7.1).

Observe from Theorem 7.3 that if $\|A\|_2 \neq 1$, the coefficient of $\varepsilon\|r_0'\|_2$ is not the square of a condition number but something of the form $\sigma_1'/\sigma_n'^2$. This is not surprising since condition numbers are typically related to $\|\Delta x'\|_2/\|x_0'\|_2$ and not to $\|\Delta x'\|_2$ alone, as shown below.

COROLLARY 7.1 (Upper bound on the relative error in the LS solution). *Consider $A_0x \approx b_0$ with* LS *solution x_0' and* $\text{rank}(A_0) = n$. *Let* (1.21) *be the* SVD *of A_0 and $\mu = \varepsilon\kappa(A_0) = \varepsilon\sigma_1'/\sigma_n' < 1$. Define $F(x_0') = \|Ax_0'\|_2/\|x_0'\|_2$, hence $\sigma_n' \leq F(x_0') \leq \sigma_1'$. Also, let θ_0 denote the angle between b_0 and $R(A_0)$. Then $\|x_0'\|_2 = \|b_0\|_2 \cos\theta_0/F(x_0')$ and $\|r_0'\|_2 = \|b_0\| \sin\theta_0$. Then for any $[\Delta A; \Delta b]$ satisfying* (7.4), *the relative error in the* LS *solution x' of the set $(A_0 + \Delta A)x \approx b_0 + \Delta b$ with $\Delta x' = x' - x_0'$, satisfies*

$$(7.16)\quad \frac{\|\Delta x'\|_2}{\|x_0'\|_2} \leq \varepsilon\left[\frac{\sigma_1'}{\sigma_n'}\frac{F(x_0')}{\sigma_n'}\frac{\tan\theta_0}{1-\mu^2} + \frac{\sigma_1'}{\sigma_n'}\frac{1}{1-\mu} + \frac{F(x_0')}{\sigma_n'}\frac{1}{\cos\theta_0}\frac{1}{1-\mu}\right].$$

Proof. For the proof see [174, p. 251]. □

For zero residual problems, the same bound (7.16) is valid if we set $\theta_0 = 0$. The upper bound (7.16) can be assessed under the same conditions (7.9) and (7.10) as given for the absolute error, but these conditions do not maximize the relative error effect. Indeed, the contribution of the *residual* to $\|\Delta x'\|_2/\|x_0'\|_2$, i.e., the first term in (7.16), is *maximized* if

$$(7.17)\qquad (7.9) \text{ is satisfied and the LS solution } x_0' = \|x_0'\|_2 v_1'.$$

The effect of the contribution of $-\Delta A x_0'$ to $\|\Delta x'\|_2/\|x_0'\|_2$, i.e., the second term in (7.16) is *maximized* if

$$(7.18)\qquad \Delta A = -\varepsilon\sigma_1' u_n' v_n'^T \in R(A_0) \text{ and } x_0' = \|x_0'\|_2 v_n',$$

while the effect of Δb to $\|\Delta x'\|_2/\|x_0'\|_2$, i.e., the third term in (7.16), is *maximized* if

$$(7.19)\qquad \Delta b = \varepsilon\|b_0\|_2 u_n' \in R(A_0) \text{ and } x_0' = \|x_0'\|_2 v_1'.$$

We immediately see that (7.17)–(7.19) *cannot* be satisfied *simultaneously*. Even for zero residual problems the conditions (7.18) and (7.19) may not be satisfied together. Hence, the *worst* possible upper bound (7.16) is *never* attained.

Observe that the effect of the residual and Δb on the relative error is maximized for $x_0'//v_1'$, while the effect of the residual on the absolute error is independent of the solution (see (7.9)) and the effect of Δb is maximal for $x_0'//v_n'$ (see (7.10)) (see Example 7.1).

From (7.16) we see that, unless $\tan\theta_0$ is large, $\kappa^2(A_0)$ plays *no* role for $\|\Delta x'\|_2/\|x_0'\|_2$ if $F(x_0')$ is a negligible fraction of σ_1', i.e., if x_0' has a *negligible* projection on all the v_i' associated with *larger* singular values. This is obviously the case for only a small fraction of all vectors x_0'. It is, however, the case for a large fraction of all vectors b_0 of given θ_0 provided that $\kappa(A_0)$ is large. Indeed, as soon as the projection of b_0 on u_n' is not small with respect to b_0, x_0' will belong approximately to the space spanned by the v_i', associated with the smaller singular values. Hence, if $\kappa(A_0)$ is large, then $\kappa^2(A_0)$ is *not realistic* with respect to the *relative error* for most least squares problems unless b_0 is nearly orthogonal to $R(A_0)$, i.e., $\theta_0 \approx \pi/2$.

From Corollary 7.1, we get a clearer picture of the importance of the various terms as functions of the *degree of incompatibility* of the set $A_0 x \approx b_0$, for which θ_0 may be taken as a measure. In particular, we see that for zero residual problems ($\theta_0 = 0$), the contribution of $-\Delta A x_0'$, i.e., the second term in (7.16)(and therefore in (7.8)) is the most important one, and this virtually remains so as long as θ_0 is small. If we let θ_0 grow, however, then already from $\theta_0 = 1/\kappa(A_0)$ on, the first term, i.e., the effect of the residual, may become the most important one, and, in fact, is the most important one if b_0 is such that $F(x_0') = \sigma_1'$, i.e., the projection of b_0 on $R(A_0)$ is parallel with the singular vector u_1', corresponding to σ_1'. Hence, if $\kappa(A_0)$ is large, already a residual corresponding to a b_0 with small θ_0 (i.e., $A_0 x \approx b_0$ is only slightly incompatible) may no longer be considered small in the sense of having no additional influence on the error. We also see that the full $\kappa^2(A_0)$ impact cannot occur until $\theta_0 \approx \pi/4$.

If A_0 is rank-deficient, i.e., $\text{rank}(A_0) = r < n$, the *minimum norm* LS solution is computed. For those cases, the worst perturbation bounds (7.9) and (7.16) must be slightly changed as shown by Hansen [77], Wedin [205], Stewart [159], and Van der Sluis and Veltkamp [175].

The worst perturbation effects on the LS solution are now compared with the upper bounds (7.8) and (7.16) in the following example.

Example 7.1.(a) Consider the following *nonzero residual* problem: If

$$(7.20)\ A_0 = \begin{bmatrix} 1 & 0 \\ 0 & 10^{-4} \\ 0 & 0 \end{bmatrix}, \Delta A = \begin{bmatrix} 0 & 0 \\ 0 & 0 \\ 0 & 10^{-8} \end{bmatrix}, b_0 = \begin{bmatrix} 1 \\ 0 \\ 1 \end{bmatrix}, \Delta b = \begin{bmatrix} 0 \\ 10^{-8} \\ 0 \end{bmatrix},$$

then

$$(7.21)\ x_0' = \begin{bmatrix} 1 \\ 0 \end{bmatrix} \quad \text{and } x_0' + \Delta x' = (A_0 + \Delta A)^\dagger (b_0 + \Delta b) = \begin{bmatrix} 1 \\ 1.0001 \end{bmatrix}.$$

Since $\kappa(A_0) = 10^4$ and $\varepsilon = 10^{-8}$, we have

$$\|\Delta x'\|_2 / \|x_0'\|_2 = \|\Delta x'\|_2 = 1.0001 \leq 1 + 10^{-4} + \sqrt{2}\, 10^{-4}.$$

These are the upper bounds given by (7.8) and (7.16). Observe that the *square* of the condition is attained since the residual is large, i.e., the set of equations is very incompatible (see also Fig. 7.3).

(b) Now consider the following *zero residual* problem:

$$(7.22) \quad A_0 = \begin{bmatrix} 1 & 0 \\ 0 & 10^{-2} \\ 0 & 0 \end{bmatrix}, \Delta A = \begin{bmatrix} 0 & 0 \\ 0 & -10^{-3} \\ 0 & 0 \end{bmatrix}, \Delta b = \begin{bmatrix} 0 \\ 10^{-3} \\ 0 \end{bmatrix}.$$

The condition $\kappa(A_0) = 10^2$ and $\varepsilon = 10^{-3}$. Observe that ΔA and Δb are worst case perturbations satisfying the conditions (7.10), (7.18) and (7.19) (when $\|b_0\|_2 = 1$). Let (1.21) be the SVD of A_0.

First, take

$$(7.23)\quad b_0 = \begin{bmatrix} 0 \\ 1 \\ 0 \end{bmatrix} // u_n', \text{ then } x_0' = \begin{bmatrix} 0 \\ 10^2 \end{bmatrix} // v_n' \text{ and } x_0' + \Delta x' = \begin{bmatrix} 0 \\ 100.1/.9 \end{bmatrix}.$$

Hence, $\|\Delta x'\|_2 / \|x_0'\|_2 = .101/.9$ and $\|\Delta x'\|_2 = 10.1/.9$. Observe that the bounds (7.8) and (7.16) are assessed exactly. Indeed

$$(7.24)\qquad \|\Delta x'\|_2 = (10^{-3} \cdot 10^2 \cdot 10^2 + 10^{-3} \cdot 1 \cdot 10^2)/.9 = 10.1/.9.$$

Now take

$$(7.25)\quad b_0 = \begin{bmatrix} 1 \\ 0 \\ 0 \end{bmatrix} // u_1', \text{ then } x_0' = \begin{bmatrix} 1 \\ 0 \end{bmatrix} // v_1' \text{ and } x_0' + \Delta x' = \begin{bmatrix} 1 \\ .1/.9 \end{bmatrix}.$$

Hence $\|\Delta x'\|_2 / \|x_0'\|_2 = \|\Delta x'\|_2 = .1/.9$. Compare this with the upper bounds (7.8) and (7.16):

$$(7.26)\quad \|\Delta x'\|_2 / \|x_0'\|_2 = \|\Delta x'\|_2 < (10^{-3} \cdot 10^2 \cdot 1 + 10^{-3} \cdot 10^2)/.9 = .2/.9$$

The upper bound on the absolute error $\|\Delta x'\|_2$, given in (7.24) with $x_0' // v_n'$, is maximal since conditions (7.10) are all satisfied but for the relative error $\|\Delta x'\|_2 / \|x_0\|_2$, larger upper bounds exist, e.g., when $x_0' // v_1'$ as shown in (7.26). Observe that the absolute and relative error upper bound can be assessed exactly, as shown in (7.24). The upper bound (7.26) is not attained. Since for zero residual problems ($\theta_0 = 0$) the effect of the contribution $-\Delta A x_0'$ on the relative error, i.e., the second term of (7.16), is the most important one, we can indeed expect larger relative errors for $x_0' // v_n'$ (maximizing the effect of $-\Delta A x_0'$) than for $x_0' // v_1'$ (which maximizes the effect of the residual and Δb), as shown in Example 7.1.

7.2.3. Upper bounds of the TLS solution perturbation. Based on the results of §7.2.2, we now deduce the *worst perturbation* effects on the TLS solution and investigate their importance. From the relationship (2.27), it is clear that a TLS solution can be considered as the LS solution of the related problem:

$$(7.27)\qquad \widetilde{A}_0 x \approx b_0 \text{ with } \widetilde{A}_0 = U'(\Sigma'^2 - \sigma_{n+1}^2 I)^{-1} \Sigma' V'^T.$$

The condition number of (7.27) is given by (7.1), i.e.,

$$(7.28)\qquad \widehat{\kappa} = \frac{\sigma_n'}{\sigma_1'} \cdot \frac{\sigma_1'^2 - \sigma_{n+1}^2}{\sigma_n'^2 - \sigma_{n+1}^2} \geq \frac{\sigma_1'}{\sigma_n'}.$$

Hence, the *same* upper bounds (7.8) and (7.16) may be used for measuring the *worst* TLS solution perturbation if we replace σ'_1 by $(\sigma_1'^2 - \sigma_{n+1}^2)/\sigma'_1$ and σ'_n by $(\sigma_n'^2 - \sigma_{n+1}^2)/\sigma'_n$.

For *zero residual* problems, i.e., $\sigma_{n+1} = 0$, (7.28) reduces to the following:

$$\widehat{\kappa} = \frac{\sigma'_1}{\sigma'_n} = \kappa(A_0). \tag{7.29}$$

Hence, the TLS and LS solution perturbations have the *same* bounds in the case of zero residual problems and attain their worst perturbation effect under the same conditions. For TLS, the *worst* perturbations ΔA and Δb occur when $R([\Delta A; \Delta b]) \subset R(A_0)$. Then σ_{n+1} remains zero. *No noise* can be *rejected* since the perturbed set is compatible. Thus the perturbations will entirely affect the solution accuracy. Under these conditions, $R([v'_1, \cdots, v'_n]) = R([v_1, \cdots, v_n])$. Hence, it follows that LS and TLS are equally affected with respect to their worst perturbations.

For *nonzero residual* problems, i.e., $\sigma_{n+1} \neq 0$, the TLS perturbation upper bounds, given by (7.8) and (7.16) in which σ'_1 and σ'_n are replaced by $(\sigma_1'^2 - \sigma_{n+1}^2)/\sigma'_1$ and $(\sigma_n'^2 - \sigma_{n+1}^2)/\sigma'_n$, respectively, are always *larger* than the LS perturbation upper bounds. This means that the nonzero residual TLS problem is *more sensitive* to errors than the corresponding nonzero residual LS problem, i.e., the TLS solution is *less stable* or robust than the LS solution. Hence, we may conclude that, in the absence of any further information about the data A, b (e.g., error distribution, orientation of b, etc.) the perturbation effects on the TLS solution may be worse than those on the LS solution. Thus considering the worst possible perturbations in the data, the accuracy of the TLS solution is more affected than that of the LS solution. As already shown in Corollary 6.2, the TLS problem becomes very sensitive and the TLS solution very *unstable* if σ_{n+1} approaches σ'_n. However, if $\sigma'_n - \sigma_{n+1}$ becomes very small, then we know from Theorem 3.11 that $v_{n+1,n+1}$ can be almost zero, i.e., b is nearly orthogonal onto u_{n+1} and $u_{n+1} \approx u'_n$. In those cases, the *nongeneric* TLS solution must be computed (see §3.4). The solution error is kept small and stability is maintained. This is illustrated in Example 7.2. Comparing $v_{n+1,n+1}$ with the perturbation level indicates whether or not b is orthogonal to u_{n+1} and whether the nongeneric TLS solution must be computed.

Example 7.2. Consider the same nonzero residual problem ($n = 2$) as in Example 7.1(a), illustrated in Fig. 7.3. Since the set of equations is highly incompatible, the nongeneric TLS problem is likely to occur. Taking the SVD (1.22) of $[A_0; b_0]$, we observe that $v_{n+1,n+1} = 0$. Hence, $b_0 \perp u'_n$ and $v_{n+1} = v'_n$ by Theorem 3.11 (see Fig. 7.3) and the nongeneric TLS solution $\widehat{x}_0$ of the unperturbed problem $A_0 x \approx b_0$ must be computed. Using Theorem 3.12, we obtain $\widehat{x}_0 = [2/((5)^{1/2} - 1), 0]^T$ from v_2.

Perturb A_0 and b_0 with ΔA and Δb as given in (7.20). Taking the SVD (1.22) of $[A; b] = [A_0 + \Delta A; b_0 + \Delta b]$, we now obtain $v_{n+1,n+1} \approx 10^{-8}$ and

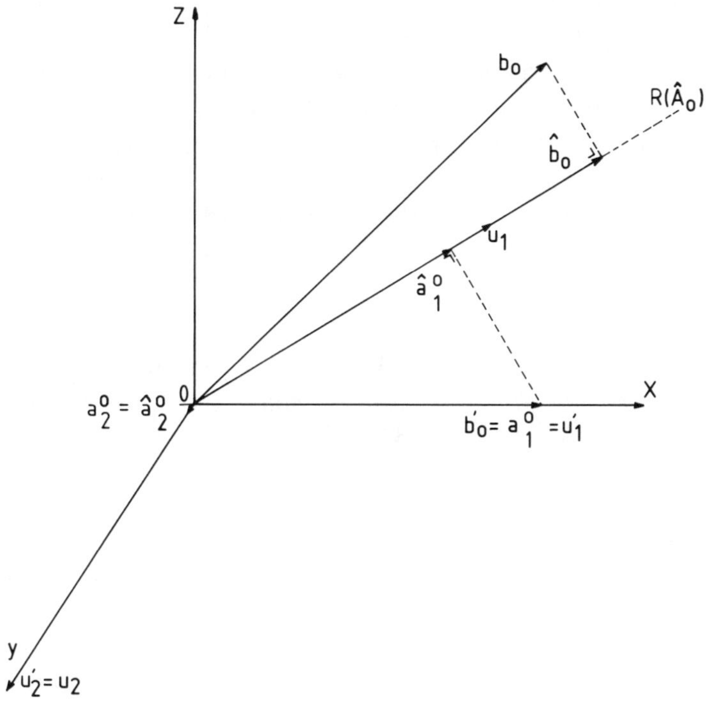

FIG. 7.3. *Geometric illustration of Examples* 7.1(a) *and* 7.2. *X, Y, and Z are the coordinate axes.* A_0 *has two columns* a_1^0 *and* a_2^0 *and* b_0 *is the observation vector. The* TLS *approximation is given by* $\widehat{a}_1^0, \widehat{a}_2^0, \widehat{b}_0$ *and* b'_0 *is the* LS *approximation.* u'_1, u'_2 *(respectively,* u_1, u_2*) are the left singular vectors of* A_0 *(respectively,* $[A_0; b_0]$*).*

$\sigma'_n - \sigma_{n+1} = \mathcal{O}(10^{-8})$. Computing the generic TLS solution (2.26) from v_{n+1} produces a very sensitive, unstable solution, given by $[1, 10^8 + 10^4]^T$. However, since the perturbation level is 10^{-8}, we may conclude that $b \perp u'_n$. Hence the nongeneric TLS solution must be computed. Using Theorem 3.12, we obtain $\widehat{x}_0 + \Delta\widehat{x} = [2/((5)^{1/2} - 1), (2 \cdot 10^{-8})/((5)^{1/2} - 3)]^T$ from v_n $(n = 2)$. Observe that $\|\Delta\widehat{x}\|_2 = 10^{-8}$ is very small and the perturbed TLS solution very stable.

This example illustrates how to stabilize the TLS solution of very unstable nonzero residual TLS problems, namely, by computing the nongeneric TLS solution. For rank-deficient problems close connections between nongeneric TLS and well-known biased regression estimators can be shown (see Chapter 9). For those problems, the nongeneric TLS estimator looks promising but the merits of the nongeneric TLS solution can be questioned when the sets of equations are more incompatible (i.e., σ_{n+1} increases). The same holds for the other regression estimators, e.g., the minimum norm LS solution of the set $A_r^0 \approx b_0$ with A_r^0 the best rank r approximation of A_0, also called the principal component (PC) estimator. For example, the presence of *outliers* in the data, i.e., large errors in the measurements, makes the generic TLS solution very unstable and deteriorates its accuracy dramatically. Then *robust* procedures that are rather insensitive to outliers should be considered, e.g., by downweighting measurement samples that give rise to high residuals (see §8.6).

Alternative upper bounds for the error in the TLS solution of $AX \approx B$, due to perturbations in the data A and B, have been proven in [68], [207] and are given below.

THEOREM 7.5. *Let* (1.21) (*respectively,* (1.22)) *be the* SVD *of* A_0 (*respectively,* $[A_0; b_0]$) *and* $\widehat{x}_0$ *be the* TLS *solution of* $A_0 x \approx b_0$. *If* $A \in \mathcal{R}^{m\times n}$ *and* $b \in \mathcal{R}^m$ *are such that* $\eta = \|[A - A_0; b - b_0]\|_F \leq \varepsilon/6$ *where* $\varepsilon = \sigma_n' - \sigma_{n+1} > 0$, *then the perturbed* TLS *problem* $Ax \approx b$ *has a unique solution* $\widehat{x} = \widehat{x}_0 + \Delta\widehat{x}$. *Moreover, if* $\widehat{x} \neq 0$, *then*

$$\frac{\|\Delta\widehat{x}\|_2}{\|\widehat{x}_0\|_2} \leq \frac{9\eta\sigma_1}{\sigma_n - \sigma_{n+1}}\left\{1 + \frac{\|b_0\|_2}{\sigma_n' - \sigma_{n+1}}\right\}\frac{1}{\|b_0\|_2 - \sigma_{n+1}}. \tag{7.30}$$

Proof. For the proof see [68, Thm. 4.4]. □

For most TLS problems, the upper bound (7.30) is too large and can never be attained. From the proof we deduce that the worst perturbation $[\Delta A; \Delta b]$ occurs if it maximizes the angle between the perturbed singular vector v_{n+1} of $[A; b]$ and the corresponding unperturbed one of $[A_0; b_0]$, associated with σ_{n+1} and if $\widehat{x}_0 // v_n'$. Observe from (7.30) that the TLS solution is less sensitive with increasing $\sigma_n' - \sigma_{n+1}$ and $\|b_0\|_2$ and decreasing σ_1.

The results of Theorem 7.5 have been generalized by Wei [207] for multidimensional TLS problems which may have more than one solution.

THEOREM 7.6. *Consider the* TLS *problem* $A_0 X \approx B_0$. *Let the* SVD *of* A_0 *and* $[A_0; B_0]$ *be given by* (1.21) *and* (1.22), *respectively, but assume that* V *is partitioned differently as in* (3.55) *in which* V_{22} *is of full rank. Furthermore, assume that for* $p \leq n$, (6.10) *or* (6.11) *hold and consider* $\widehat{X}_0 = -V_{12}V_{22}^{\dagger} = (V_{11}^T)^{\dagger}V_{21}^T$ *as the minimum norm* TLS *solution of* $A_0 X \approx B_0$. *Perturb* $[A_0; B_0]$ *such that* $[A; B] = [A_0; B_0] + [\Delta A; \Delta B]$. *If* $\widetilde{U}\widetilde{\Sigma}\widetilde{V}^T$ *is the* SVD *of* $[A; B]$ *and* $\widetilde{V}$ *is partitioned conformally with* V *as in* (3.55) (*replace* V_{ij} *by* $\widetilde{V}_{ij}$), *then the minimum norm* TLS *solution of the problem* $AX \approx B$ *is given by* $\widehat{X} = -\widetilde{V}_{12}\widetilde{V}_{22}^{\dagger} = (\widetilde{V}_{11}^T)^{\dagger}\widetilde{V}_{21}^T$. *Moreover, if* $\eta = \|[\Delta A; \Delta B]\|_2 \leq \frac{1}{6}(\sigma_p' - \sigma_{n+1})$, *we have*

$$\begin{aligned}\|\Delta\widehat{X}\|_2 = \|\widehat{X}_0 - \widehat{X}\|_2 &\leq \frac{\eta + \sigma_{n+1}}{\sigma_p' - \sigma_{n+1}}(3 + 5\|\widehat{X}_0\|_2) \leq 6\frac{\eta + \sigma_{n+1}}{\sigma_p' - \sigma_{n+1}}\sqrt{1 + \|\widehat{X}_0\|_2^2} \\ &\leq 12\frac{\eta + \sigma_{n+1}}{\sigma_p' - \sigma_{n+1}}\frac{\sigma_1}{\|B_0\|_2 - \sigma_{n+1}}\|\widehat{X}_0\|_2.\end{aligned}$$

The last inequality holds for $\widehat{X}_0 \neq 0$.

Proof. For the proof see [207, Thm. 4.1]. □

In particular, assume that $p = n$ and $\text{rank}(\widehat{A}) = n$, i.e., the multidimensional TLS problem with unique solution is considered (see §3.2). Then we have the following result.

COROLLARY 7.2. *If, in Theorem 7.6, $p = n$, the following inequalities hold:*

$$\begin{aligned}\|\Delta \widehat{X}\|_2 = \|\widehat{X}_0 - \widehat{X}\|_2 &\leq 3\frac{\eta + \sigma_{n+1}}{\sigma_n' - \sigma_{n+1}}(1 + \|\widehat{X}_0\|_2) \\ &\leq \frac{9(\eta + \sigma_{n+1})}{2(\sigma_n' - \sigma_{n+1})}\sqrt{1 + \|\widehat{X}_0\|_2^2} \\ &\leq 9\frac{\eta + \sigma_{n+1}}{\sigma_n' - \sigma_{n+1}} \frac{\sigma_1}{\|B_0\|_2 - \sigma_{n+1}}\|\widehat{X}_0\|_2.\end{aligned}$$

The last inequality holds for $\widehat{X}_0 \neq 0$.

Proof. For the proof see [207, Cor. 4.1]. □

If $\|[\Delta A; \Delta B]\|_2 \leq \frac{1}{6}(\sigma_p' - \sigma_{n+1})$, then Theorem 7.6 and Corollary 7.2 show that the minimum norm solution $\widehat{X} = -V_{12}V_{22}^{\dagger}$, as computed by Algorithm 3.1, is a good approximation of the true minimum norm TLS solution $\widehat{X}_0$ of the corresponding unperturbed problem $A_0X \approx B_0$, provided $\sigma_p' >> \sigma_{n+1}$ and the correct rank p can be determined. In perturbed rank-deficient TLS problems, it is, however, possible that the rank of A_0 is incorrectly estimated by some integer q, $n \geq q > p$. In this case, the perturbed minimum norm solution is not close to $\widehat{X}_0$. It is, however, close to another TLS solution of the corresponding unperturbed TLS problem $A_0X \approx B_0$, as proved in [207, Thm. 4.2]. Hence, if we just look for a TLS solution, not necessarily of minimal norm, then it is not necessary to find the correct rank p.

From the previous discussion, it can be concluded that TLS is *not better* than LS and is even worse for nonzero residual problems. However, it must be stressed that we only investigated the *worst* perturbations, which are only attained for a *special choice* of B_0 and the perturbations ΔA and ΔB. Under these conditions only, the given upper bounds, e.g., (7.8) and (7.16), are realistic for LS and generic TLS solutions. However, if further *information* about the *errors* ΔA and ΔB is given, e.g., their distribution function or covariance matrix, those upper bounds may be unrealistic. Other more important properties making TLS superior to LS from a *practical* point of view can be deduced (see §7.4 and Chapter 8).

7.3. Sensitivity properties of the SVD

In §7.2, upper bounds for the TLS and LS solution errors, due to perturbation effects on the data A_0 and B_0, were derived. These upper bounds correspond with a special orientation of B_0 with respect to $R(A_0)$ and may not be realistic for the other orientations. In this section, the *perturbation* effects on the *singular values* and associated *singular subspaces* of a given matrix C_0, e.g., $[A_0; B_0]$, due to perturbations on its matrix entries, are investigated, as well as their impact on the accuracy of the TLS and LS problem.

Let us start with a *general* description of the perturbation effects. Consider an $m \times p$ matrix C_0 of rank $r, m \geq p \geq r$, whose SVD is partitioned as:

$$(7.31) \qquad C_0 = U_0 \Sigma_0 V_0^T = [\underbrace{U_1^0}_{r} ; \underbrace{U_2^0}_{m-r}] \left[\Sigma_1^0 0 0 \Sigma_2^0 \right] [\underbrace{V_1^0}_{r} ; \underbrace{V_2^0}_{p-r}]^T = \sum_{i=1}^{r} \sigma_i^0 u_i^0 v_i^{0T},$$

with $U_0^T U_0 = I_m, V_0^T V_0 = I_p$, $\sigma_1^0 \geq \cdots \geq \sigma_r^0 > 0$, $\Sigma_1^0 = \text{diag}(\sigma_1^0, \cdots, \sigma_r^0) \in \mathcal{R}^{r \times r}$ and $\Sigma_2^0 = 0_{(m-r)\times(p-r)}$. Add perturbations ΔC to C_0 and let the SVD of $C = C_0 + \Delta C$ be partitioned analogously to (7.31):

$$(7.32)\ C = C_0 + \Delta C = U \Sigma V^T = [U_1; U_2] \begin{bmatrix} \Sigma_1^0 & 0 \\ 0 & \Sigma_2^0 \end{bmatrix} [V_1; V_2]^T = \sum_{i=1}^{p} \sigma_i u_i v_i^T,$$

with $U^T U = I_m, V^T V = I_p$, $\sigma_1 \geq \cdots \geq \sigma_p \geq 0$, $\Sigma_1 = \text{diag}(\sigma_1, \cdots, \sigma_r) \in \mathcal{R}^{r \times r}$ and $\Sigma_2 = \text{diag}(\sigma_{r+1}, \cdots, \sigma_p) \in \mathcal{R}^{(m-r)\times(p-r)}$. Then the relationship between C, C_0, and their SVD is given by

$$(7.33) \qquad C_0 + \Delta C = U \Sigma V^T = (U_0 G)(\Sigma_0 + \Delta\Sigma) H^T V_0^T,$$

where $G = U_0^T U$ and $H = V_0^T V$ are orthonormal and

$$(7.34) \qquad \|\Delta C\|_F = \|G(\Sigma_0 + \Delta\Sigma)H^T - \Sigma_0\|_F.$$

Equation (7.33) says that the perturbation ΔC of C_0 causes a *variation* $\Delta\Sigma$ in the singular value spectrum of C_0 and *a rotation* G (respectively, H) of its left (respectively, right) singular vectors and subspaces. G and H can be considered as *products of elementary Givens rotations* [69, §5.1.8], which describe the simultaneous rotation of two singular vectors over a certain angle within their own range.

Bounds on the *perturbation* effects $\Delta\Sigma$ of the *singular value spectrum* have been given in [156], [165], and [69, §8.3]. Using (7.31) and (7.32), we have

$$(7.35) \qquad \|\Delta\Sigma\|_F \leq \|\Delta C\|_F \text{ and } |\sigma_i - \sigma_i^0| \leq \|\Delta C\|_2 \quad i = 1, \cdots, p.$$

Equation (7.35) says that each singular value is "*perfectly conditioned*," i.e., the absolute change in a singular value is not more than the absolute change in the matrix.

The perturbation effects G and H on the *singular subspaces* of C_0, e.g., $R(U_1^0)$ and $R(V_1^0)$, are more delicate, as shown below. First, note that the singular values γ_i, $i = 1, \cdots, r$ of $G_{11} = U_1^{0T} U_1$ (respectively, $H_{11} = V_1^{0T} V_1$) are the *cosine* of the canonical angles θ_i between the ith left (respectively, right) principal singular subspaces of C_0 and C, i.e., the subspaces generated by their first i singular vectors.

Bounds on the perturbation effects related to singular subspaces have been given by Stewart [158], [165], Davis and Kahan [34], Staar [156], and others. Most interesting for the sensitivity study of the TLS problem are the bounds given by Wedin [204]. Here, we use the important property that the *distance* between any two subspaces is just the *sine* of the *largest canonical angle.* Then, the following interesting *upper bounds* for the distance between perturbed singular subspaces and their corresponding unperturbed ones can be derived.

THEOREM 7.7 (Perturbation of singular subspaces). *Let the* SVD *of the m by p matrix $C_0 (m \geq p)$ be given by* (7.31). *Call $C_1^0 = U_1^0 \Sigma_1^0 V_1^{0T}$ and $C_2^0 = U_2^0 \Sigma_2^0 V_2^{0T}$. Add perturbations ΔC to C_0 and let the* SVD *of $C = C_0 + \Delta C$ be given by* (7.32). *Call $C_1 = U_1 \Sigma_1 V_1^T$ and $C_2 = U_2 \Sigma_2 V_2^T$. If $\delta_1 = \sigma_r - \sigma_{r+1}^0 > 0$ and $\varepsilon_1 = \max(\|\Delta C V_1\|_2, \|\Delta C^T U_1\|_2)$, then*

$$\text{dist}(R(C_1), R(C_1^0)) \leq \frac{\varepsilon_1}{\delta_1}, \tag{7.36}$$

$$\text{dist}(R(C_1^T), R(C_1^{0T})) = \text{dist}(R(C_2^T), R(C_2^{0T})) \leq \frac{\varepsilon_1}{\delta_1}. \tag{7.37}$$

Also, if $\delta_2 = \sigma_r^0 - \sigma_{r+1} > 0$ and $\varepsilon_2 = \max(\|\Delta C V_2\|_2, \|\Delta C^T U_2\|_2)$, then

$$\text{dist}(R(C_1), R(C_1^0)) \leq \frac{\varepsilon_2}{\delta_2}, \tag{7.38}$$

$$\text{dist}(R(C_1^T), R(C_1^{0T})) = \text{dist}(R(C_2^T), R(C_2^{0T})) \leq \frac{\varepsilon_2}{\delta_2}, \tag{7.39}$$

$$\text{dist}(R(C_2), R(C_2^0)) \leq \sqrt{2}\,\frac{\varepsilon_2}{\delta_2}. \tag{7.40}$$

Theorem 7.7 is a slightly modified version of the generalized $\sin\theta$-theorem and the $\sin\theta$-theorem with complementary residuals in Wedin [204]. It is restricted to $m \times p$ matrices, $m \geq p$ where p equals the dimension of $R(C_1) \cup R(C_2)$ and the 2-norm is used. Since the subspaces C_1^T and C_2^T are orthogonal complements, we can make the following statement:

$$\text{dist}(R(C_2^T), R(C_2^{0T})) = \text{dist}(N(C_1), N(C_1^0)) = \text{dist}(R(C_1^T), R(C_1^{0T})) \tag{7.41}$$

and use the upper bounds (7.37) or (7.39). The subspaces $R(C_1^0)$ and $R(C_2^0)$ are not orthogonal complements. To estimate $\text{dist}(R(C_2), R(C_2^0))$, complementary residuals were defined to make a similar theorem possible. Note that the upper bounds in Theorem 7.7 depend on the *perturbation effects* $\|\Delta C\|_2$ and *the gap* between the smallest nonzero singular value of C_1 and the largest singular value of C_2^0. This means that the sensitivity of the singular subspaces grows when the *perturbations* are *larger* and when the *separation* between the concerned singular values *narrows*. In other words, Theorem 7.7 says that ε changes in C_0 can alter a singular subspace by an amount ε/δ where δ measures the separation of the involved singular values.

Theorem 7.7 has important consequences for the perturbation of the *singular subspaces* related to the TLS and LS problem.

Let (1.21) (respectively, (1.22)) be the SVD of A (respectively, $[A; B]$). To make the set of equations $AX \approx B$ compatible, LS projects the data A, B into the singular subspaces $R([u'_1, \cdots, u'_r])$ and $R([v'_1, \cdots, v'_r])$ of the matrix A of numerical rank $r \leq n$, while TLS projects A, B into the singular subspaces $R([u_1, \cdots, u_r])$ and $R([v_1, \cdots, v_r])$ of $[A; B]$. Indeed, from their respective definitions, the LS solution X' is the minimum norm solution of

$$A_r X = B' \text{ with } A_r = \sum_{i=1}^{r} \sigma'_i u'_i v'^T_i \text{ and } B' = \sum_{i=1}^{r} u'_i u'^T_i B, \tag{7.42}$$

and the TLS solution $\widehat{X}$ is the minimum norm solution of

$$\widehat{A} X = \widehat{B} \text{ with } [\widehat{A}; \widehat{B}] = \sum_{i=1}^{r} \sigma_i u_i v_i^T. \tag{7.43}$$

The differences in sensitivity of these singular subspaces explain the better performance of TLS in the presence of perturbations, as shown below.

First consider the unperturbed TLS *problem* $A_0 X \approx B_0$. Let the $m \times (n+d)$ matrix $[A_0; B_0]$ be perturbed by $[\Delta A; \Delta B]$ such that $[A; B] = [A_0; B_0] + [\Delta A; \Delta B]$. Denote by $r\ (\leq n)$ the rank of the approximation $[\widehat{A}_0; \widehat{B}_0]$ (defined by (7.43)) of $[A_0; B_0]$. Denote the SVD of $[A_0; B_0] = \sum_{i=1}^{n+d} \sigma_i^0 u_i^0 v_i^{0T}$. Applying Theorem 7.7 to $[A_0; B_0]$ and $[A; B]$, we obtain

$$\begin{aligned} \text{dist}(R(\widehat{A}), R(\widehat{A}_0)) &= \text{dist}(R([\widehat{A}; \widehat{B}]), R([\widehat{A}_0; \widehat{B}_0])) \\ &\leq \frac{\|[\Delta A; \Delta B]\|_2}{\sigma_r - \sigma_{r+1}^0} \end{aligned} \tag{7.44}$$

$$\begin{aligned} \text{dist}(R([\widehat{A}; \widehat{B}]^T), R([\widehat{A}_0; \widehat{B}_0]^T)) &= \text{dist}(N([\widehat{A}; \widehat{B}]), N([\widehat{A}_0; \widehat{B}_0])) \\ &\leq \frac{\|[\Delta A; \Delta B]\|_2}{\sigma_r - \sigma_{r+1}^0}. \end{aligned} \tag{7.45}$$

For *zero residual* problems, $A_0 X = B_0$ represents an *exact* relationship. In this case, $[\widehat{A}_0; \widehat{B}_0] = [A_0; B_0]$ and thus $\sigma_{r+1}^0 = 0$ in (7.44) and (7.45). Usually, A_0 has *maximal rank* n. Then (7.44) and (7.45) reduce to

$$\text{dist}(R(\widehat{A}), R(A_0)) \leq \frac{\|[\Delta A; \Delta B]\|_2}{\sigma_n}, \tag{7.46}$$

$$\text{dist}(N([\widehat{A}; \widehat{B}]), N([A_0; B_0])) \leq \frac{\|[\Delta A; \Delta B]\|_2}{\sigma_n}. \tag{7.47}$$

This means that the TLS *solution space approaches* the *exact* solution space if the *perturbations* $\|[\Delta A; \Delta B]\|_2$ are kept *small* and if the *gap* between the rth and $(r+1)$th singular value of $[A; B]$ is *large*.

Now consider LS *problems*. Let the $m \times n$ matrix A_0 of rank r $(r \leq n)$ be perturbed by ΔA such that $A = A_0 + \Delta A$. Denote the SVD of $A_0 = \sum_{i=1}^{r} \sigma_i^{0'} u_i^{0'} v_i^{0'T}$. Applying Theorem 7.1 to A_0 and A, we obtain

$$\text{dist}(R(A), R(A_0)) \leq \frac{\|\Delta A\|_2}{\sigma_r'}, \tag{7.48}$$

$$\text{dist}(R(A^T), R(A_0^T)) \leq \frac{\|\Delta A\|_2}{\sigma_r'}. \tag{7.49}$$

Usually, A_0 has *maximal rank* n. Then (7.48) reduces to

$$\text{dist}(R(A), R(A_0)) \leq \frac{\|\Delta A\|_2}{\sigma_n'}. \tag{7.50}$$

Comparing the sensitivity of the singular subspaces associated with the LS and TLS problem from the expressions above, the following important observations are made.

- First assume that the data A and B are measured samples of a zero residual problem $A_0 X = B_0$. The measurement errors ΔA and ΔB may be arbitrarily distributed provided $\|\Delta A\|_2 \approx \|[\Delta A; \Delta B]\|_2$ (e.g., when all data are equally perturbed). Then, using the interlacing Theorem 2.4 for singular values, it is concluded that the singular subspaces related to the TLS problem are *less noise sensitive* than those related to the LS problem. This implies that the former are expected to be "closer" to their corresponding unperturbed subspaces. Hence, the TLS solution is expected to be *more accurate* than the LS solution. This is also the case for nonzero residual problems $A_0 X \approx B_0$ as long as $\sigma_r - \sigma_{r+1}^0 > \sigma_r'$ in (7.44) and (7.48). In other words, the corresponding *unperturbed* problem $A_0 X \approx B_0$ must be *sufficiently compatible*.

- Since the *differences in sensitivity* depend on the ratio $(\sigma_r - \sigma_{r+1}^0)/\sigma_r'$, the *better accuracy* of TLS as opposed to LS will be *more* pronounced when this *ratio* is *larger*. This ratio is mainly influenced by the length (Frobenius norm), the column dimension, and the orientation of the observation matrix B_0 (or equivalently, of the solution X_0) with respect to the singular vectors of A_0. These observations are confirmed by simulations in §7.4.

Furthermore, let us investigate which *subspace perturbations maximally affect the accuracy of the singular subspaces* associated with the TLS *problem*. From Chapter 3, we know that the TLS solution of $AX \approx B$ is entirely deduced from the right singular subspace, associated with the smallest singular values of the matrix $[A; B]$. Hence, the worst subspace perturbations for TLS occur by rotating away this subspace as far as possible without affecting the others. Let us first consider simultaneous left and right singular subspace rotations.

The largest angle θ over which two singular vectors in the left and right singular subspace can rotate simultaneously within their own range is given in the next theorem.

THEOREM 7.8 (Left and right singular subspace rotation of maximal sensitivity). *Let the* SVD *of* C_0 *be given by* (7.31). *The largest angle* θ *over which a pair of right singular vectors* v_i^0 *and* v_j^0 *and left singular vectors* u_i^0 *and* u_j^0 *of* C_0 *can rotate simultaneously for a fixed variation* $\|\Delta C\|_F$ *is given by*

$$\theta = \arcsin \frac{\|\Delta C\|_F}{\sqrt{2}\,|\sigma_i^0 - \sigma_j^0|}. \tag{7.51}$$

Proof. For the proof see [156, Thm. 3-3]. □

In fact, we obtain a maximal forward effect on the matrices U_0 and V_0 for a minimal effect on C_0. Therefore, we use all the energy of ΔC to rotate away as far as possible one pair of singular vectors, without affecting the others. It is clear from (7.51) that bending a singular vector in the direction of another singular vector is much easier when the involved singular values are close.

The left and right singular subspace perturbation, which *maximally affect* the *accuracy* of the TLS problem is obtained by applying Theorem 7.8 to the singular vectors associated with σ_n^0 and σ_{n+1}^0 of $[A_0; B_0]$ or ($\angle$ denotes angle):

$$\theta = \max\{\gamma | \gamma = \angle(u_{n+1}^0, u_{n+1}) = \angle(v_{n+1}^0, v_{n+1})\} = \arcsin \frac{\|[\Delta A; \Delta B]\|_F}{\sqrt{2}|\sigma_n^0 - \sigma_{n+1}^0|}. \tag{7.52}$$

If we allow $\angle(u_{n+1}^0, u_{n+1})$ to be *different* from $\angle(v_{n+1}^0, v_{n+1})$, then *worse* perturbations than (7.52) with respect to the accuracy of the TLS solution are possible. Indeed, since the TLS solution is computed from the right singular subspace of $[A_0; B_0]$, we can concentrate the variation $[\Delta A; \Delta B]$ on $[A_0; B_0]$ entirely in the rotation of the two right singular vectors associated with the singular values σ_n^0 and σ_{n+1}^0. The maximal angle θ over which these vectors can rotate within their own range for a given $\|[\Delta A; \Delta B]\|_F$, and without affecting the corresponding left singular vectors, is derived in the following theorem.

THEOREM 7.9 (Right singular subspace rotation of maximal sensitivity). *Let the* SVD *of* C_0 *be given by* (7.31). *The largest angle* θ *over which a pair of right singular vectors* v_i^0 *and* v_j^0 *of* C_0 *can rotate for a fixed variation* $\|\Delta C\|_F$, *without affecting the corresponding left singular vectors* u_i^0 *and* u_j^0, *is given by*

$$\theta = 2 \arcsin \frac{\|\Delta C\|_F}{2\sqrt{\sigma_i^{02} + \sigma_j^{02}}}. \tag{7.53}$$

Proof. To obtain a maximal forward effect on V for a minimal ΔC, the whole variation ΔC should be used to rotate away one pair of right singular vectors as far as possible without affecting the others, i.e., the variation ΔC on $C_0 = U_0 \Sigma_0 V_0^T$ is concentrated in the singular vectors of two of its triplets only. Using the dyadic decomposition, the proof merely requires the verification of the following identity:

$$\begin{aligned} &\left\| \begin{bmatrix} \sigma_i^0 & 0 \\ 0 & \sigma_j^0 \end{bmatrix} \begin{bmatrix} \cos z & -\sin z \\ \sin z & \cos z \end{bmatrix} - \begin{bmatrix} \sigma_i^0 & 0 \\ 0 & \sigma_j^0 \end{bmatrix} \right\|_F^2 \\ &\qquad = \|\Delta C\|_F^2 = 4(\sigma_i^{02} + \sigma_j^{02}) \sin^2 \frac{\theta}{2}. \end{aligned}$$ □

For a given $\|\Delta C\|_F$, the angle θ will be maximal for the smallest singular values. Applying Theorem 7.9 to the zero residual problem $A_0 X = B_0$, i.e., $\sigma_{n+1}^0 = 0$, and taking $\Delta C = [\Delta A; \Delta B], j = n$, and $i = n+1$, we obtain

$$\theta = 2 \arcsin \frac{\|[\Delta A; \Delta B]\|_F}{2\sigma_n^0}. \tag{7.54}$$

Equation (7.54) gives the *worst singular subspace rotation for zero residual* TLS *problems*, i.e., the accuracy of the TLS solution is maximally affected. However, if $\sigma_{n+1}^0 \neq 0$, it is no longer evident which subspace rotation is the worst with respect to the TLS solution accuracy. Indeed, if σ_{n+1}^0 approaches σ_n^0, the angle given in (7.52) may be much larger than the angle given in (7.53). For each fixed variation $\|\Delta C\|_F$, there must exist an angle $\angle(u_{n+1}^0, u_{n+1})$ and an angle $\angle(v_{n+1}^0, v_{n+1})$ such that the accuracy of the TLS solution is most affected. This problem has not yet been solved.

Finally, it is shown how the perturbation of the singular vector v_{n+1} (from which the solution of the one-dimensional TLS problem is computed) affects the accuracy of the TLS solution.

THEOREM 7.10 (Singular vector perturbation effect on the TLS solution accuracy). *Let σ_{n+1}^0 and σ_{n+1} be the minimal singular values of $[A_0; b_0]$ and $[A; b]$, respectively, and v_{n+1}^0 and v_{n+1} their respective corresponding right singular vectors. Denote by σ_n' the minimal singular value of A and let $\widehat{x}_0$ and $\widehat{x}$ be the solution of the (un)perturbed* TLS *problem $A_0 x \approx b_0$ and $Ax \approx b$, respectively. If $\widehat{x}_0^T \widehat{x} \geq 0$, then*

$$\cos \angle(\widehat{x}_0, \widehat{x}) = \frac{\sqrt{(\|\widehat{x}_0\|_2^2 + 1)(\|\widehat{x}\|_2^2 + 1)} \cos \angle(v_{n+1}^0, v_{n+1}) - 1}{\|\widehat{x}_0\|_2 \|\widehat{x}\|_2}. \tag{7.55}$$

Proof. From (2.26) we obtain, using the cos-formula and assuming $\widehat{x}_0^T \widehat{x} > 0$,

$$\begin{aligned} \cos \angle\left(\begin{bmatrix} \widehat{x}_0 \\ -1 \end{bmatrix}, \begin{bmatrix} \widehat{x} \\ -1 \end{bmatrix} \right) &= \frac{\widehat{x}_0^T \widehat{x} + 1}{\sqrt{\|\widehat{x}_0\|_2^2 + 1}\sqrt{\|\widehat{x}\|_2^2 + 1}} \\ &= \cos \angle(v_{n+1}^0, v_{n+1}) = v_{n+1}^T v_{n+1}^0. \end{aligned} \tag{7.56}$$

Substituting $\cos \angle(\widehat{x}_0, \widehat{x}) = \widehat{x}_0^T \widehat{x}/(\|\widehat{x}_0\|_2 \|\widehat{x}\|_2)$ into (7.56) gives (7.55). □

Theorem 7.10 has important consequences. If $\|\widehat{x}_0\|_2 = \|\widehat{x}\|_2$, Theorem 7.10 reduces to

$$\cos \angle(\widehat{x}_0, \widehat{x}) = \cos \angle(v_{n+1}^0, v_{n+1}) - \frac{1 - \cos \angle(v_{n+1}^0, v_{n+1})}{\|\widehat{x}_0\|_2^2}. \tag{7.57}$$

This means that, if $\|\widehat{x}_0\|_2 = \|\widehat{x}\|_2$, the angle between $\widehat{x}_0$ and $\widehat{x}$ *approaches* the angle between v_{n+1}^0 and v_{n+1} more and more with *increasing* $\|\widehat{x}_0\|_2$ and *smaller angle* $\angle(v_{n+1}^0, v_{n+1})$ but always remains larger, e.g., if $\|\widehat{x}_0\|_2 = 1$ then

$$\begin{aligned} \angle(\widehat{x}_0, \widehat{x}) = 14.16^\circ \quad &\text{if} \quad \angle(v_{n+1}^0, v_{n+1}) = 10^\circ \\ \angle(\widehat{x}_0, \widehat{x}) = 43^\circ \quad &\text{if} \quad \angle(v_{n+1}^0, v_{n+1}) = 30^\circ \\ \angle(\widehat{x}_0, \widehat{x}) = 90^\circ \quad &\text{if} \quad \angle(v_{n+1}^0, v_{n+1}) = 60^\circ (\text{TLS solution error is 100 percent}). \end{aligned}$$

If $\|\widehat{x}_0\|_2 \neq \|\widehat{x}\|_2$ then

$$\cos \angle(\widehat{x}_0, \widehat{x}) \begin{cases} < \text{expression (7.57) if } \|\widehat{x}_0\|_2 > \|\widehat{x}\|_2 \\ > \text{expression (7.57) if } \|\widehat{x}_0\|_2 < \|\widehat{x}\|_2 \end{cases}.$$

This implies that for a given $\angle(v_{n+1}^0, v_{n+1})$, the deviations between the angle $\angle(\widehat{x}_0, \widehat{x})$ and the angle $\angle(v_{n+1}^0, v_{n+1})$ decrease (respectively, increase) with respect to the case $\|\widehat{x}_0\|_2 = \|\widehat{x}\|_2$ if the length of the perturbed solution $\widehat{x}$ is larger (respectively, smaller) than the length of the unperturbed solution $\widehat{x}_0$.

7.4. Simulation Results

In this section we describe a set of simulations confirming the main results of the sensitivity analysis, discussed in the previous sections. The *expected differences in accuracy* between the TLS and LS solution in the presence of *perturbations* are shown. Only *zero residual* problems are considered.

Simulation procedure. We start from an exact, overdetermined set of m equations $A_0 X = B_0$ in $n \times d$ unknowns X with d observation vectors B_0 and A_0 of full rank. Denote by X_0 the exact solution of $A_0 X = B_0$. Random errors Δa_{ij} and Δb_{ij}, normally distributed with zero mean and variance σ_ν^2, are then added to the entries of the exact data matrix $[A_0; B_0]$. Call the perturbed matrix $[A; B] = [A_0; B_0] + [\Delta A; \Delta B]$. We then solve the perturbed set of equations $AX \approx B$ to obtain the TLS solution $\widehat{X}$ and the LS solution X'. The relative errors

$$\|X' - X_0\|_F / \|X_0\|_F \quad \text{and} \quad \|\hat{X} - X_0\|_F / \|X_0\|_F \tag{7.58}$$

are averaged over 100 sets of equations affected by random errors with zero mean and equal variance σ_ν^2. We repeat the simulation for progressively larger noise variances σ_ν^2. The average relative errors are then plotted versus the

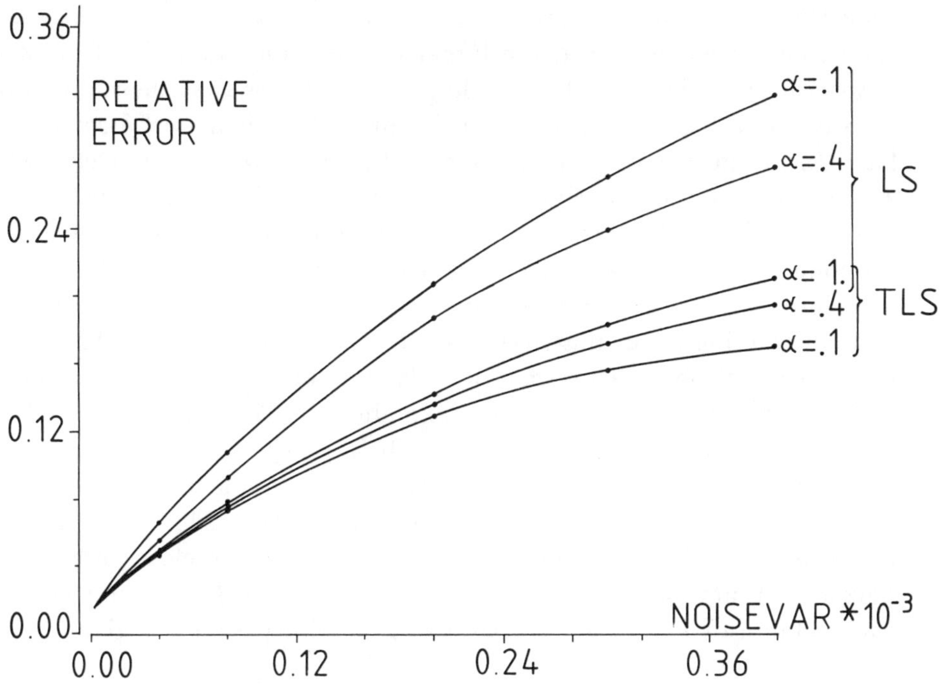

FIG. 7.4. *Comparison of the average relative error of the* LS *and* TLS *solution versus the variance of the noise added to the set* $(A_0)_{18\times 5}x = b_0$ *with* $\sigma'(A_0) = \{10, 8, 4, 1, .1\}$ *and* $b_0 = \sum_{i=1}^{5} \beta_i u_i'(A_0)$ *for different coefficient ratios* $\alpha = \max_{i=1,..,4} \frac{\beta_i/\sigma_i'}{\beta_5/\sigma_5'}$. *The values* 1, .4, *and* .1 *of* α *correspond with a ratio* $\sigma_5([A_0; b_0])/\sigma_5'(A_0)$ *equal to* 1.1, 1.6, *and* 3.6, *respectively.*

noise variance σ_ν^2 (see Figs. 7.4–7.6). The numerical rank $r \leq n$ of the matrix A (respectively, $[A; B]$) in LS (respectively, TLS) problems is estimated, using the rank determinator (3.124), i.e., $\sigma_r'^2 \geq 2\max\{m, n\}\sigma_\nu^2 > \sigma_{r+1}'^2$ (respectively, $\sigma_r^2 \geq 2\max\{m, n+d\}\sigma_\nu^2 > \sigma_{r+1}^2$). If $r < n$, the minimum norm solution is computed.

Let (1.21) (respectively, (1.22)) be the SVD of A_0 (respectively, $[A_0; B_0]$). From the previous discussion and since $A_0X = B_0$ (i.e., $\sigma_{n+1}^0 = 0$), it is clear that the difference in accuracy depends on the ratio of the nth singular values of $[A; B]$ and A or equivalently, on the *ratio* σ_n/σ_n'. Let us investigate the parameters that *influence* this ratio.

- First, the *orientation of the observation matrix* B_0 with respect to the lowest left singular vector u_n', or, equivalently, the orientation of the solution X_0 with respect to the lowest right singular vector v_n' of A_0, plays an important role. From the interlacing Theorem 2.4 for singular values, it is clear that the singular values of $[A_0; B_0]$ are always larger than the corresponding ones of A_0, i.e., $\sigma_i \geq \sigma_i'$ $i = 1, \cdots, n$. However, the extent to which the σ_i are larger depends on the orientation of B_0 with respect to the u_i' of A_0. It is not difficult to see that those directions u_i' of A_0 onto which B_0 has a larger projection will be "*favored*" and have the largest ratio σ_i/σ_i'. Hence, the ratio σ_n/σ_n' increases when B_0 becomes closer and closer to the direction of u_n'. As illustrated in Fig. 7.4, the better accuracy of TLS as opposed to LS is then more pronounced. In this case, X_0 has a negligible projection on all v_i' of A_0, associated with larger singular values. This is obviously the case for only a small fraction of all matrices X_0. It is, however, the case for a *large* fraction of all observation matrices B_0 provided the *condition number* $\kappa(A_0) = \sigma_1'/\sigma_n'$ of A_0 *is large* (as explained in §7.2.2). Indeed, a large condition number implies a small σ_n'. If $b_0 = \sum_{i=1}^n \beta_i u_i'$, then the exact solution is $x_0 = \sum_{i=1}^n (\beta_i/\sigma_i')v_i'$. Hence, the component β_n/σ_n' of x_0 in the direction of v_n' is large whenever σ_n' is small (provided the projection β_n of b_0 onto u_n' is not too small) and hence, x_0 is close to v_n'. The better accuracy of TLS with respect to LS is most pronounced when each column of B_0 is *proportional* to u_n'. In this case, σ_n is given by

$$\sigma_n = \min\left\{\sigma_{n-1}', \sqrt{\sigma_n'^2 + \|B_0\|_F^2}\right\}, \tag{7.59}$$

and the ratio σ_n/σ_n' reaches its maximal value for a given $\|B_0\|_F$. Also, when $B_0 // u_n'$, the LS solution is most sensitive to noise since the conditions (7.10) and (7.18) are then satisfied (see §7.2.2). This is also evident from Theorem 7.7, applied to the perturbation of singular vectors u_i' or v_i' (take $C_2 = \sigma_i' u_i' v_i'^T$). Indeed, by adding noise, the smallest singular values of A_0 and their associated singular vectors are most sensitive to noise. Hence, it reduces the accuracy of the LS solution, especially when B_0 is close to u_n'.

- Second, the *length of each observation vector* in B_0 influences the ratio σ_n/σ_n'. This is clear from (7.59). Making $\|B_0\|_F$ larger increases the ratio σ_n/σ_n' if $B_0//u_n'$ (as long as $\sigma_{n-1} > \sqrt{\sigma_n'^2 + \|B_0\|_F^2}$) and hence, the better accuracy of the TLS solution with respect to the LS solution will be more pronounced, as shown in Fig. 7.5.

- Third, the ratio σ_n/σ_n' depends on the *number d of observation vectors* in B_0. Indeed, if all d vectors of B_0 (instead of one) are now close to the lowest singular vector u_n' of A_0, the singular value σ_n of $[A_0; B_0]$ increases. Because d observation vectors together have a larger norm than one vector, the ratio σ_n/σ_n' increases with increasing number d of

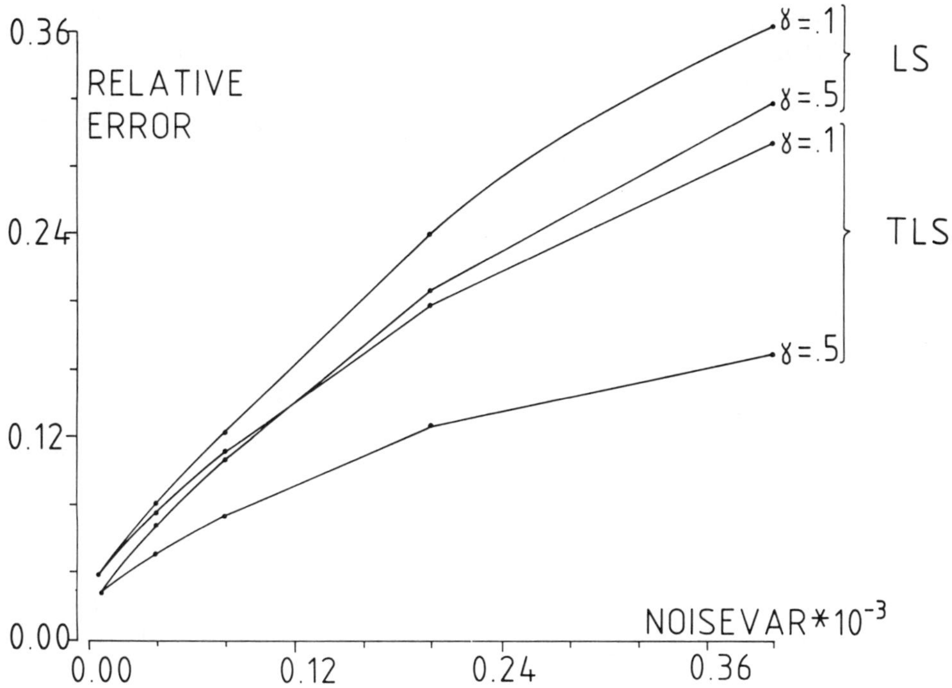

FIG. 7.5. *Comparison of the average relative error of the* LS *and* TLS *solution versus the variance of the noise added to the set* $(A_0)_{18\times5}x = b_0$ *with* $\sigma'(A_0) = \{10, 8, 4, 1, .1\}$ *and* $b_0 = \gamma u'_5$ *for different lengths* γ *of* b_0. *The values* .1 *and* .5 *of* γ *correspond with a ratio* $\sigma_5([A_0; b_0])/\sigma'_5(A_0)$ *equal to* 1.4 *and* 5.1, *respectively.*

observation vectors. In the extreme case, where $B_0//u'_n, \sigma_n$ is given by (7.59) and the ratio σ_n/σ'_n is maximal. The better accuracy of the TLS solution as opposed to LS is most pronounced, not only if all observation vectors are equally oriented along the lowest singular vector u'_n of A_0, but also if the condition number of A is large and the projection of B_0 onto u'_n is not too small (as explained before). This is illustrated in Fig. 7.6. Observe that the accuracy of the TLS solution improves with increasing number d of observation vectors while the accuracy of LS is independent of d.

- Furthermore, the difference in accuracy between TLS and LS also depends on the *number m of equations*, i.e., the degree of overdetermination. This will be statistically proven in Chapter 8 and confirmed by simulations (see Fig. 8.1).

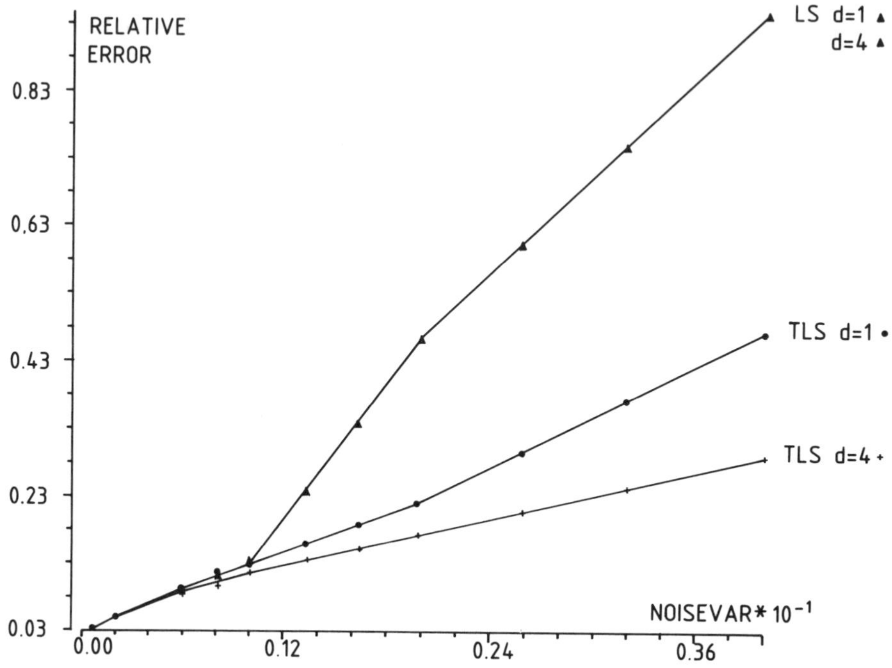

FIG. 7.6. *Comparison of the average relative error of the* LS *and* TLS *solution versus the variance of the noise added to the set* $(A_0)_{14\times 4}X = B_0$ *with* $\sigma'(A_0) = \{264, 9.7, 2.7, .75\}$ *for different numbers d of observation vectors in* B_0. *All d columns of* B_0 *are equal to* $\sum_{i=1}^{4} u_i'(A_0)$. *The values* 1 *and* 4 *of d correspond with ratio* $\sigma_4([A_0; B_0])/\sigma_4'(A_0)$ *equal to* 1.6 *and* 2.2, *respectively.*

It is illustrated here that the TLS solution is *expected* to have *better accuracy* than LS especially when B_0 is close to u_n' and all data are equally perturbed. Hence, the observation made in §7.2 that the TLS solution error may equal or even be worse than the LS solution error for a particular set $AX \approx B$ is generally not the case. The upper bounds given in §7.2, are only realistic for the worst type of perturbations $[\Delta A; \Delta b]$ of the data and the worst orientation of B_0. They are rather pessimistic if more information about the perturbations is available (see Chapter 8).

7.5. Concluding Remarks

This chapter examines how perturbations in all data A and B affect the accuracy of the TLS and LS problem. This is done by evaluating the relative changes in their solutions and related subspaces with respect to the *unperturbed* problem $A_0X \approx B_0$.

Provided the perturbations are *sufficiently small*, the unperturbed problem $A_0X \approx B_0$ is zero residual and *no* a priori information about the data is available (e.g., error distribution, orientation of B_0, etc.), the minimum norm solutions of all compatible sets $\widetilde{A}X = \widetilde{B}$, where $[\widetilde{A}; \widetilde{B}]$ is an orthogonal projection of $[A; B]$ onto $R(\widetilde{A})$, are shown to differ mutually only by second order terms. Hence, each of these solutions, including TLS and LS, is an adequate estimator of the true solution X_0 of $A_0X \approx B_0$ (§7.2.1).

Furthermore, the *worst perturbation* effects on the LS and TLS *solution* of zero and nonzero residual problems are investigated (§7.2). The importance of this notion of worst perturbation is that in the absence of further information, the worst perturbation should be considered as an attainable perturbation of the solution. Upper bounds for the absolute and relative errors are given and it is shown that they are maximized for different conditions of perturbation and orientation of the data (§7.2.2). In case of *zero residual* problems the TLS and LS solution perturbations attain the *same* upper bounds under the same conditions, but for *nonzero* residual problems the perturbation effects on the TLS solution may be even *worse* (§7.2.3). However, it must be stressed that those worst perturbations are only attained for a special choice of B_0 and the perturbations on the data. They are quite unrealistic for other conditions.

Second, the *sensitivity* properties of the SVD are investigated (§7.3). The perturbation results for singular values and their associated singular subspaces clearly show the smaller sensitivity-to-noise of the singular subspaces related to the TLS problem and explain the better performance of TLS with respect to LS in any zero residual problem and sufficiently compatible nonzero residual problems. This better accuracy of TLS as opposed to LS is *more pronounced* if B_0, or, equivalently, the solution X_0, is "*close*" to the singular vector u'_n or v'_n of A_0, associated with its *smallest* singular value, and if the *number of observation vectors* in B_0 and their *length* increase, as confirmed by simulations (§7.4). When $B_0 // u'_n$, the difference in expected accuracy of the TLS solution as opposed to LS *is maximal.*

Furthermore, subspace perturbations that maximally affect the accuracy of the TLS problem are computed, as well as the effect of singular subspace perturbations on the TLS solution accuracy for the one-dimensional problem (§7.3).

Chapter 8

Statistical Properties of the Total Least Squares Problem

8.1. Introduction

In the previous chapters, the TLS technique has been studied mainly from the viewpoint of a numerical analyst. No assumptions are made about the distribution of errors added to the data. If some *information* about the *error distribution* is *available* now, e.g., its covariance matrix, then statistical properties of the solution can be derived. These prove whether a certain solution technique is *expected* to be better and more accurate than others. In this chapter, it is assumed (unless stated otherwise) that *all errors* in the data matrix $[A; B]$ of the set $AX \approx B$ are row-wise *independently* and *identically distributed* with *zero mean* and *common covariance matrix* of the form $\sigma_\nu^2 I$. Under these assumptions, the TLS technique offers the *best* estimate and is more accurate than the LS solution, as statistically proved and confirmed by simulations.

The statistical properties in this chapter clearly show the reader the *expected quality* of the TLS solution of $AX \approx B$ as an estimate of the unknown parameters of linear relations (see, e.g., (1.1)) present in the data for any number m of observations. Note, however, that, in contrast to Chapter 7, these properties do not inform the reader about the quality of the TLS estimator for a *particular* choice of the error matrix. Instead, they judge the merit of the TLS solution, as parameter estimator, by the *distribution* of estimates to which it gives rise. Additionally, estimators of other parameters in the model, such as the error variance σ_ν^2 or an intercept, are derived from the TLS results and provide very valuable information to the user. Moreover, all the properties given in this chapter also extend to problems whose common error covariance matrix has the more general form $\sigma_\nu^2 C_0$, C_0 known and positive definite.

The statistical properties of a solution technique, called an estimator, are always related to a model whose parameters must be estimated. In order to appreciate the statistical properties of the TLS problem we must first of all derive its most suitable model. The most successful approach, directly related to the TLS concept, is the "*Errors-In-Variables*" (EIV) model, which considers all observations as coming from some unknown true values plus measurement errors. Knowledge of the error covariance matrix (or an estimate of it), up to a factor of proportionality, is required. Section 8.2 clarifies the essential differences between a regression model and the EIV model. At first sight, estimation of the parameters in the latter model looks like a problem in regression analysis, and indeed, this resemblance has given rise to much confusion. Next, in §8.3 estimators of various parameters of the EIV model are derived from the TLS results. In §8.4, it is shown that these TLS estimators are *strongly consistent* estimators of the true model parameters in both intercept and no-intercept models. The strong consistency results are obtained without any assumptions about the distribution of the errors Δ beyond those stated above. In particular, it is not assumed that the rows of Δ have a multivariate normal distribution. Strong consistency results (which assert convergence with probability one) provide deeper knowledge of the large-sample properties of estimators than do weak consistency results (which assert only convergence in probability) [59]. Section 8.5 considers *asymptotic distributional* results of the TLS estimators in both the intercept and no-intercept models. Finally, §8.6 discusses the *practical relevance* of the assumptions needed to carry through the asymptotic theory and presents some simulation results. The properties of the TLS estimator are also discussed when the assumptions of the model are violated.

8.2. Regression versus errors-in-variables model

The differences between the well-known classical regression model and the Errors-In-Variables (EIV) model are best illustrated by considering the *univariate*, one-dimensional case of the more *general regression model* with *errors in the variables*:

$$(8.1) \qquad b_0 = \alpha + a_0 x_0 + \varepsilon\,; \qquad b = b_0 + \Delta b \text{ and } a = a_0 + \Delta a;$$

a_0 and b_0 are the true but unobservable variables, Δa and Δb are their measurement or observation errors, and a and b are the corresponding observable variables. α is the intercept if any else zero. In addition to the errors in the variables Δa and Δb, there is an equation error, which summarizes all the factors that have an influence on b_0 but are not represented by explicit observable variables in the equation. ε is due to the inherent structure of the relationship between the variables. For example, body weight varies with height in an intrinsic way quite unconnected with any other errors of measurement. It is assumed that the errors $\Delta a, \Delta b, \varepsilon$ are uncorrelated with a_0 and b_0, their variances $\sigma^2_{\Delta a}$ and $\sigma^2_{\Delta b}$ and their covariance $\sigma_{\Delta a \Delta b}$ (which is zero in the case of independence) are known, while σ^2_ε is an unknown parameter to be estimated along with α and x_0. Model (8.1) typically describes *nonzero residual* problems (see §7.2.1).

If we assume ε to be *zero* in (8.1), we obtain the *classical* EIV model. There exists an exact linear relation between a_0 and b_0 in which we are interested. This model thus typically describes *zero residual* problems. a_0 and hence b_0 may be either random or nonrandom variables. If a_0 is a random variable, the relation $b_0 = \alpha + a_0 x_0$ is called a *structural* relation. Otherwise, if a_0 is fixed, the relation is called a *functional* relation [110]. Note that this situation is *symmetric* in that the expression $a_0 = (b_0 - \alpha)/x_0$ is an equally meaningful way of writing the relation $b_0 = \alpha + a_0 x_0$ in this context.

In a *classical regression* situation, however, we are concerned with the dependence of b_0 upon a_0 that is *not* subject to error: the error variable Δa in (8.1) is identically zero in value, so that $\sigma^2_{\Delta a} = 0$. Thus the classical regression model is essentially a *special case* of the more *general* EIV regression model (8.1) but may *not* be considered as a special case of the classical EIV model. Indeed, in contrast to the latter model, the *variation* of the dependent variable b_0 in regression analysis is *not* necessarily, or even usually, due to the error Δb alone, i.e., $\varepsilon \neq 0$. Hence, one important property of the regression situation when a_0 is a random variable is its *asymmetry*, i.e., the expression $a_0 = (b_0 - \alpha)/x_0$ is not a meaningful relation in this context. Also note that when we write $b = b_0 + \Delta b + \varepsilon$, we must distinguish between Δb and ε. The variable Δb is an error of observation which we presumably may eliminate by making finer and finer observations, whereas even if we get rid of Δb, we could never eliminate ε. The variable ε is inextricably tied to the distribution of b_0 given a_0. The variable Δb is called a "*fluctuation*" while ε is called the "*individual part*" of an observed quantity [101].

We may easily convince ourselves that the existence of errors in *both* a_0 and b_0 poses a problem quite *distinct* from that of regression. A useful consequence of (8.1) is

(8.2) $$b = \alpha + a x_0 + u \quad \text{with} \quad u = \Delta b + \varepsilon - \Delta a x_0.$$

This is not a classical regression situation: a is a random variable and it is correlated with the error term u [88]. Finally, we should note that, in contrast to EIV models, it makes no difference in regression analysis whether a_0 is fixed or random, for in either case, a_0 is not treated as a random variable in considering the linear relationship $b_0 = \alpha + a_0 x_0$ of interest.

Consider now the *multivariate* ($n > 1$), multidimensional ($d > 1$) EIV model and assume that a_0 is fixed, not random. A vector b_i^0 of unobservable values of d dependent variables and a vector a_i^0 of unobservable values of n independent variables generated by some phenomenon of interest for $i = 1, \cdots, m$ are related by the following linear model:

(8.3) $$b_i^{0T} = \alpha^T + a_i^{0T} X_0 \qquad i = 1, \cdots, m.$$

The d-dimensional vector α and the $n \times d$ matrix X_0 are the true, unknown parameters to be estimated. α is the vector of intercepts and X_0 is the matrix of slopes. Actually we can only observe b_i^0 and a_i^0 with errors, i.e.,

$$a_i = a_i^0 + \Delta a_i \text{ and } b_i = b_i^0 + \Delta b_i \quad i = 1, \cdots, m. \tag{8.4}$$

$\Delta_i = [\Delta a_i^T ; \Delta b_i^T]^T$ is a vector of random errors whose covariance matrix $\mathcal{C} = \mathcal{E}(\Delta_i \Delta_i^T)$ is known, up to a factor of proportionality. To condense the notation, let A, A_0, and ΔA be the $m \times n$ matrices whose rows are a_i^T, a_i^{0T}, and $\Delta a_i^T, i = 1, \cdots, m$, respectively. Let B, B_0 and ΔB be the $m \times d$ matrices whose rows are b_i^T, b_i^{0T}, and Δb_i^T, $i = 1, \cdots, m$, respectively. In terms of these matrices, (8.3) and (8.4) can be reformulated as follows.

DEFINITION 8.1 (The multivariate linear errors-in-variables model).

$$\begin{aligned} B_0 &= 1_m \alpha^T + A_0 X_0 \qquad\qquad B_0 \in \mathcal{R}^{m \times d}, A_0 \in \mathcal{R}^{m \times n} \\ A &= A_0 + \Delta A \\ B &= B_0 + \Delta B \end{aligned} \tag{8.5}$$

where $1_m = [1, \cdots, 1]^T$. X_0 are the true but unknown parameters to be estimated. The intercept vector α is either known to be zero (no-intercept model) or else is unknown (intercept model) and must be estimated. The observations $[A; B]$ of the unknown fixed values $[A_0; B_0]$ contain measurement errors $[\Delta A; \Delta B]$ such that the rows of $[\Delta A; \Delta B]$ are independently and identically distributed with zero mean and positive definite covariance matrix $\mathcal{C}$, known up to a factor of proportionality.

If we additionally assume that

(8.6) *the rows of* $[\Delta A; \Delta B]$ *in* (8.5) *are independently and identically distributed with common zero mean vector and common covariance matrix* $\mathcal{C} = \sigma_\nu^2 I_{n+d}$ *where* $\sigma_\nu^2 > 0$ *is unknown,*

then the TLS method is able to compute "*strongly consistent*" estimates of the unknown parameters X_0, A_0, α, and σ_ν of the EIV model (8.5) (see §8.4).

There are basically three situations in which EIV models are *useful*. First, the EIV model is useful when the primary goal is to estimate the *true parameters* of the model generating the data rather than prediction and if we cannot a priori be sure that the observations are error-free. However, if we wish to predict b_i^0 in (8.3) given a_i, ordinary regression of b_i on a_i, i.e., LS, should be used [8].

A second application of the EIV model is due to the immediate connection between TLS and *eigenvalue-eigenvector analysis* or, equivalently, SVD. TLS gives the hyperplane that passes through the intercept if any else zero, and is parallel to the plane spanned by the first n right singular vectors of the data matrix, see, e.g., [150].

Third, there are many situations in modeling in which it is important to treat the variables a_i and b_i *symmetrically*. In science we are frequently interested only in the parameters of a model and not in predicting one variable from other variables. There are no "independent" and "dependent" variables. Instead, the variables are jointly distributed random variables and should be treated symmetrically, e.g., when estimating the constant in the adiabatic Boyle's law. TLS automatically treats the variables symmetrically. A symmetric formulation of model (8.3) is

$$a_i^{0T} X_0 = \alpha^T \quad i = 1, \cdots, m,$$

where a_i^0 is an $(n+d) \times 1$ vector whose last d components are $-b_i^0$ and X_0 is an $(n+d) \times d$ matrix.

Finally, it is interesting to note that more *general* (although less considered) EIV model formulations exist than the one described in (8.5), e.g., [55].

DEFINITION 8.2 (*The general errors-in-variables model).*

$$\begin{array}{rcll} B_0 = A_0 X_0 & = & A_1 X_1^0 + A_2^0 X_2^0 & A_1 \in \mathcal{R}^{m \times n_1},\ X_1^0 \in \mathcal{R}^{n_1 \times d},\ X_2^0 \in \mathcal{R}^{n_2 \times d} \\ (8.7) \qquad A_2 & = & A_2^0 + \Delta A_2 & \in \mathcal{R}^{m \times n_2}, \quad n = n_1 + n_2, \\ B & = & B_0 + \Delta B & \in \mathcal{R}^{m \times d}. \end{array}$$

$X_0 = [X_1^{0T}; X_2^{0T}]^T$ are the true but unknown parameters to be estimated. $A_0 = [A_1; A_2^0]$ consists of constants as well as B_0. A_1 is known but A_2^0 and B_0 are not. The observations A_2 and B of the unknown values A_2^0 and B_0 contain measurement errors ΔA_2 and ΔB such that the rows of $[\Delta A_2; \Delta B]$ are independently and identically distributed with zero mean and known positive definite covariance matrix C, up to a factor of proportionality σ_ν^2.

Allowing the covariance matrix C to be singular provides a unified and even more general treatment of the models (8.5) and (8.7) that contain both explanatory variables measured without error and explanatory variables measured with error [53]. Observe that model (8.5) is just a special case of model (8.7). Indeed, model (8.5) assumes that only one variable, namely, the intercept α if any, is observed without error. By setting $n_1 = 1$, $n_2 = n$, $A_1 = 1_m$, $A_2^0 = A_0$, $X_1^0 = \alpha^T$, and $X_2^0 = X_0$, model (8.7) reduces to model (8.5). Based on model (8.7), the classical TLS problem formulation, given in §§2.3 and 3.2, has been generalized and the solution of this so-called *generalized* TLS problem is shown to be a consistent estimate of the true parameters X_1^0, X_2^0 of this model. For more details, the reader is referred to [188] (see also Chapter 10).

8.3. Parameter estimators

It is well known in regression analysis that the *ordinary* LS solution X' of (8.7) is generally an *inconsistent* estimate of the true parameters X_0. Indeed, assume $d = 1$ and partition C as

$$\begin{bmatrix} C_a & C_{ab} \\ C_{ab}^T & \sigma_\nu^2 \end{bmatrix} \begin{matrix} n \\ 1 \end{matrix} .$$
$$\quad n \quad\quad 1$$

Then, applying [138, Lemma 30] to [54, Eq. (1.2.2)], it is easily demonstrated that under quite general conditions $x' = (A^TA)^{-1}A^Tb$ does not converge to x_0 but ("plim" means probability limit):

$$\text{(8.8)} \qquad \underset{m\to\infty}{\text{plim}}\ x' = [I_n - (\frac{1}{m}A_0^TA_0 + C_a)^{-1}C_a][x_0 + (\frac{1}{m}A_0^TA_0)^{-1}C_{ab}].$$

Assuming that (8.6) holds, i.e., $C = \sigma_\nu^2 I_{n+1}$, (8.8) reduces to the following:

$$\text{(8.9)}\ \underset{m\to\infty}{\text{plim}}\ x' - x_0 = -\sigma_\nu^2(\frac{1}{m}A_0^TA_0 + \sigma_\nu^2 I_n)^{-1}x_0 = -\sigma_\nu^2(\underset{m\to\infty}{\text{plim}}\ \frac{1}{m}A^TA)^{-1}x_0.$$

This implies that LS is asymptotically *biased.* Observe from (8.8) and (8.9) that the size of the measurement errors and the "spread" of the true values A_0 especially indicate whether or not x' is acceptable. Large errors (large C, σ_ν), ill-conditioned A_0 (i.e., $\kappa(A_0) = \sigma_1'/\sigma_n'$ large) and x_0 oriented close to the lowest right singular vector v_n' of A_0, increase the bias and make the LS estimate more and more inaccurate, as shown in §§7.4 and 8.5 (see also [163]).

If the error covariance matrix C, partitioned as above, is known, the asymptotic bias can be removed and a consistent estimator, called *Corrected Least Squares* (CLS), can be derived [140], [91], [53]:

$$\text{(8.10)} \qquad x_{\text{CLS}} = (A^TA - mC_a)^{-1}(A^Tb - mC_{ab}).$$

This is clearly a modification of the ordinary LS estimate. It corrects for the extra sources of error in an errors-in-variables model and can be considered as a "method-of-moments" estimator in the following sense:

$$\text{(8.11)} \qquad \mathcal{E}(A^TA - mC_a) = A_0^TA_0$$
$$\text{(8.12)} \qquad \mathcal{E}(A^Tb - mC_{ab}) = A_0^TA_0x_0$$

Let (1.22) be the SVD of $[A; b]$. Since, under the assumption (8.6), $C = \sigma_\nu^2 I_{n+1}$, and $\lim_{m\to\infty}(1/m)\sigma_{n+1}^2 = \sigma_\nu^2$ with probability one (see §8.4), we immediately see from (2.27) and (8.10) that CLS and TLS asymptotically yield the same consistent estimator $\hat{x}$ of the true parameters x_0 in (8.5). Using the same assumption, Gleser [59] proved consistency of the TLS solution $\hat{X}$ for the multidimensional problem ($d > 1$), as well as consistency of the estimators of other parameters α and σ_ν in (8.5) (see §8.4). These estimators are derived from the TLS results, as shown below.

Because the analyses of the no-intercept and intercept form of the model (8.5) involve similar steps, these two forms can be treated simultaneously. To this end, we premultiply the data $[A; B]$ with a matrix G, given by:

$$\begin{aligned} G &= I_m && \text{for the no-intercept model} \\ &= I_m - \frac{1}{m} 1_m 1_m^T && \text{for the intercept model with } 1_m = [1, \cdots, 1]^T. \end{aligned} \tag{8.13}$$

This means that the variables of the intercept model are *centered.* This is done by subtracting from each entry in A and B the arithmetic mean of its corresponding column. The columns of the transformed data $G[A; B]$ then have zero mean. Now compute the SVD (1.22) of $G[A; B]$ and assume that V_{22} and V_{11} are nonsingular. We then consider the following estimators $\widehat{X}, \widehat{A}, \widehat{\alpha}$ and $\widehat{\sigma}$ of the respective parameters X_0, A_0, α and σ_ν in (8.5):

- $\widehat{X}$ is the TLS solution given by (3.13) or (3.10).
- $\widehat{A}$ is the TLS approximation of A, given by $U_1 \Sigma_1 V_{11}^T$ in (3.11).
- $\widehat{\alpha}^T$ $= 0$ for the no-intercept model

(8.14) $= [\bar{b}_1, \cdots, \bar{b}_d] - [\bar{a}_1, \cdots, \bar{a}_n]\widehat{X}$ for the intercept model with $\bar{a}_i$ and $\bar{b}_i$ the arithmetic means of the ith column in A and B respectively.

(8.15) • $\widehat{\sigma}^2 = \dfrac{1}{mt} \|\Sigma_2\|_F^2 = \dfrac{1}{mt} \sum_{i=1}^{t} \sigma_{n+i}^2$ with $t = \min\{m-n, d\}$

If $m \geq n + d$, (8.6) holds and the common distribution of the rows of $[\Delta A; \Delta B]$ is multivariate normal, then the estimators $\widehat{X}, \widehat{\alpha}, \widehat{A}$ and $(d/(n+d))\widehat{\sigma}^2$ defined above, are the unique (with probability one) *maximum likelihood* estimators of X_0, α, A_0, and σ_ν^2 in (8.5), respectively, [59]. Alternative derivations of the maximum likelihood estimators for model (8.5), where the covariance matrix $\mathcal{C}$ is allowed to be singular, are given in [53].

It is interesting to note that the estimators $\widehat{X}$ and $\widehat{\alpha}$ of the intercept model can be computed straightforwardly from the original data $[A; B]$ (no centering is required). Indeed, $[\begin{smallmatrix} \widehat{\alpha}^T \\ \widehat{X} \end{smallmatrix}]$ is the TLS solution of the following set in which the first column is exactly known:

$$\left[\begin{array}{c|c} 0 & 0 \cdots 0 \\ \hline 1_m & [A; B] \end{array} \right] \left[\begin{array}{c} \alpha^T \\ X \\ -I_d \end{array} \right] \approx 0. \tag{8.16}$$

Applying the second step of Algorithm 3.2 yields:

$$(8.17)\qquad \begin{bmatrix} -\sqrt{m} & \vert & -\frac{1}{\sqrt{m}}1_m^T[A;B] \\ -- & - & ------- \\ 0 & \vert & \\ \vdots & \vert & G[A;B] \\ 0 & \vert & \end{bmatrix} \begin{bmatrix} \alpha^T \\ X \\ -I_d \end{bmatrix} \approx 0$$

and thus $\widehat{X}$ is the TLS solution of $G[A;B][{}_{-I_d}^{X}] \approx 0$ while $\widehat{\alpha}^T = -(1/m)1_m^T[A;B][{}_{-I_d}^{\widehat{X}}]$, equals (8.14). In practice, we should not add the extra zero row in (8.16) but compute $\widehat{\alpha}$ and $\widehat{X}$ by applying Algorithm 3.2 to $[1_m; A; B]$. According to [188, Thm. 5], both TLS solutions must coincide.

8.4. Consistency properties

We begin our investigation of the asymptotic properties of the estimators $\widehat{\alpha}$, $\widehat{X}$, and $\widehat{\sigma}^2$ of the parameters in model (8.5) with a consideration of consistency. If an estimator can be shown to converge in probability to the true value as $m \to \infty$, then in practice we can be confident that the estimate is likely to be near the true value if sufficient observations are available. Conditions on the data A_0 and the errors $[\Delta A; \Delta B]$, which ensure consistency, are considered here. No conditions are imposed on the distribution of the rows of $[\Delta A; \Delta B]$ beyond those appearing in assumption (8.6). Reference is made to the following assumptions on A_0 given in (8.5):

$$(8.18)\qquad \lim_{m\to\infty} \frac{1}{m} A_0^T A_0 \quad \text{exists.}$$

$$(8.19)\qquad \text{the matrix} \quad \Omega = \lim_{m\to\infty} \frac{1}{m} A_0^T G A_0 \quad \text{is positive definite.}$$

Assumption (8.18) also implies the existence of an *asymptotic mean* μ, given by

$$(8.20)\qquad \mu = \lim_{m\to\infty} \frac{1}{m}[A_0; B_0]^T 1_m = \begin{bmatrix} \mu_1 \\ \alpha + X_0^T \mu_1 \end{bmatrix} \quad \text{with} \quad \mu_1 = \lim_{m\to\infty} \frac{1}{m} A_0^T 1_m$$

$$(8.21)\qquad \text{and} \quad \lim_{m\to\infty} \frac{1}{m}[A_0; B_0]^T G [A_0; B_0] = \begin{bmatrix} I_n \\ X_0^T \end{bmatrix} \Omega [\, I_n \,;\, X_0 \,]$$

Using the notation defined above, the following strong consistency properties are proved:

THEOREM 8.1 (Strong consistency of parameter estimators in the EIV model). *Consider the* EIV *model* (8.5). *Let* Ω *and* μ *be defined by* (8.19) *and*

(8.20), *respectively, and let* (1.22) *be the* SVD *of* $G[A;B]$. *If assumptions* (8.6) *and* (8.18) *hold, then* (w.p.1 *means "with probability one"*):

(8.22)
$$\lim_{m\to\infty}\frac{1}{m}[A;B]^TG[A;B]=\Theta = \sigma_\nu^2 I_{n+d}+\lim_{m\to\infty}\frac{1}{m}[A_0;B_0]^TG[A_0;B_0] \text{ w.p.1}$$

(8.23)
$$\lim_{m\to\infty}\frac{1}{m}[A;B]^T 1_m=\mu \qquad \text{w.p.1}$$

(8.24)
$$\lim_{m\to\infty}\frac{1}{m}\Sigma^T\Sigma=\operatorname{diag}(\theta_1,\cdots,\theta_{n+d}) = \operatorname{diag}(\sigma_\nu^2+\gamma_1,\cdots,\sigma_\nu^2+\gamma_n,\sigma_\nu^2,\cdots,\sigma_\nu^2) \text{ w.p.1,}$$

where θ_i *and* γ_i *are the (ordered) eigenvalues of* Θ *in* (8.22) *and of*

$$\Omega(I_n+X_0X_0^T),$$

respectively,

$$\lim_{m\to\infty}\frac{1}{m}\Sigma_2^T\Sigma_2 = \sigma_\nu^2 I_d \qquad \text{w.p.1} \tag{8.25}$$

$$\lim_{m\to\infty}\hat{\sigma}^2 = \sigma_\nu^2 \qquad \text{w.p.1.} \tag{8.26}$$

Furthermore, if the assumptions (8.6), (8.18), *and* (8.19) *hold, then:*

$$\lim_{m\to\infty}\hat{X} = X_0 \qquad \text{w.p.1,} \tag{8.27}$$

$$\lim_{m\to\infty}\hat{\alpha} = \alpha \qquad \text{w.p.1,} \tag{8.28}$$

$$\lim_{m\to\infty}\frac{1}{m}V_{11}(\Sigma_1^T\Sigma_1-m\hat{\sigma}^2I)V_{11}^T = \Omega \qquad \text{w.p.1,} \tag{8.29}$$

$$\lim_{m\to\infty}V_{11}V_{11}^T = (I_n+X_0X_0^T)^{-1} \qquad \text{w.p.1,} \tag{8.30}$$

$$\lim_{m\to\infty}V_{22}V_{22}^T = (I_d+X_0^TX_0)^{-1} \qquad \text{w.p.1,} \tag{8.31}$$

$$\lim_{m\to\infty}\frac{1}{m}(B-A\hat{X})^T(B-A\hat{X}) = \sigma_\nu^2(I_d+X_0^TX_0) \qquad \text{w.p.1,} \tag{8.32}$$

$$\lim_{m\to\infty}\frac{1}{m}\Delta\hat{B}^T\Delta\hat{B} = \sigma_\nu^2(I_d+X_0^TX_0)^{-1} \qquad \text{w.p.1.} \tag{8.33}$$

Proof. For the proof see [59, §3]. □

The following comments are in order:

- Equation (8.22) means that the error matrix $[\Delta A;\Delta B]$ becomes *uncorrelated* with the *exact* data $[A_0;B_0]$ with increasing *overdetermination.* By subtracting $\sigma_\nu^2 I_{n+d}$ from the observations $[A;B]^TG[A;B]$ we obtain a strongly consistent estimate of the exact data covariance matrix $[A_0;B_0]^TG[A_0;B_0]$.

- Equation (8.23) implies that the arithmetic mean of each column in A and B is a strongly consistent estimator of the asymptotic mean of the corresponding unobservable column in A_0 and B_0.

- From (8.22), (8.24), and (8.25), it follows that the first n squared singular values σ_i^2 of $G[A; B]$ are strongly consistent estimators of $(\sigma_i^0)^2 + m\sigma_\nu^2$ with $(\sigma_i^0)^2$ the corresponding squared singular values of the exact data matrix $G[A_0; B_0]$. The smallest squared singular values σ_{n+i}^2, $i = 1, \cdots, d$ of $G[A; B]$ are strongly consistent estimators of $m\sigma_\nu^2$ or

$$\begin{aligned} &\lim_{m\to\infty} \frac{1}{m}\sigma_i^2 &&= \tfrac{1}{m}(\sigma_i^0)^2 + \sigma_\nu^2 \quad &&\text{w.p.1} \quad \text{for } i = 1, \cdots, n \\ (8.34) \qquad &\lim_{m\to\infty} \frac{1}{m}\sigma_{n+i}^2 &&= \sigma_\nu^2 &&\text{w.p.1} \quad \text{for } i = 1, \cdots, d. \end{aligned}$$

- The estimator $\widehat{\sigma}^2$, given by the *mean square* of the t smallest singular values of $[A; B]$ divided by m (see (8.15)), yields a strongly consistent estimate of the unknown *variance* σ_ν^2 of the *errors* in the data $[A; B]$ (see (8.26)). Thus, provided that (8.6) and (8.18) hold, TLS is able to estimate the covariance matrix of the errors $[\Delta A; \Delta B]$ in the data.

- Furthermore, (8.27) and (8.28) prove that the TLS solution $\widehat{X}$ and $\widehat{\alpha}$ are strongly consistent estimators of the exact solution X_0 and the intercept α of model (8.5). This means that the TLS *solution* $\widehat{X}$ and the *intercept* $\widehat{\alpha}$ converge to the *true parameters* X_0 and α of (8.5) with increasing degree of overdetermination.

- From (8.19), it might be conjectured that $(1/m)\widehat{A}^T G\widehat{A} = (1/m)V_{11}\Sigma_1^T\Sigma_1 V_{11}^T$ defines a consistent sequence of estimators for Ω. However, from (8.26), (8.29), and (8.30) we have

$$(8.35) \qquad \lim_{m\to\infty} \frac{1}{m}\widehat{A}^T G\widehat{A} = \Omega + \sigma_\nu^2(I_n + X_0X_0^T)^{-1} \qquad \text{w.p.1}$$

  so that this conjecture is false. This implies that the *exact* data covariance matrix $(1/m)A_0^T GA_0$ *cannot* be consistently estimated by the TLS estimator $(1/m)\widehat{A}^T G\widehat{A}$.

- Equations (8.30)–(8.33) follow directly from the expressions (3.27), (3.28), (6.16), and (6.42), respectively, by taking the asymptotic limits.

Assumptions (8.6), (8.18), and (8.19) seem somewhat *restrictive*. Indeed, loosely speaking, assumption (8.6) requires that all measurements in A, B are affected by errors and moreover, these errors must be uncorrelated and equally sized. If these conditions are not satisfied, e.g., when considering the general EIV model (8.7), the classical TLS solution of $AX \approx B$ is no longer a consistent estimate of the model parameters. However, provided that the error covariance matrix $\mathcal{C}$ of the perturbed data is known up to a factor of proportionality, the classical TLS Algorithms 3.1 and 3.2 can still be used for computing consistent estimates. We can transform the data $[A; B]$ explicitly to new data $[A^\star; B^\star] = [A; B]C^{-1}$, where C is a square root of $\mathcal{C}$, such that the error covariance matrix of the transformed data is now diagonal with equal error variances (i.e., assumption (8.6) is satisfied). The classical TLS algorithm can then be used to solve this transformed set of equations and finally the solution of the transformed system must be converted to a solution of the original set of equations. Such transformation procedures are described in [58], [59] for the case that $n_1 = 0$ and $\mathcal{C}$ in (8.7) is positive definite. This approach is, however, not recommended in general since computing $[A; B]C^{-1}$ (with $\mathcal{C} = C^T C$) would usually lead to unnecessarily large numerical errors if C is ill-conditioned with respect to the solution of equations.

To avoid the explicit transformation of the data A, B it is better to use the generalized SVD (GSVD) [69, §8.7.3]. The great advantage of the GSVD is that it replaces these transformation procedures by *one* that is numerically reliable and can more easily handle (nearly) *singular* covariance matrices $\mathcal{C}$ in (8.7). Moreover, the GSVD reveals the *structure* of the general "errors-in-variables" model more clearly than the usual transformation procedures. Using this GSVD, Van Huffel and Vandewalle [188] have developed a numerically reliable algorithm, called Generalized TLS (GTLS), which is able to compute consistent estimates of the parameters in any general EIV model of the form (8.7). For more details, see [188], [215].

Thus the basic question that must be raised about the *applicability* of using assumption (8.6) is whether or not the form of the *error covariance* matrix $\mathcal{C}$ can be *known* up to a constant scalar multiple. This may still appear too restrictive. Indeed, this type of information is not always available to an experimenter. Nevertheless, there is no other more general assumption on $\mathcal{C}$ that is sufficient to maximize the likelihood. If in a given situation we do not possess enough information to produce a reasonable $\mathcal{C}$, then different types of procedures must be found. For example, assume that independent repeated observations are available for each variable observed with error (typically, perhaps, each is measured twice); then, this type of replication provides enough information about the error covariance matrix to derive consistent unbiased estimates of $\mathcal{C}$ [55], [53]. Using these consistent estimates instead of the true covariance matrix $\mathcal{C}$ does not change the consistency properties of the parameter estimators for model (8.5), as proved by Fuller [53].

Let us now consider assumptions (8.18) and (8.19). Loosely speaking, assumption (8.19) requires that the row vectors a_i^{0T} of A_0 sample all directions in the n-dimensional space with relative frequencies of commensurate magnitude. Certainly, any well-designed experiment would require that the observations on the vector of independent variables, i.e., the rows of A_0, have this property.

Mathematically, assumption (8.19) implies that the *rank* of Ω must equal the dimension of the space spanned by the independent variables. Indeed, consider the EIV model (8.5) for the case $n = d = 3$. If one dimension in the three-dimensional space $R([A_0; B_0])$ is under-sampled, the resulting estimates of X_0 bear little resemblance to the true value. On the other hand, when all dimensions are nearly equally sampled, the estimates of X_0 are reasonably accurate even in moderately small samples, as demonstrated in [60].

Nevertheless, assumptions (8.18) and (8.19) are somewhat restrictive. For example, (8.18) does not hold in a univariate linear EIV model (8.5) with $n = d = 1$, if the values of the independent variable vary linearly with the sample size. Indeed, when $a_i^0 = i$ for $i = 1, \cdots, m$, then $\sum_{i=1}^m i^2 = \mathcal{O}(m^3)$ as $m \to \infty$.

These assumptions are, moreover, much stronger than conditions that have been shown to be sufficient for consistency of the usual linear regression estimate. On this matter, results of increasing strength and generality have been produced. Conditions on the errors vary somewhat but the condition on A_0 that is crucial to all of these is

$$\lambda_{\min}(\ A_0^T A_0\ ) \to \infty \quad \text{as } m \to \infty, \tag{8.36}$$

where $\lambda_{\min}$ denotes the minimal eigenvalue of A_0 as defined in model (8.7). This condition is much weaker than the assumptions (8.18) and (8.19). Equation (8.36) requires only that the independent variables are not highly correlated among each other and that the values of these variables "spread out" fast enough. For example, in the univariate errors-in-variables intercept model (8.5) with $n = d = 1$, (8.36) requires that $\sum_{i=1}^m (a_i^0 - \bar{a}_0)^2 \to \infty$ as $m \to \infty$ where $\bar{a}_0$ is the mean value, i.e., the values a_i^0 may not be clustered around a particular value.

Now asssuming conditions "intermediate" between (8.36) and assumptions (8.18) and (8.19), Gallo proved *weak consistency* of the TLS estimate.

THEOREM 8.2 (*Consistency of the TLS solution*). *Consider the one-dimensional* ($d = 1$) *errors-in-variables model* (8.7) *with* $n_2 = n$; $n_1 = 1$ *if an intercept is assumed, otherwise* $n_1 = 0$. *Assume that* (8.6) *holds. Now if the following conditions on* A_0 *are satisfied*

$$\frac{1}{\sqrt{m}} \lambda_{\min}(A_0^T A_0) \quad \to \infty \quad as \quad m \to \infty, \tag{8.37}$$

$$\frac{\lambda_{\min}^2(A_0^T A_0)}{\lambda_{\max}(A_0^T A_0)} \quad \to \infty \quad as \quad m \to \infty \tag{8.38}$$

with $\lambda_{\min}$ (*respectively,* $\lambda_{\max}$) *the minimal* (*respectively, maximal*) *eigenvalue, and the joint distribution of the errors* $[\Delta A_2; \Delta b]$ *possesses finite fourth moment, then the* TLS *solution* $\hat{x}$ *is a weakly consistent estimate of* x_0, *i.e.,*

$$\operatorname*{plim}_{m \to \infty} \hat{x} = x_0.$$

Proof. For the proof see [54, Thm. 2.1] or [55, Thm. 2]. □

Assumptions (8.37) and (8.38) are *intermediate* in the following sense: either one implies (8.36) while both are implied by assumptions (8.18) and (8.19). Condition (8.37) requires that the variables "spread out" fast enough at a faster rate than does (8.36). Condition (8.38) also holds much more generally than assumptions (8.18) and (8.19). It is satisfied, for example, if (8.37) holds and the independent variables are bounded. Finally, note that Gallo needs to assume finite fourth moments of the errors for consistency. However, he could weaken this assumption provided (8.37) is strengthened [54], [55].

As shown by (8.46), the LS estimate x' is generally *inconsistent* in model (8.5). Nevertheless, it is worth considering whether or not x' ever tends to x_0 since any consistent EIV estimate of the form (8.10), e.g., the TLS estimate $\widehat{x}$, requires knowledge of the error covariance matrix C and is not always applicable. The following consistency result holds.

THEOREM 8.3 (Consistency of the LS solution). *Consider the* EIV *model* (8.7) *with* $d = 1$. *If* (8.38) *holds and*

$$\frac{1}{m}\lambda_{\min}(A_0^T A_0) \to \infty \quad as \quad m \to \infty \tag{8.39}$$

then

$$\operatorname*{plim}_{m\to\infty} x' = x_0$$

Proof. For the proof see [54, Thm. 2.7]. □

This result might seem a bit disconcerting at first. However, it should be clear that x' can be badly inconsistent if (8.39) does not hold, which occurs, for example, if the independent variables are bounded. The main reason, though, that Theorem 8.3 is not particularly important is evidenced by some results of Anderson [9]. Whether or not (8.39) holds for moderate sample sizes x' is generally not as likely to be close to x_0 as the errors-in-variables estimates, e.g., TLS when (8.6) holds. Indeed, the distributions of the latter estimates are "centered" around the true values while the distribution of x' is not, even if (8.39) holds. By further restricting the behavior of A_0, we have the following theorem.

THEOREM 8.4 (Consistency of the LS solution). *Consider the* EIV *model* (8.7), $d = 1$ *and let* C *be partitioned as*

$$\begin{array}{cc} \left[\begin{array}{cc} C_a & C_{ab} \\ C_{ab}^T & \sigma_\nu^2 \end{array}\right] & \begin{array}{c} n_2 \\ 1 \end{array} \\ \begin{array}{cc} n_2 & \quad 1 \end{array} & \end{array}.$$

If there exist M_1 *and* M_2 *such that for all* $m \geq n$,

$$0 < M_1 \leq \frac{1}{m}\lambda_{\min}(\, A_0^T\, A_0\,) \leq \frac{1}{m}\lambda_{\max}(\, A_0^T\, A_0\,) \leq M_2 < \infty \tag{8.40}$$

holds, then

$$\operatorname*{plim}_{m\to\infty} x' = x_0 \Longleftrightarrow \mathcal{C}_{ab} = \mathcal{C}_a x_2^0.$$

Proof. For the proof see [54, Thm. 2.8]. □

Theorem 8.4 implies that, if Δb is independent of ΔA, then x_0 is estimated consistently by LS if and only if $x_0 = 0$! This is the case when $\mathcal{C} = \sigma_\nu^2 I$, as assumed in models (8.5) in which the TLS solution $\widehat{x}$ is a consistent estimate. This theorem provides more evidence to *discourage* the use of LS in EIV models. Nevertheless, the LS estimate can have some merits in estimating consistently certain linear functions of the elements of x_0 in some special classes of problems [25].

Finally, it is interesting to note that EIV models with a *special structure* also occur in *system identification* and *parameter estimation*. For instance, the transfer function model, arising in the identification of systems modeled by autoregressive moving average (ARMA) difference equations, has the following form:

$$y(t) = -a_1 y(t-1) - \cdots - a_{n_a} y(t-n_a) + b_1 u(t-1) + \cdots + b_{n_b} u(t-n_b). \tag{8.41}$$

$\{u(t)\}$ and $\{y(t)\}$ are the input and output sequences, respectively, of the system and $\{a_j\}$ and $\{b_j\}$ are the unknown constant parameters to be estimated. The *observations* at the input and output are assumed to be perturbed by mutually independent *white noise* sequences (i.e., i.i.d. random variables) with *zero mean* and *equal variances.* If sufficient observations are taken, (8.41) gives rise to an overdetermined set of equations $AX \approx B$, where the corresponding error covariance matrix on the data has the form $\sigma_\nu^2 I_{n+d}$. The matrix $[A; B]$ consists of Toeplitz or Hankel expansions of the input and/or output sequences. A well-known estimator of the parameters in these models is the eigenvector method [98] or Koopmans–Levin method [49] whose statistical properties have been studied by Aoki and Yue [12]. Since this estimator coincides with the TLS solution, their results also apply to the TLS problem. Necessary and sufficient conditions for consistency, similar to those given by Gleser, have been derived. See [12], [188], [191] for more details.

8.5. Asymptotic distributions

The consistency properties of the previous section are important but for purposes of statistical inference, some knowledge of the distributions of the estimates is necessary. Since fixed sample size distribution theory is extremely unwieldy even in the simplest EIV model formulations, only asymptotic distributions are considered. Assume that assumptions (8.6) and (8.18) hold and that the common distribution of the rows of $\Delta = [\Delta A; \Delta B]$ has finite fourth moments. Gleser [59] then proved that the elements of $1/\sqrt{m}([A;B]^T G[A;B] - \mathcal{E}([A;B]^T G[A;B]))$, together with the elements of $1/\sqrt{m}([A;B]^T 1_m - [A_0;B_0]^T 1_m)$ in case of an intercept model, have a symmetric, asymptotic multivariate normal distribution with zero means. The *asymptotic distributions* of $\sqrt{m}(\widehat{X} - X_0)$ and $\sqrt{m}(\widehat{\sigma}^2 - \sigma_\nu^2)$ in both the no-intercept and intercept model, and the asymptotic distribution of $\sqrt{m}(\widehat{\alpha} - \alpha)$ in the intercept model are all *(multivariate) normal* with *zero* means. However, the covariance matrices of these distributions involve third-order and fourth-order cross moments, and thus are both complicated in form and hard to estimate from data. To obtain *concise* expressions for the asymptotic covariances, *additional assumptions* are imposed (which also permit the construction of large-sample approximate confidence regions for the parameters σ_ν, X_0 and α of model (8.5) [59]).

THEOREM 8.5 (Asymptotic distribution of the TLS solution). *Let the estimators* $\widehat{X}, \widehat{\alpha}$ *and* $\widehat{\sigma}$ *be given by* (3.13), (8.14), *and* (8.15), *respectively. Let* Ω *and* μ_1 *be given by* (8.19) *and* (8.20), *respectively. Assume that* (8.6), (8.18), *and* (8.19) *hold and assume further that the cross moments of the common distribution of the rows of the error matrix* $[\Delta A\ ;\ \Delta B]$ *in* (8.5) *are identical, up to and including moments of order four, to the corresponding moments of the multivariate normal distribution with zero mean and covariance matrix* $\sigma_\nu^2 I_{n+d}$.

Under both the intercept and no-intercept models, we have that

1. $\sqrt{m}(\widehat{\sigma}^2 - \sigma_\nu^2)$ *has an asymptotic normal distribution with zero mean and variance* $(2/d)\sigma_\nu^4$.

2. *The elements of* $\sqrt{m}(\widehat{X} - X_0)$ *have an asymptotic* $n \times d$*-variate normal distribution with zero means and covariance between the* (i,j)*th and* (k,l)*th elements given by*

$$\sigma_\nu^2[\sigma_\nu^2\Omega^{-1}(I_n + X_0X_0^T)^{-1}\Omega^{-1} + \Omega^{-1}]_{ik}\ [I_d + X_0^TX_0]_{jl}. \tag{8.42}$$

3. *In the intercept model,* $\sqrt{m}(\widehat{\alpha} - \alpha)$ *has an asymptotic multivariate normal distribution with zero mean and covariance matrix*

$$\sigma_\nu^2\{1 + \mu_1^T[\sigma_\nu^2\Omega^{-1}(I_n + X_0X_0^T)^{-1}\Omega^{-1} + \Omega^{-1}]\mu_1\}(I_d + X_0^TX_0), \tag{8.43}$$

and, furthermore, the elements of $\sqrt{m}(\widehat{X} - X_0)$ *and* $\sqrt{m}(\widehat{\alpha} - \alpha)^T$ *have an asymptotic* $(n+1) \times d$*-variate normal distribution where the asymptotic covariance between the* (i,j)*th element of* $\sqrt{m}(\widehat{X} - X_0)$ *and the* l*th element of* $\sqrt{m}(\widehat{\alpha} - \alpha)^T$ *is*

$$\sigma_\nu^2(I_d + X_0^TX_0)_{jl}\{[\sigma_\nu^2\Omega^{-1}(I_n + X_0X_0^T)^{-1}\Omega^{-1} + \Omega^{-1}]\mu_1\}_i. \tag{8.44}$$

Proof. For the proof see [59, Thm. 4.2]. □

However, as pointed out by Gallo [54], there is a flaw in Gleser's argument since his conclusions depend directly on an erroneous lemma. The conclusions of Gleser's lemma, and hence Theorem 8.5, can be shown to be true if we require the errors to have finite moments of order more than four. In addition,

Gallo proved the following asymptotic distribution (first assume for simplicity that the intercept is zero).

THEOREM 8.6 (Asymptotic distribution of the TLS solution). *Consider the one-dimensional* EIV *no-intercept model* (8.5) *and assume that* (8.6) *holds. If* $\lim_{m\to\infty}(1/m)A_0^TA_0 = \Omega$ *exists and is positive definite and if the distribution of the rows of* $[\Delta A; \Delta b]$ *possesses finite fourth moment, then* $\sqrt{m}(\widehat{x} - x_0)$ *has an asymptotic zero-mean multivariate normal distribution as* $m \to \infty$; *if the third and fourth moments of the distribution of the rows of* $[\Delta A; \Delta b]$ *are the same as those of the normal distribution, then* $\sqrt{m}(\widehat{x} - x_0)$ *is asymptotically normally distributed with zero mean and covariance matrix*:

$$(8.45)\quad \sigma_\nu^2(1 + x_0^Tx_0)\{\Omega^{-1} + \Omega^{-1}M\Omega^{-1}\} \quad \textit{where} \quad M = \sigma_\nu^2(I_n + x_0x_0^T)^{-1}.$$

Proof. For the proof see [54, Thm. 3.3]. □

The assumption that the error moments are those of the normal distribution is not necessary for asymptotic normality of $\widehat{x}$. Nevertheless, the asymptotic variance depends on the third and fourth moments since this yields a concise expression more easily compared with those of other estimates. Other asymptotic distributions of the TLS estimators for model (8.5) are proved in [53], [140]. In particular, Schneeweiss considered several other special cases of noise distribution, as well as models including trend-like variables.

When the amount of *additive noise* is *low* and (8.6) holds, the covariance matrix of the TLS estimator $\widehat{x}$ can be approximated by (neglect the second term in (8.45)):

$$\operatorname{cov}(\widehat{x}) \approx \frac{1}{m}[x_0^T\,;\,-1]\mathcal{C}[x_0^T\,;\,-1]^T(\frac{1}{m}A_0^TA_0)^{-1} = (1+\|\,x_0\,\|_2^2)\sigma_\nu^2(A_0^TA_0)^{-1}.$$
(8.46)

For finite sample size m, $\operatorname{cov}(\widehat{x})$ can be approximated by the estimates $\widehat{x}, \sigma_{n+1}^2/m$ and $A^TA - \sigma_{n+1}^2I_n$ of the true parameters x_0, σ_ν^2, and $A_0^TA_0$, respectively:

$$(8.47)\quad \operatorname{cov}(\widehat{x}) \approx (1+\|\,\widehat{x}\,\|_2^2)\sigma_\nu^2(A^TA - m\sigma_\nu^2 I)^{-1} \quad \text{with} \quad \sigma_\nu^2 \approx \frac{\sigma_{n+1}^2}{m}.$$

This approximate covariance matrix is found to be reasonably accurate [146].

Note that the asymptotic variance of $\widehat{x}$ must exceed the asymptotic variance of the consistent LS estimator x' in the usual regression case (i.e., A is error-free), given by

$$(8.48)\qquad \operatorname{cov}(x') = \sigma_\nu^2(A_0^TA_0)^{-1},$$

where σ_ν is the variance of the errors on b.

Theorem 8.6 and its simplifications (8.46) and (8.47) also apply to the *intercept* model (8.5) provided $\sqrt{m}(\widehat{x} - x_0), \operatorname{cov}(\widehat{x}), \operatorname{cov}(x'), A_0, M$ and $A^TA - m\sigma_\nu^2 I$ are replaced by $\sqrt{m}[\begin{smallmatrix}\widehat{\alpha}-\alpha\\ \widehat{x}-x_0\end{smallmatrix}]$, $\operatorname{cov}([\begin{smallmatrix}\widehat{\alpha}\\ \widehat{x}\end{smallmatrix}])$, $\operatorname{cov}([\begin{smallmatrix}\alpha'\\ x'\end{smallmatrix}])$, $[1_m; A_0]$, $[\begin{smallmatrix}0 & 0\\ 0 & M\end{smallmatrix}]$ and $[1_m; A]^T[1_m; A] - m\sigma_\nu^2[\begin{smallmatrix}0 & 0\\ 0 & I_n\end{smallmatrix}]$, respectively.

Observe from (8.46) and (8.48) that the *covariance matrix* of the TLS estimator $\widehat{x}$ is *larger* than the covariance matrix of the LS estimator x', even if A is noisy. This is confirmed by our simulations in the next section and has been proved in [91] for the univariate case. Hence, despite its asymptotic *bias* owing to observational errors, *ordinary* LS estimation is thus *not* necessarily inferior to TLS estimation since the asymptotic distribution of the TLS estimator is characterized by a larger variance. This is clear when we consider the *asymptotic mean squared error* (AMSE), which is the sum of the asymptotic variance and the squared asymptotic bias. The differences between the AMSE of $\widehat{x}$ and x' are only significant for sufficiently large noise variances σ_ν^2 when the bias of LS becomes more important (see Fig. 8.1).

Based on the asymptotic MSE, Ketellapper [91] investigated the choice between ordinary LS and corrected least squares (CLS) estimation (see (8.10)) of parameters in a univariate linear EIV model under conditions relevant in practice. He found that CLS, which equals TLS under the assumption (8.6), is in general the preferable estimation procedure.

Furthermore, it has been established that for *Gaussian noise*, the *bias* of the TLS estimator $\widehat{x}$ is *negligible* compared to its standard deviation, and that (8.46) equals the *covariance* matrix given by the *Cramér–Rao lower bound* for joint unbiased estimates [98]. This means that the TLS solution $\widehat{x}$ has minimum variance with respect to all joint unbiased estimates when the amount of additive noise is low and Gaussian.

Finally, note that we generally assume here that A_0 and B_0 in model (8.5) and (8.7) are *fixed* (i.e., the *functional* equations model). Kelly [87] considers the case in which these elements are *random* (called the *structural* equations model). More specifically, Kelly assumes that the rows of $[A_0; B_0]$ are independently and identically distributed with common mean vector and covariance matrix. Additionally, these rows must be independent from the errors, i.e., the rows of $[\Delta A \; ; \; \Delta B]$. Similar properties can be derived for the structural equations model. In particular, Kelly derives an explicit expression for the asymptotic covariance matrix of the TLS estimator in these models [87].

8.6. Practical relevance and simulation results

In the previous sections, asymptotic properties have been proved to evaluate the quality of estimates for the parameters of the errors-in-variables model. Hence, our analysis and conclusions are valid for *large-sample* models whereas applications often involve *moderate* sample sizes. However, there is evidence based on Monte Carlo simulation experiments, that the moments of the asymptotic *distributions* of the LS and CLS (which equals TLS under the assumption (8.6)) estimators are good approximations for the corresponding simulated finite-sample moments at least for samples having 20 observations or more [90].

Similar conclusions hold for the *consistency* properties of the TLS solution that are fairly well approximated in applications of moderate sample size, as confirmed by simulations now. We therefore compare the accuracy of the TLS and LS solution with respect to their squared bias, total variance, and mean squared error (MSE). These quantities are related as follows [88]:

$$
\begin{aligned}
&(8.49) \\
\mathrm{MSE}(\widehat{X}) &= \mathcal{E}\{(\widehat{X}-X_0)^T(\widehat{X}-X_0)\} \\
(8.50) \qquad &= \mathcal{E}\{(\widehat{X}-\mathcal{E}(\widehat{X}))^T(\widehat{X}-\mathcal{E}(\widehat{X}))\}+(\mathcal{E}(\widehat{X})-X_0)^T\mathcal{E}((\widehat{X})-X_0) \\
&= \mathrm{trace}\{\ \mathrm{cov}(\widehat{X})\}+\ \mathrm{bias}^2(\widehat{X}) \\
&= \text{total variance}(\widehat{X})+\ \text{squared bias}(\widehat{X})
\end{aligned}
$$

with X_0 the exact solution of $A_0X = B_0$. We use the same simulation procedure as explained in §7.4 but instead of relative errors, the squared bias, total variance, and MSE of the solution are estimated. The expected values above are estimated by the sample mean, using the solution of 100 sets of equations affected by noise with the same variance. We divide those estimated values of the squared bias, total variance and MSE by $\| X \|_F^2$ and then plot those values in a log-log diagram versus the noise variance.

Fig. 8.1 illustrates the difference in *accuracy* between TLS and LS with *increasing degree m of overdetermination.* A linear plot of the same example comparing the TLS and LS relative errors with increasing m, has been given in [181]. From Fig. 8.1 the following observations are made.

- The *bias* of TLS is much *smaller* than the bias of LS (given by (8.9)) and decreases with increasing m, conformal to the asymptotic unbiasedness property (8.27) of TLS. These differences in bias are more pronounced when the noise variance or m increases, when A_0 is more ill-conditioned or when x_0 is more oriented in the direction of the lowest singular vector v_n' of A_0, according to (8.9) and (8.27). Observe also that the squared bias of TLS does not exceed the noise variance for $m = 22$.

- The *total variance* of TLS is *larger* than that of LS. This has been proved for the one-dimensional univariate model [91] and can be deduced for the multivariate model from (8.46) and (8.48). The decrease in variance for the case $m = 10$ and $m = 14$ that occurs at larger noise variances, are due to *rank reduction* of the data matrix. Indeed, in the simulations the rank of the data matrix is reduced as soon as its squared minimal singular value is smaller than the used rank determinator (3.124) given by m times the noise variance. In this case, the minimum norm solution is computed. As is well known in principal component analysis [72], this rank reduction causes an increase in bias but a decrease in variance of the solution error.

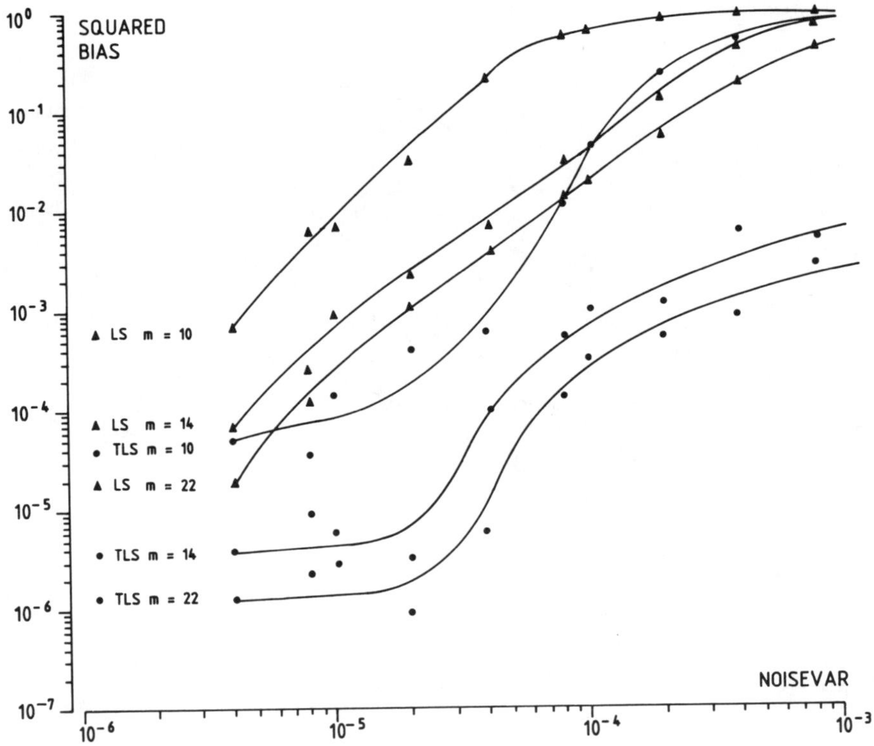

(a)

FIG. 8.1. *Comparison of* (a) *the squared bias,* (b) *the total variance, and* (c) *the mean squared error* (MSE) *of the* LS *and* TLS *solution (all values divided by* $\| x \|_2^2$) *versus the variance of the noise added to the set* $(A_0)_{m\times 5}x = b_0$ *for different row dimensions* $m = 10, 14,$ *and* 22. *The* 22×5 *matrix* A_0 *has singular values* $10, 8, 4, 1, .1,$ *and* $b_0 = .5u_5'$, *where* u_5' *is the left singular vector of* A_0 *corresponding to* .1. *By deleting the last rows in this* 22×6 *matrix* $[A_0; b_0]$, *the other* $m\times 6$ *matrices* $[A_0; b_0]$ *are derived.*

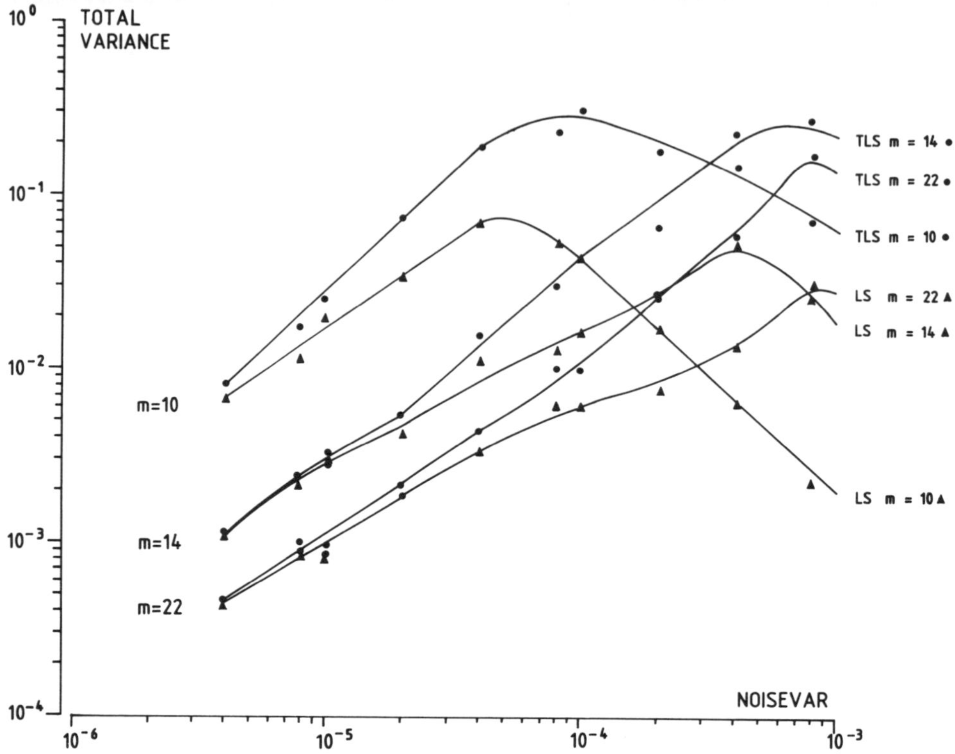
TOTAL
VARIANCE
TLS m = 14 •
TLS m = 22 •
TLS m = 10 •
LS m = 22 ▲
LS m = 14 ▲
LS m = 10 ▲
m=10
m=14
m=22
NOISEVAR
10^{0}
10^{-1}
10^{-2}
10^{-3}
10^{-4}
10^{-6}
10^{-5}
10^{-4}
10^{-3}

(b)

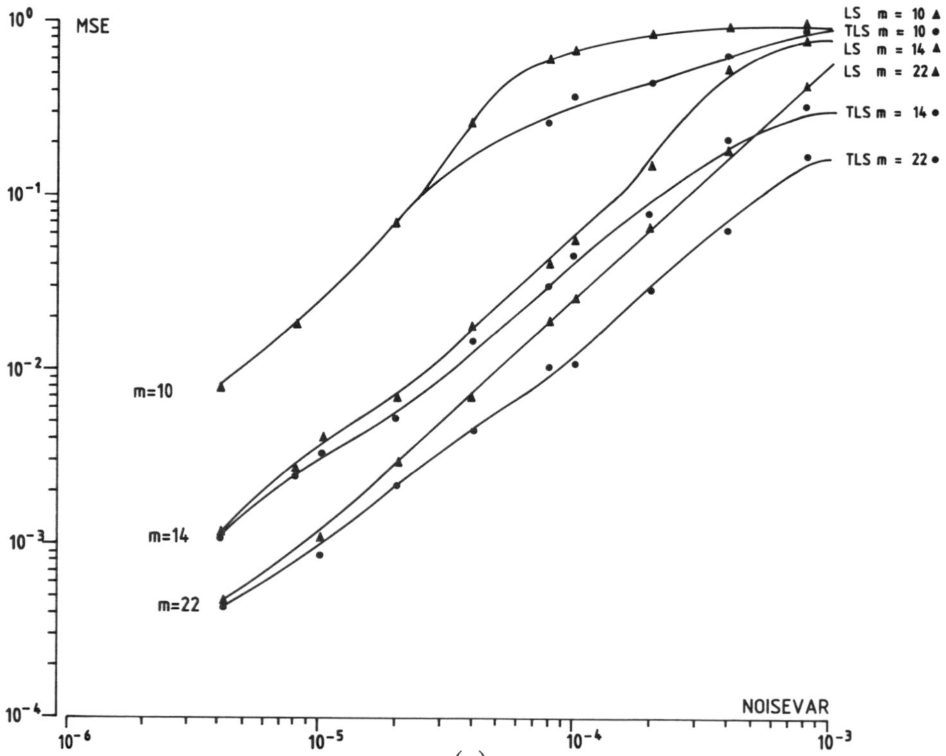
MSE
LS m = 10 ▲
TLS m = 10 •
LS m = 14 ▲
LS m = 22 ▲
TLS m = 14 •
TLS m = 22 •
m=10
m=14
m=22
NOISEVAR
10^{0}
10^{-1}
10^{-2}
10^{-3}
10^{-4}
10^{-6}
10^{-5}
10^{-4}
10^{-3}

(c)

- And finally, the MSE *combines* the effects of *bias* and *variance.* At the smallest noise variances, the LS bias is rather low (and usually substantially smaller than the variance) and the variance of TLS is somewhat larger than the LS variance. These two effects annihilate each other, producing comparable MSE for TLS and LS. By increasing the noise in the data, the bias of LS becomes more important and even the dominating term in (8.50). Hence, the differences in MSE are more pronounced, showing the better performance of TLS. It is evident that the TLS solution will be more *accurate* especially when the sets of equations are more *overdetermined.* Indeed, the TLS estimator then approaches its asymptotic properties. Fig. 8.1 confirms that these properties are already fairly well approximated for *moderate* sample size m.

Similar, though more detailed, conclusions are given in [7], [8], [211]. Ammann and Van Ness performed several Monte Carlo simulations on both TLS and LS and their robust versions to study robustness and efficiency of these estimates in EIV models satisfying assumption (8.6). An estimator is called "*efficient*" if the ratio of its estimates' MSE to the MSE of the maximum likelihood estimator, having minimum variance in large samples, approaches one. As said before, TLS is the maximum likelihood estimator if the errors in the data, satisfying the EIV model, are *Gaussian* distributed. *Robustness* of the various estimates is estimated when the data deviate from the assumed model in various ways and is measured by computing the variance of the MSE estimate. The results of these simulations are based on 400 replications of the data set and the rows of all A_0 are zero-mean Gaussian distributed.

It is concluded that TLS gives *better* estimates of the *true* model parameters x_0 than does LS in the "*uncontaminated*" EIV model (8.5) ($d = 1$), i.e., the data satisfy the assumed model. By "better" it is meant that the MSE is smaller. This is true, even if the errors are *not* Gaussian but exponentially distributed or t-distributed with three degrees of freedom.

For *small* sample sizes m, there is not much difference between LS and TLS. Moreover, the TLS estimates are less robust and have larger variances than the LS estimates. However, when the sample size m is *increased*, the TLS estimates clearly "*beat*" LS in accuracy in the sense that their MSE is much smaller. The improvement in using TLS over LS in EIV models is much more pronounced for estimating the *slope* parameters x_0 than for estimating the intercept α.

Also, *heavy-tailed* errors, i.e., errors having a heavy-tailed distribution function, have some effect. The performance of TLS deteriorates as the error tails get heavier and the clear dominance of TLS over LS is only apparent at larger sample sizes m.

Furthermore, the *size* of the errors in the measured data relative to the variance of the data A_0 influences the relative differences in accuracy between LS and TLS. Large errors affect more the accuracy of the LS estimates more than that of the TLS estimates, especially when estimating the slope parameters x_0. Moreover, if contamination in the measurements A of A_0 is more severe than in the measurements b of b_0, the better accuracy of TLS compared to LS is more pronounced [211].

Finally, *increasing* x_0 has a detrimental effect on any estimate (including LS and TLS) of both α and x_0. This does not happen when LS is applied on data satisfying the ordinary regression model. However, for the EIV model it is predicted by the orthogonal regression theory [87] (see also (8.46)) that the asymptotic variances of the estimates of all the parameters depend on all the components of x_0. This effect is more complex in higher dimensions.

Summarizing, these results support the *use of* TLS *over* LS for estimating the parameters of the *uncontaminated* EIV model. However, for small sample sizes, small errors, or small x_0 there is not much difference between LS and TLS: the MSE of the LS estimates may be slightly larger than the MSE of the TLS estimates but, on the other hand, LS is *more robust* than TLS, which is characterized by larger variances, especially when the sample size m gets too small relative to the number n of parameters to be estimated or when the errors are too large relative to the variance of the data A_0.

This instability or *nonrobustness* has quite serious implications for the accuracy of the TLS estimator in the presence of *outliers* in the data, i.e., large errors in the measurements. This has been theoretically proved by Kelly [87] and Cheng and Van Ness [30] and is confirmed by Monte Carlo results in [7], [8], [211]. Ammann and Van Ness contaminated the data by 10 percent outliers. The outliers themselves are simulated either by slipping the decimal point one position to the right (multiplication by 10) and/or by misrecording the sign. When outliers of this type are present in the data, the TLS estimates are very *unstable* and a dramatic *deterioration* of the TLS estimates is shown. Also, the LS estimates encounter serious stability problems, although less dramatically, and efficient *robust* procedures should be considered, which are quite efficient and rather insensitive to outliers, e.g., by downweighting measurement samples that have high residual and/or high leverage.

An excellent survey of the theory of robust *ordinary* regression can be found in [76, Chap. 6]. This includes the M-estimation method of Huber [82] which "robustifies" a regression estimator as follows. Rather than minimizing the sum of squared deviations, Huber proposes minimizing the sum of a less rapidly increasing convex function of the deviations. However, M-estimators are only the first step in the "robustification" of a regression estimator since they do not have bounded influence. Generalized M-estimators are introduced to overcome this problem and provide bounded influence estimators, see, e.g., [94]. Little theoretical work has been done on *robust* EIV *regression*. Gallo [54] was the first to extend M-estimation to EIV models. Orthogonal regression analogues of M-estimators of regression have been presented in [211], [8]. The theoretical properties of these estimates, as well as their robustness, have been investigated in problems where the assumptions of the model are violated. Procedures for computing robust orthogonal regression estimators are presented in [30], [211] and seem to work quite well in general. If no outliers are present and the data have a Gaussian distribution, TLS outperforms robust orthogonal regression. However, in the face of contamination by outliers, TLS is disturbingly unstable while robust orthogonal regression can improve the *efficiency* and *robustness* of the TLS estimates considerably, as shown in [7], [8].

8.7. Concluding remarks

As soon as some *prior* information about the *distribution of the errors* in the data is available, statistical properties of the TLS problem can be derived, and they prove the *expected* quality of TLS in estimating the parameters of a model. The most suitable model, directly related to the TLS concept, is the "*errors-in-variables*" (EIV) model. This model assumes an unknown but *exact* linear relation between the true variables that can only be observed with errors. The linear *regression* model, however, assumes the independent variables to be exactly observable ($A = A_0$) but, in addition, allows the observations B of the dependent variable to be influenced by other factors than observation errors alone (§8.2). If the errors in the observations $[A; B]$ of $A_0 X = B_0$ are independently and identically distributed with zero mean and common covariance matrix $C = \sigma_\nu^2 I$ and if $\lim_{m\to\infty}(1/m)A_0^T A_0$ exists and is positive definite, it is shown that the TLS solution of $AX \approx B$ is a *strongly consistent* estimate of the true but unknown parameters X_0 of the EIV model. As is well known in linear regression, the *ordinary* LS solution is *inconsistent* in this case.

Given the frequency of practical situations in which the "independent" variables are recorded with error, TLS should prove to be quite a useful tool to data analysts. TLS gives better estimates of the parameters of the EIV model than does LS. However, TLS does not usually provide a better predictor of B_0 given the observations A. Moreover, TLS is very closely related to eigenvalue-eigenvector analysis or SVD and can be very useful in building models, where all the variables should be treated symmetrically.

Beyond an estimate for the parameters X_0, TLS also provides estimators for the intercept of the model and the variance σ_ν^2 of the errors (§8.3). Under reasonable assumptions, all these estimators are shown to be *strongly consistent* estimators of the true model parameters, regardless of the common distribution of the errors in the data (§8.4). The *restrictive* nature of the *assumptions* used is evaluated. In particular, *weak* consistency of the TLS solution is proved under less restrictive conditions. Moreover, the error covariance matrix C need not have the form $\sigma_\nu^2 I$ to compute consistent estimates of the parameters X_0 in the EIV model. Indeed, provided that C is known up to a factor of proportionality, TLS still computes consistent estimates if the data are transformed appropriately. However, if C is (nearly) singular, these transformations must be performed implicitly by using the generalized TLS algorithm based on the generalized SVD.

Furthermore, the TLS estimators in both the intercept and no-intercept model are shown to be *asymptotically normally distributed* and expressions of their asymptotic covariance matrix are given (§8.5).

Finally, the *practical relevance* of the asymptotic properties given here is investigated, as well as the performance of the TLS estimator in the case that the assumed model is violated (§8.6).

As long as the data satisfy the assumed EIV model, TLS gives *better* estimates of the true model parameters X_0 than does LS, regardless of the common error distribution and should be preferred to LS. For *small* sample sizes, small errors, or small X_0, there is not much difference between TLS and LS: the mean squared error (MSE) of the LS estimates may be slightly larger than the MSE of the TLS estimates but on the other hand, TLS is *less robust* than LS and has larger variances. However, when the sample size is increased, the TLS estimates clearly "beat" LS in accuracy. These observations are confirmed by simulations, proving at the same time that the asymptotic properties given in this chapter are fairly well approximated in applications of *moderate* sample sizes.

However, when the data significantly *violate* the assumptions of the model, e.g., when *outliers* are present, the TLS estimates are very *unstable* and their accuracy is strongly affected. Also LS encounters stability problems, although less dramatically. In these situations, *robust* procedures, which are quite efficient and rather insensitive to outliers, should be applied.

Chapter 9

Algebraic Connections Between Total Least Squares Estimation and Classical Linear Regression in Multicollinearity Problems

9.1. Introduction

Multivariate linear least squares regression (LS) has been in use for a long time as the major statistical technique for estimating the regression parameters of a linear regression model. However, the full implications, limitations, and inherent problems associated with it have only recently been treated in the literature. Despite possessing the very desirable property of having minimum variance among all linear unbiased estimators under the usual conditions imposed on the model, the LS estimators can, nevertheless, have extremely large variances when the data are multicollinear, i.e., nearly linearly dependent. Most of the modifications of the ordinary least squares approach are attempts to deal with the problem of *multicollinearity* and many *biased estimators* have been proposed: principal component regression (PC) [47], latent root regression (LR) [203], shrinkage [157], ridge regression (RR) [47], [81], etc.

Despite their better performance in the presence of a multicollinearity, all the techniques cited above assume the *independent* variables to be *free of error.* They are no longer adequate when all data are observed with errors. New techniques have been introduced for estimating the parameters of the linear regression model with *errors in the variables.*

If the errors in all data are independently and identically distributed with zero mean and common covariance matrix $\sigma_\nu^2 I$, then under mild conditions the generic TLS solution is proved to be a strongly consistent estimate of the parameters in the linear errors-in-variables (EIV) model (8.5), regardless of the common distribution of the errors (see §8.4).

Several other and more general approaches to EIV regression have led to many other EIV estimators for the linear as well as for the nonlinear case (see [141], [127] for a list of references). However, all these estimators assume that the independent variables in A_0 are mutually sufficiently independent. If this is not the case, e.g., when multicollinearities are present in the data, then the coefficient estimates become very sensitive and unstable and have very large variances. The stronger the multicollinearities are, the more damaging are their effects.

One attempt to deal with *multicollinearities in* EIV regression is *minimum norm (non)generic* TLS (see §§3.3 and 3.4). This estimator extends the method of biased estimation to EIV models: by introducing a small bias, the coefficients of the generic TLS solution (3.13) are stabilized and its variance is greatly reduced. Despite the absence of exact distributional theory and a thorough statistical analysis, nongeneric TLS has been shown to perform better than generic TLS in the presence of multicollinearities (see §7.2.3).

It is the aim of this chapter to present *algebraic equivalences* and interrelations between the classical *linear regression* estimators and *(non)generic* TLS estimation in the presence of *multicollinearities.* Section 9.2 briefly describes the classical linear regression techniques: least squares (LS), principal component (PC), ridge regression (RR), and latent root regression (LR). The SVD technique, proved to be an excellent tool in regression analysis [102], is used to elucidate their equivalences and relation with TLS. Next, in §9.3, the algebraic relationships between TLS and the classical regression techniques are presented. Geometric concepts clarify the difference in approach of classical regression versus TLS estimation. The equivalence between PC and LR in the presence of exactly defined multicollinearities is proved and the differences between LR and (non)generic TLS estimation are investigated.

9.2. Unified SVD approach of classical linear regression

9.2.1. Multicollinearities. The *most general linear regression model* of the form (8.1) ($A = A_0$) is

$$b = \alpha 1_m + A x_0 + \varepsilon, \tag{9.1}$$

where b is an m-dimensional vector of response variables; 1_m is an m-dimensional vector of ones. $A = [a_1, \cdots, a_n]$ is a full column rank matrix of nonstochastic *error-free* regressor variables and is usually *standardized*, i.e., all columns of A have zero mean and unit length; ε is an m-dimensional vector of unobservable random error variables that are normally distributed with zero mean and covariance $\sigma_\varepsilon^2 I$. α is zero for no-intercept models and an unknown constant to be estimated in intercept models. x_0 is an n-dimensional vector of unknown regression coefficients.

The major statistical technique for estimating x_0 has been LS estimation. However, in spite of the highly desirable estimator properties of unbiasedness, minimum variance (among unbiased estimators of regression coefficients), and known probability distributions, multicollinearities can produce extremely *large* variances and thus *destroy* the effectiveness of LS estimation.

Multicollinearities are *approximate* linear *dependencies* between the columns of A and are essentially constant for all responses b (zero, when expressed in terms of standardized variables). They are a consequence of the duplication of information that is provided by the variables. More formally, if there exist k linearly independent nonzero vectors $w_j = [w_{1j}, \cdots, w_{nj}]^T$ such that

$$\sum_{q=1}^{n} w_{qj} a_q \approx 0 \quad j = 1, \cdots, k, \tag{9.2}$$

then the columns a_q of A are said to be "multicollinear." The existence of a linear relation such as (9.2) is, however, usually referred to as "collinearity" (which is thus used as a generic term, including a generalization of the more limited case of points on the same line) [102].

The closer the linear combinations in (9.2) are to zero, the stronger are the multicollinearities and the more damaging are their effects on the LS estimator. This has prompted research into *biased* regression estimators. These estimators attempt to introduce a *small bias* into the regression estimator while greatly *reducing* the *variance* appearing in the LS estimator. Despite their relative paucity of exact distributional theory [73], biased estimators can provide more realistic parameter estimates. They are, in general, more accurate when the data are multicollinear.

9.2.2. Biased regression estimators. Our discussion is facilitated by expressing each of the various estimators (except latent root) as a linear combination of the right singular vectors of A. Denote the singular value decomposition of A as in (1.21). Small singular values and the corresponding right singular vectors identify multicollinearities. This is because

$$\sigma_j'^2 = v_j'^T A^T A v_j' = (Av_j')^T (Av_j') \quad j = 1, \cdots, n. \tag{9.3}$$

So if the smallest $n - r$ singular values of A are suitably small, this relation is

$$0 \approx (Av_j')^T (Av_j') \quad \text{and hence} \quad Av_j' \approx 0 \quad j = r+1, \cdots, n. \tag{9.4}$$

The elements of the *right singular* vectors can thus be used to determine the coefficients used in (9.2). The general form of the estimators is

$$x_e = \sum_{j=1}^{n} \nu_j \tau_j v_j'. \tag{9.5}$$

The subscript e denotes the used regression technique. ν_j depends on the particular estimator and $\tau_j = \sigma_j' u_j'^T b$ is the same for all estimators. Thus the estimators differ only in the values of ν_j used in (9.5). Although the latent root estimator cannot be expressed as a linear combination of the v_j', its form is similar to (9.5).

The *least squares* (LS) estimator of x_0 in (9.1) can be written as

$$x_{\rm LS} = \sum_{j=1}^{n} \frac{\tau_j}{\sigma_j'^2} v_j'. \tag{9.6}$$

This form of the estimator requires that $\nu_j = 1/\sigma_j'^2$ in (9.5), thereby tending to place large weights on singular vectors associated with small singular values of A.

The *principal component* (PC) estimator of x_0 in (9.1), more commonly known as the minimum norm LS solution or truncated SVD solution [28] in numerical analysis, is given by

$$x_{\text{PC}} = \sum_{j=1}^{r} \frac{\tau_j}{\sigma_j'^2} v_j'. \tag{9.7}$$

This is equivalent to setting $\nu_j = 0, \quad j = r+1, \cdots, n$, and $\nu_j = 1/\sigma_j'^2$ for $j = 1, \cdots, r$ in (9.5). PC estimators eliminate multicollinearities from the LS estimator, thereby greatly reducing estimator variances while attempting to introduce only a small amount of bias.

The *ridge regression* (RR) estimator of x_0 in (9.1), proposed by Hoerl and Kennard [81], can be written using (9.5) with $\nu_j = 1/(\sigma_j'^2 + k)$ for some $k > 0$, i.e.,

$$x_{\text{RR}} = \sum_{j=1}^{n} \frac{\tau_j}{\sigma_j'^2 + k} v_j'. \tag{9.8}$$

Ridge regression adjusts the LS estimator by reducing the influence of multicollinearities on the parameter estimates. Unlike PC and LR regression, no singular vectors v_j' defining multicollinearities are eliminated from the resulting estimator, but their effect is greatly diminished. Relatively, the largest reductions occur with terms having the smallest singular values. The major controversy surrounding the use of this estimator concerns the appropriate method for choosing k.

The *latent root* (LR) *regression* estimator was developed by Hawkins [78] and Webster, Gunst, and Mason [203] as an alternative for the PC estimator. Two major differences exist between PC and LR regression: first, the matrices from which the eigenvalues and eigenvectors are extracted differ, and second, the techniques for deciding whether to delete multicollinear vectors from the respective estimators differ. Let $[A; b]$ be the $m \times (n+1)$ matrix whose first n columns contain the (standardized) regressor variables and whose last column contains the standardized values of the response variable. x_{LR} is then calculated from the eigenvalues and eigenvectors of $[A; b]^T[A; b]$, the "correlation" matrix of regressor and response variables or, similarly, from the singular values and right singular vectors of $[A; b]$. Through geometrical arguments the developers show that the response variable aids in the determination of the final estimator and can reveal when LS should be preferred to LR. To be more precise, denote the singular values and corresponding right singular vectors of $[A; b]$ as in (1.22) and let $v_j^T =$

$[v_j^{1T}; v_{n+1,j}]$. Then the LR estimator of x_0 in (9.1) is given by

$$x_{\mathrm{LR}} = \sum_{j=1}^{n+1-s} f_j v_j^1, \tag{9.9}$$

where

$$f_j = -\eta \frac{v_{n+1,j}\sigma_j^{-2}}{\sum_{q=1}^{n+1-s} v_{n+1,q}^2 \sigma_q^{-2}},$$

s is the number of nonpredictive multicollinearities (defined below) and $\eta^2 = \sum_{i=1}^m (b_i - \bar{b})^2$ with $\bar{b} = (1/m)\sum_{i=1}^m b_i$ ($\eta = 1$ if all columns in $[A; b]$ are standardized).

The form (9.9) is similar to (9.5) but the vectors v_j^1 are not the right singular vectors v_j'. LR regression *extends* PC regression by utilizing an assessment of whether multicollinearities have predictive value and should be retained in the resulting estimator (9.9). When σ_j is nearly zero but $|v_{n+1,j}|$ differs significantly from zero, the multicollinearity is referred to as a *predictive multicollinearity.* However, when both $|v_{n+1,j}|$ and the associated singular value σ_j are nearly zero, the multicollinearity is referred to as a *nonpredictive multicollinearity,* i.e., a multicollinearity among the predictor variables in A that is of little value in predicting the response variable b. The LR estimator (9.9) *eliminates all nonpredictive* multicollinearities, corresponding to the s terms deleted in (9.9). To accomplish this, the coefficients f_j in (9.9) are determined by first setting $f_j = 0$ if v_j identifies a nonpredictive multicollinearity and then selecting the remaining values to minimize the residual sum of squares. If $s = 0$ then $x_{\mathrm{LR}} = x_{\mathrm{LS}}$. Gunst, Webster, and Mason [74] report that in many cases x_{LR} behaves similarly to x_{PC} so that $\mathrm{MSE}(x_{\mathrm{LR}}) \approx \mathrm{MSE}(x_{\mathrm{PC}})$ for these situations. This result agrees well with our Theorem 9.2, proved in the next section.

Many other biased estimators, e.g., a shrunken estimator of the form ϕx_{LS}, where $0 < \phi < 1$ [157], have been proposed, but they are not relevant for our comparison with the (non)generic TLS estimator (see [79], [80] for a list of references).

9.3. Algebraic comparison between classical linear regression and (non)generic TLS estimation

In the comparison of the estimators, the presence or absence of (non)predictive multicollinearities is crucial. To allow an adequate algebraic comparison, let us consider the *exact* case throughout this section. This means that nonpredictive multicollinearities are defined by vectors v_j whose $(n+1)$th component $v_{n+1,j}$ is exactly zero and the associated singular value σ_j is suitably small. Predictive multicollinearities are defined by vectors v_j, associated with zero singular values σ_j and $v_{n+1,j} \neq 0$. Let (1.21) (respectively, (1.22)) be the SVD of A (respectively, $[A;b]$). Throughout this section, s denotes the number of nonpredictive multicollinearities and p denotes the total number of multicollinearities (predictive and nonpredictive).

If $[A;b]$ has rank $n+1-p$, then its p smallest singular values are zero and thus, $[A;b]$ has p multicollinearities. Hence, if at least one multicollinearity is predictive, the TLS *estimator* is directly obtained from (3.35):

(9.10)

$$x_{\text{TLS}} = \sum_{j=n+2-p}^{n+1} \lambda_j v_j^1 \text{ with } \lambda_j = \frac{-v_{n+1,j}}{\sum_{i=n+2-p}^{n+1} v_{n+1,i}^2} \text{ and } v_j^1 = [v_{1,j}, \cdots, v_{n,j}]^T.$$

If, however, all s multicollinearities that appear in $[A;b]$ are nonpredictive ($s \geq 1$), then all $v_{n+1,i} = 0$ for $i = n+2-s, \cdots, n+1$, i.e., the TLS problem is nongeneric. Using Theorem 3.11, this means that the vector b is orthogonal onto the singular vectors u'_j of A with $j = n+1-s, \cdots, n$. Nongeneric TLS then approximates $[A;b]$ with $[\widehat{A};\widehat{b}]$ obtained by making the next larger singular value σ_{n+1-s} zero (see §3.41). This means that v_{n+1-s} is made predictive if $v_{n+1,n+1-s} \neq 0$. Provided $\sigma_{n-s} > \sigma_{n+1-s}$, this nongeneric TLS estimator is also expressed by

$$x_{\text{TLS}} = \sum_{j=1}^{n-s} \frac{\sigma'_j}{\sigma_j'^2 - \sigma_{n+1-s}^2} (u_j'^T b) v'_j. \tag{9.11}$$

Equation (9.11) is obtained from (3.83), using the dyadic decomposition $A = \sum_{i=1}^n u'_i \sigma'_i v_i'^T$, and s is the number of nonpredictive multicollinearities. If $s = 0$ (respectively, $s > 0$), (9.11) yields the generic (respectively, nongeneric) TLS estimator (2.27) (respectively, (3.83)).

Because we approximate $[A;b]$ with $[\widehat{A};\widehat{b}]$ by making v_{n+1-s} predictive, the (non)generic TLS estimator can also be expressed by (9.10), provided p equals the number of multicollinearities of the TLS approximation $[\widehat{A};\widehat{b}]$, namely $p = s+1$.

Equation (9.11) allows us to compare, at least formally, the expressions of the PC, LS, and RR *estimators with* TLS *estimation.* From (9.8) and (9.11) we observe that $x_{\text{TLS}} = x_{\text{RR}}$ if we take $k = -\sigma_{n+1}^2$ and assume no multicollinearities. Ridge regression is, in fact, a way of regularizing the solution to an ill-conditioned LS problem. TLS is a deregularizing procedure, a kind of *reverse* ridge regression. This implies that TLS can only be useful in EIV problems (8.5), for which it can be proved that the expected value of the perturbed $\sigma_i'^2 - \sigma_{n+1}^2$ equals the ith singular value of the error-free matrix A_0 corresponding to A. In those cases, the TLS solution can estimate the solution x_0 of the error-free problem consistently (see §8.4).

If we subtract σ_{n+1-s}^2 from each squared singular value $\sigma_i'^2$ of A, we transform the LS estimator (respectively, PC estimator) into the generic (respectively, nongeneric) TLS estimator if the number s of nonpredictive multicollinearities is zero (respectively, > 0). Hence, the TLS estimator (9.11) differs from the PC (respectively, LS) estimator (9.7) (respectively, (9.6)), depending on the size of the singular value σ_{n+1-s} of $[A;b]$, corresponding to v_{n+1-s}, which is made predictive by TLS. They coincide if $\sigma_{n+1-s} = 0$, i.e., if $[A;b] = [\widehat{A};\widehat{b}]$ has one predictive multicollinearity (see Theorem 9.1 below). Observe also from (9.7) and (9.11) that PC and nongeneric TLS eliminate the same multicollinearities, thereby greatly reducing the variances of the generic TLS and LS estimator, respectively. This is evident from Theorems 3.11 and 3.12.

The comparison between LR and TLS *estimation* is based on (9.10). Assuming s ($s < p$) nonpredictive multicollinearities, we observe from (9.9) and (9.10) the strong, although different, relation between LR and TLS estimation in contrast to the assertion of Stewart [160]. The relation between LR and (non)generic TLS depends primarily on the size of the singular value σ_{n+1-s}, corresponding to v_{n+1-s}, which is predictive or made predictive by TLS ($v_{n+1,n+1-s} \neq 0$). Indeed, $\sigma_{n+1-s} = 0$ implies that $[A;b]$ has one predictive multicollinearity and thus, the estimates of generic TLS and LR coincide (see Theorem 9.1). Otherwise, TLS makes v_{n+1-s} predictive by making σ_{n+1-s} zero in its approximation $[\widehat{A};\widehat{b}]$, to obtain its estimate (9.10) with $p = s+1$ (which is nongeneric if $s > 0$). LR, however, minimizes the residual sum of squares (like LS), after eliminating the s nonpredictive multicollinearities, to obtain its estimate (9.9). Observe that in (9.9) the contribution of v_{n+1-s} will dominate and thus x_{LR} approaches x_{TLS}, depending on the size of the corresponding singular value σ_{n+1-s}.

Investigation of the properties of the nongeneric TLS estimate (see Theorem 3.12) shows that LR and nongeneric TLS impose the same additional constraints on their estimate x, i.e., $\left[\begin{smallmatrix} x \\ -1 \end{smallmatrix}\right] \perp$ all nonpredictive v_j. Moreover, Theorem 3.11 proves the equivalence between these nonpredictive v_j with $v_{n+1,j} = 0$ and corresponding multicollinearities v_i' of A. This implies that LR, PC, and nongeneric TLS eliminate the *same* multicollinearities. In this sense, one could thus say that *nongeneric* TLS *stabilizes* the coefficients of the *generic* TLS estimate in EIV models (8.5), just as PC and LR *stabilize* the coefficients of the LS estimate in *regression* models (9.1), when *multicollinearities* are present in the data A_0.

In generalizing the relations between all estimators, the presence or absence of (non)predictive multicollinearities (see the definition above) in $[A;b]$ is crucial. First, we can prove that, in case there is an *exact solution*, it is found by *all* methods.

THEOREM 9.1. *If $[A;b]$ has one or more predictive multicollinearities, then*

$$x_{\text{LS}} = x_{\text{PC}} = x_{\text{LR}} = x_{\text{TLS}} = x_0, \tag{9.12}$$

with x_0 the minimum norm exact solution of $Ax = b$.

Proof. Let (1.21) and (1.22) be the SVD of A and $[A;b]$, respectively. As soon as $[A;b]$ has one predictive multicollinearity, assume v_{n+1}, there exists a linear relation between the columns of A and b, namely, $[A;b]v_{n+1} = 0$. As $v_{n+1,n+1} \neq 0$ it follows immediately that $b \in R(A)$ and $R(A) = R([A;b])$. This means that $Ax = b$ has an exact solution x_0. Hence the minimal residual norm must be zero. As σ_{n+1} is also zero, we obtain (9.12) from the definitions of the respective estimators (9.6), (9.7), (9.9), and (9.10).

If $[A;b]$ has p predictive multicollinearities $v_{n-p+2}, \cdots, v_{n+1}$ then we can always transform these $p = s+1$ multicollinearities to s nonpredictive multicollinearities $[{Y \atop 0}]_1^n$ and one predictive multicollinearity $[{z \atop \gamma}]$, $\gamma \neq 0$, as done in (3.32). The TLS solution $\hat{x} = -z/\gamma$, which can also be expressed by (9.10) and (9.11), then follows immediately and has minimal norm as proved in Theorem 3.7.

Moreover, A must also have s multicollinearities $v'_{n-s+1}, \cdots, v'_n$ or equivalently, s zero singular values according to the interlacing Theorem 2.4 for singular values. Indeed,

$$\sigma'_{n-s+1} = \cdots = \sigma'_n = \sigma_{n-s+1} = \cdots = \sigma_{n+1} = 0.$$

On the other hand, $Av'_i = 0$ implies that $u_i'^T b = 0$ for $i = n-s+1, \cdots, n$ since $u'_i \perp R(A)$ and $b \in R(A)$ is assumed. Substituting these expressions in (9.6), (9.7), and (9.11) proves $x_{\text{LS}} = x_{\text{PC}} = x_{\text{TLS}}$. $x_{\text{LR}} = x_{\text{TLS}}$ follows from equality of the expressions (9.9) and (9.10) for $\sigma_i \to 0 \quad i = n-p+2, \cdots, n+1$. □

The *difference* between the estimators becomes apparent when the model does *not* have an *exact* solution, i.e., $R(A) \neq R([A;b])$ and $b \notin R(A)$. This implies that the number of predictive multicollinearities equals zero. In this case, all multicollinearities, if any, are nonpredictive. Using the interesting Theorem 3.11, we can prove the following important relation between LR *regression* and PC.

THEOREM 9.2. *If there are only nonpredictive multicollinearities, then*

$$x_{\text{PC}} = x_{\text{LR}}. \tag{9.13}$$

Proof. Assume s nonpredictive multicollinearities corresponding to isolated singular values. By Theorem 3.11(b) the s smallest singular triplets in the SVD (1.21) of A and (1.22) of $[A;b]$ then coincide, i.e.,

$$u_{j+1} = \pm u'_j, \sigma_{j+1} = \sigma'_j \quad \text{and} \quad [\pm v_j'^T; 0]^T = v_{j+1} = [v_{j+1}^{1T}; 0]^T \qquad \text{for } j = n-s+1, \cdots, n. \tag{9.14}$$

If $s = 0$, LS, PC, and LR all minimize the residual norm $\|b - Ax\|_2$ and thus $x_{\text{LS}} = x_{\text{PC}} = x_{\text{LR}}$. If $s > 0$ then $v_{n+1,j} = 0$ for $j = n-s+2, \cdots, n+1$. Using Theorem 3.11 this implies that $u_j'^T b = 0$ for $j = n-s+1, \cdots, n$. As $v'_j = v^1_{j+1}$

for $j = n - s + 1, \cdots, n$ from (9.14), PC and LR will delete the same terms in the expression of their respective estimators (9.7) and (9.9). Hence $x_{\text{PC}} = x_{\text{LR}}$. In the case of nonpredictive multicollinearities corresponding to singular values with multiplicity $r > 1$, the proof still holds if we deal with the corresponding r-dimensional singular subspaces instead of the singular vectors. □

Using Theorem 3.11(e), we can also immediately prove that $x_{\text{LS}} = x_{\text{PC}}$ in the case of s nonpredictive multicollinearities.

Geometrically, Theorem 9.2 implies that b is orthogonal onto the vectors u'_j of A for $j = n-s+1, \cdots, n$. To find a nonzero approximation b' such that b' belongs to the column space $R(A)$ of A, b must be projected orthogonally onto the $(n-s)$-dimensional space $R(A_{n-s}) = R([u'_1, \cdots, u'_{n-s}])$ (cf. Fig. 2.1(a)). Hence, $b' = b''$ with b'' the orthogonal projection of b onto $R(A_{n-s})$. As such, LS, PC, and LR obtain their estimate of x_0 in (9.1), which minimizes the residual norm $\|A_{n-s}x - b\|_2$.

Now combining Theorems 9.1 and 9.2 we immediately conclude:

If $[A; b]$ *has* s *nonpredictive multicollinearities, then*

$$x_{\text{LS}} = x_{\text{PC}} = x_{\text{LR}} = \sum_{j=1}^{n-s} \frac{1}{\sigma'_j}(u_j'^T b)v'_j. \tag{9.15}$$

Comparing (9.11) with (9.15) we observe that $x_{\text{TLS}} \neq x_{\text{PC}}$ and $x_{\text{TLS}} \neq x_{\text{LR}}$ if $\sigma_{n-s+1} > 0$. This *difference in estimation* depends on the *size* of the *singular value* σ_{n-s+1} and can be best understood when we present the problem geometrically. Like LS, PC, and LR, TLS also tries first to find an approximation $[\widetilde{A}; \widetilde{b}]$ orthogonal to the vectors u_{j+1} for $j = n - s + 1, \cdots, n$ to make $\widetilde{b} \in R(\widetilde{A})$. Because b is orthogonal to $u_{j+1} = u'_j$ for $j = n - s + 1, \cdots, n$, according to Theorem 3.11, b remains unchanged after the orthogonal projection. Hence $\widetilde{b}$ still $\notin R(A_{n-s}) = R(\widetilde{A})$ and the TLS problem is nongeneric provided $\sigma_{n-s+1} \neq 0$. To obtain a solution, nongeneric TLS is used and makes the TLS approximation $[\widehat{A}; \widehat{b}]$ of $[A; b]$ orthogonal onto u_{n-s+1} (cf. Fig. 2.2(b)) by making the next larger singular value σ_{n-s+1} of $[A; b]$ zero, i.e., v_{n-s+1} is made predictive. If now $v_{n+1,n-s+1} \neq 0$, $\widehat{b} \in R(\widehat{A})$. As such, the corrections applied to the data $\|[\Delta\widehat{A}; \Delta\widehat{b}]\|_F$, equal to σ_{n-s+1}, are minimized such that $\widehat{b} \in R(\widehat{A})$ and $[\begin{smallmatrix}\widehat{x}\\-1\end{smallmatrix}] \perp$ all nonpredictive v_j. The nongeneric TLS estimate x_{TLS} of x_0 in (8.5) is obtained from (3.78). Observe that σ_{n+1-s} measures the *degree of incompatibility* of the set of *linear equations* $Ax \approx b$ and thus, indicates *how closely* the data A, b *fit* the linear model (9.1) or (8.5).

We now summarize the main *conclusions.*

- If $b \in R(A)$, $[A; b]$ has at least *one predictive* multicollinearity and the *exact solution* of $Ax = b$ is found by LS, PC, LR, and TLS estimation.
- If $b \notin R(A)$, *all* multicollinearities of $[A; b]$ must be *nonpredictive.*

Assume that this number is equal to s ($s > 0$). Observe that the TLS problem is generic (respectively, nongeneric) if $s = 0$ (respectively, $s > 0$). The *difference in estimation* becomes now apparent and is proportional with the size of the next larger singular value σ_{n+1-s} of $[A; b]$, measuring how closely the data fit the model (9.1) or (8.5). LS, PC, and LR take the orthogonal projection b' of b onto $R(A)$ to make $b' \in R(A)$ and obtain their estimate (9.15) by solving $Ax = b'$. As such, the residual norm $\|Ax - b\|_2$ is minimized. TLS approximates $[A; b]$ with $[\widehat{A}; \widehat{b}]$ by making σ_{n+1-s} zero, i.e., TLS makes one v_{n+1-s} predictive such that $[\widehat{A}; \widehat{b}]$ now has one predictive multicollinearity. In this way, $\widehat{b}$ is made $\in R(\widehat{A})$, while $[{\widehat{x} \atop -1}] \perp$ all nonpredictive v_j and TLS obtains its estimate (9.11) by solving $\widehat{A}x = \widehat{b}$. As such, the corrections $\|[A\ ;\ b] - [\widehat{A}\ ;\ \widehat{b}]\|_F$, applied to the data such that $\widehat{b} \in R(\widehat{A})$ and $[{\widehat{x} \atop -1}] \perp v_j$, for all $j = n - s + 2, \cdots, n + 1$, are minimized and given by σ_{n+1-s}.

- The *residual* norm is *minimized with* LS, PC, *or* LR, i.e., $\|Ax_{\text{TLS}} - b\|_2 \geq \|Ax_{LS} - b\|_2$ but the *corrections* applied to the *data* are *minimal with* TLS, i.e., $\|[A; b] - [\widehat{A}; \widehat{b}]\|_F \leq \|b - b'\|_2$, as proved in §3.4.1.

Note that in practice the multicollinearities are not exact, i.e., numerically we have $\sigma_j \approx 0$ and $v_{n+1,j} \approx 0$. Depending on the strength of the multicollinearities, the same relations hold approximately for the inexact cases.

Finally, we mention one *chemical example in multivariate regression* [209]. In this example, the ordinary LS, the PC, the RR, the shrunken estimator, and the minimum norm generic TLS estimator are compared with respect to their prediction accuracy and the stability of their estimates, in the presence of multicollinearities among the independent variables A. In many chemical applications of multivariate regression, such as the present example of relationships between chemical structure and biological activity, the *predictive* properties of the model are of prime importance and the regression estimates therefore often need to be *stabilized.* The example [209] shows that TLS gives a solution to the multivariate regression problem that is stabilized in comparison with the LS solution and that has (at least in the example investigated) minimal prediction error. Stabilization is performed by reducing the matrix $[A; b]$ to a matrix of smaller rank. All variables remain in the model. In applied work, the multicollinearity problem is often handled by selecting a subset of variables such that the size of the subset equals the rank of the matrix $[A; b]$ [182].

9.4. Concluding remarks

In this chapter, the TLS estimator of the parameters in a linear errors-in-variables (EIV) model is compared with the classical regression estimators of the parameters in a linear regression model. Special attention is given to *multicollinearities* in the data. The major statistical technique for estimating the parameters of a linear regression model has been LS estimation. However, despite possessing the very desirable property of having minimum variance among all unbiased estimators, it can nevertheless have extremely large variances, when the data are multicollinear (§9.2.1). *Biased* estimators, which introduce a small bias while greatly reducing the variance from that of LS, are shown to perform *better* in regression models when multicollinearities are present in the data. Three biased regression estimators are described using a SVD approach: *principal component* (PC) regression, *ridge* regression (RR), and *latent root* (LR) regression (§9.2).

Multicollinearities also occur in EIV models but, so far, this problem has not been considered. *Minimum norm (non)generic* TLS is one attempt to deal with this problem. This estimator extends the method of biased estimation to EIV models and greatly reduces the variance of the generic TLS solution in the presence of multicollinearities.

Algebraic equivalences and interrelations between these regression estimators and (non)generic TLS estimation in the presence of *exactly* defined multicollinearities are presented, based on the SVD and geometric concepts. It is shown that LR and PC estimators eliminate the same multicollinearities from the LS estimator as does the nongeneric TLS estimator from the generic TLS estimator, thereby greatly reducing the estimator variances. In this sense one could thus say that nongeneric TLS stabilizes the coefficients of the generic TLS solution in EIV models, just as PC and LR stabilize the coefficients of the LS estimate in regression models, when multicollinearities are present in the data. Furthermore, the *equivalence* between PC and LR is proved and the *difference* between LR and TLS is clarified (§9.3).

Chapter 10

Conclusions

10.1. Introduction

The research reported in this book is clearly stimulated by the need for a more global approach to the problem of *linear parameter estimation* in cases where all variables that characterize the system cannot be observed without error. Such problems arise in a broad class of scientific disciplines, such as signal processing, automatic control, system identification, and, in general, engineering, statistics, physics, economics, biology, medicine, etc. The classical least squares (LS) approach, which assumes that all but one variable can be measured without error, is no longer appropriate. New fitting techniques must be devised to compensate for data errors. The total least squares (TLS) approach analyzed throughout this book is such a method of *fitting* that is appropriate when *errors* occur in *all* data and amounts to fitting a best subspace to the data. It was introduced into the field of numerical analysis by Golub and Van Loan and is based on the powerful singular value decomposition (SVD). Because of its nice numerical and geometrical properties and its potential use in a wide range of scientific problems, we were encouraged to study the computational aspects and analyze the TLS problem from different points of view (numerical, statistical, etc.). This is the main *objective* of this book. Therefore, different TLS algorithms are presented and evaluated (Chapters 2–5). Additionally, the validity of the TLS solution, i.e., its domain of applicability, is analyzed and delineated (Chapters 6–9).

In the following section, the obtained results are reviewed. Generalizations of the TLS problem, as well as other TLS related problems, are briefly described in §10.3 and, finally, a number of suggestions for further research related to the obtained results are made.

10.2. Review of the main results on TLS

Every linear parameter estimation problem gives rise to an overdetermined set of linear equations $Ax \approx b$. Whenever both A and b are noisy, the TLS method may be recommended for solving this set to estimate the parameters x. The basic principle of TLS is that the noisy data $[A; b]$, while not satisfying a linear relation, are modified with minimal effort, as measured by the Frobenius norm, in a "nearby" matrix $[\widehat{A}; \widehat{b}]$, which is rank-deficient so that the set $\widehat{A}x = \widehat{b}$ is compatible. In contrast, the LS approach consists in modifying only the right-hand side vector b with minimal effort, as measured by the Euclidean norm, in a "nearby" vector b' such that $Ax = b'$ is compatible.

In the introductive Chapter 1, a simple example illustrates the effects of the use of TLS as opposed to LS. Additionally, the improved results obtained so far in many applications in very diverse fields are reviewed and confirm the practical use of TLS.

In the two chapters that follow, different formulations of the TLS problem are presented and fully analyzed.

In Chapter 2, the main principle and solution of the LS and basic TLS problem are presented and compared geometrically. By "basic" we mean that only one right-hand side vector b is considered (one-dimensional) and that the TLS problem is solvable (generic) and has a unique solution.

In Chapter 3, the basic TLS problem is *extended* to multidimensional TLS problems $AX \approx B$ with more than one right-hand side vector and to mixed LS-TLS problems where the data matrix A contains some error-free columns. It is shown how to compute the TLS solution in each case. In particular, if the solution of the TLS problem is not unique, the TLS solution with minimal Frobenius norm is singled out. If the set of equations $AX \approx B$ is highly incompatible or if the data matrix A is (nearly) rank-deficient, the generic TLS problem is very ill-conditioned or even lacks a finite solution. To stabilize the TLS solution, additional constraints, which eliminate all nonpredictive multicollinearities from the TLS solution space, are imposed and the algorithm of Golub and Van Loan is generalized to solve these so-called *nongeneric* TLS problems. Finally, the TLS computations are summarized in one algorithm that includes all the extensions described above.

Since the TLS solution is deduced from *only one* right singular vector or, in general, *a basis* of the right singular subspace associated with the smallest singular values of the data matrix $[A; B]$, its computational cost can be considerably reduced by computing the SVD only partially. An improved *partial* SVD algorithm PSVD is presented in Chapter 4, computing the *singular subspace* of a matrix, associated with its *smallest* singular values. By analyzing the operation counts and presenting computational results, its relative efficiency with respect to classical SVD is demonstrated. Depending on the gap, the desired numerical accuracy, and the dimension of the desired subspaces, PSVD can be *three times faster* than the classical SVD algorithm, while the same accuracy can be maintained. By incorporating the PSVD algorithm into the classical TLS computations, an improved *partial* TLS algorithm PTLS is obtained. Its *higher efficiency* with respect to the classical

TLS algorithm is demonstrated and confirmed with computational results. Typically, PTLS reduces the computation time with a factor 2. All these algorithms are implemented in Fortran 77 and fully documented. Their code can be obtained through netlib [46].

If a priori *information* about the solution of the problem to be solved is available, e.g., when solving slowly varying TLS problems, the TLS computations can be speeded up in an *iterative* way. Various iterative methods are investigated in Chapter 5. In particular, inverse iteration, ordinary Chebyshev iteration, and inverse Chebyshev iteration are discussed. Also, Rayleigh quotient iteration and the Lanczos methods are briefly described. By analyzing the convergence properties of these methods, it is shown that inverse iteration is generally the *best* choice in terms of efficiency and ease-of-use provided the ratio between the singular values, corresponding to the desired and undesired singular subspace of the matrix $C = [A; B]$, is high. If this ratio is small, say less than one order of magnitude, convergence can be *accelerated* by applying the Chebyshev polynomials instead of the inverse power functions to the cross-product matrix C^TC (respectively, $(C^TC)^{-1}$). This method is called *ordinary* (respectively, *inverse*) Chebyshev iteration. Inverse Chebyshev iteration is proved to always converge *faster*, mostly considerably, to the desired basis than ordinary Chebyshev iteration. Moreover, this method always converges *faster* than inverse iteration provided a lower bound sufficiently close to the undesired squared singular value spectrum is known. The smaller the gap, the larger the gain in speed. Ordinary Chebyshev iteration is only efficient in problems characterized by a very dense singular value spectrum. The (inverse) Lanczos method is shown to have a similar convergence behavior as (inverse) Chebyshev iteration with optimal bounds. Moreover, no bounds on the singular values need be known. Note, however, that roundoff errors make the Lanczos methods difficult to use in practice. Additional computations are required that reduce their computational efficiency and question their practical use in moderately sized TLS problems.

Based on the convergence rate and the number of operations in each iteration step, the efficiency of iterative methods can be compared with that of the direct computation methods: classical and partial SVD. It is concluded that iterative methods are most efficient when only *one* singular vector is to be computed, their convergence rate is sufficiently *high*, the desired accuracy is rather low, and a good start vector is available. Iterative methods are, therefore, particularly attractive in solving slowly varying TLS problems because then the solution of a previous set is mostly a good initial guess for the solution of the next set. Simulated examples, as well as a real-life problem concerning the estimation of frequency response functions, illustrate the efficiency of the iterative methods compared to that of classical and partial TLS.

In the following chapters, the properties of the TLS solution are analyzed to delineate its domain of applicability and evaluate the practical significance of the TLS method.

In Chapter 6, *algebraic connections* between the TLS and LS problem are proved with respect to their solutions, their residuals, their corrections applied to data fitting, and their approximate subspaces. They all clearly reveal that the *same* parameters mainly determine the correspondences and differences between both approaches. As shown in Chapter 7, these parameters also determine the relative differences in sensitivity between TLS and LS. In particular, it is shown how the differences between both approaches increase when the equations $AX \approx B$ are less compatible, when the norm of B or the solution X is growing, or when A tends to be rank-deficient. They are most pronounced when the orthogonal projection B' of B onto $R(A)$ is parallel with the singular vector of A, associated with its smallest singular value. Furthermore, it is shown how TLS is led to a weighted LS problem and assumptions about the underlying perturbation model of both techniques are deduced. They show that many perturbation models correspond with the same TLS solution.

Chapter 7 examines how perturbations in all data A and B affect the accuracy of the TLS and LS problem. The relative changes in their solutions and related subspaces are evaluated with respect to the unperturbed problem $A_0X \approx B_0$. Assume first that no a priori information about the data (e.g., error distribution, orientation of B_0, etc.) is available. Then the following conclusions and *practical guidelines* hold.

Provided that the perturbations are *sufficiently small* and the corresponding unperturbed problem is zero residual (i.e., $A_0X = B_0$ is compatible), the minimum norm solutions of all consistent sets $\widetilde{A}X = \widetilde{B}$, where $[\widetilde{A}; \widetilde{B}]$ is an orthogonal projection of $[A; B]$ onto $R(\widetilde{A})$, differ mutually only by second-order terms. Hence, there is no reason to prefer one method to the other and thus, any of these solutions, including TLS and LS, may be used to estimate the true solution X_0 of $A_0X = B_0$.

In the absence of any information on the data, we can only investigate the *worst* perturbation effects in the LS and TLS solution. These effects are quantified by the condition number of the corresponding unperturbed problems. It is shown that zero residual TLS problems are equally well-conditioned as LS problems but nonzero residual TLS problems are always more ill-conditioned than the corresponding LS problems. This means that for zero residual problems, the TLS and LS solution are equally affected in the presence of worst-case perturbations but for nonzero residual problems, the worst perturbation effects on the TLS solution are *larger*. However, these worst perturbation effects are only obtained for a *special choice* of the perturbations of the data and of the orientation of the observation vector. Hence, they are quite unrealistic for other circumstances.

Now assume that some more a priori information about the data A, B and their errors $\Delta A, \Delta B$ is known.

Provided that $\|\Delta A\|_2 \approx \|[\Delta A; \Delta B]\|_2$, the perturbation results for singular values and their associated singular subspaces clearly show the smaller sensitivity-to-noise of the singular subspaces related to the TLS problem in any zero residual problem and sufficiently compatible nonzero residual problems. This means that in these problems the TLS solution is expected to be more accurate than the LS solution. This *better accuracy* of the TLS solution as opposed to LS is *more* pronounced and already significant at progressively smaller perturbations when the solution X_0 gets closer and closer to the *lowest* singular vector v'_n of A_0 associated with its *smallest* singular value. This not only occurs when the observation vectors in B_0 are close to u'_n of A_0 (assume that (1.21) is the SVD of A_0) but also when the condition number of A_0, given by σ'_1/σ'_n, is large and $|u'^T_n B_0|$ is not too small. Increasing at the same time the *length* and the *number* of observation vectors in B_0, i.e., enlarging $\|B_0\|_F$, further improves the expected accuracy of TLS with respect to LS. When $X_0 // v'_n$ or, equivalently, $B_0 // u'_n$, the expected accuracy of the TLS solution of $AX \approx B$ as opposed to the LS solution is at its best. In these problems we should clearly prefer TLS to LS. However, for progressively more incompatible nonzero residual problems the expected advantages in accuracy of TLS as opposed to LS are progressively less pronounced and may be considered less important than the effects of ill conditioning of the TLS problem. In these cases the user might prefer LS or another more robust regression technique.

Now if some more information about the *distribution of the errors* in the data is available, *statistical* properties can be derived to prove the expected accuracy of the TLS solution in these conditions. Throughout Chapter 8, it is assumed that the data A, B satisfy the following *errors-in-variables* (EIV) model: the corresponding unperturbed problem $A_0 X = B_0$ is zero residual and the errors $[\Delta A; \Delta B]$ in the observations $[A; B]$ are row-wise independently and identically distributed with zero mean and common covariance matrix $C = \sigma_\nu^2 I, \sigma_\nu$ unknown. Roughly speaking, this means that the errors have *zero mean* and are mutually *uncorrelated* and *equally sized.* Under these fairly reasonable assumptions, the TLS solution $\widehat{X}$ is shown to be a *strongly consistent* estimate of the true but unknown parameters X_0 of the model when $\lim_{m \to \infty} \frac{1}{m} A_0^T A_0$ exists and is positive definite, i.e., $\widehat{X}$ converges to X_0 as the number of equations goes to infinity. The LS solution is generally inconsistent in this case.

Also, strongly consistent estimators for the intercept, if any, of a linear relation present in the data, and for the error variance σ_ν^2 are deduced from the TLS results. Under less restrictive conditions, weak consistency of the TLS solution is proved. Moreover, the error covariance matrix C need not have the form $\sigma_\nu^2 I$. Provided that C is known up to a factor of proportionality, TLS still computes consistent estimates if the data are transformed appropriately. However, if C is (nearly) singular, these transformations must be performed implicitly by applying the generalized TLS algorithm based on the generalized SVD.

As long as the data satisfy the assumed EIV model, TLS gives better estimates of the true model parameters X_0 than does LS, regardless of the common error distribution, and should be preferred to LS. For small sample sizes, small errors, or small X_0, there is not much difference between TLS and LS: the mean squared error of the LS estimates may be slightly larger than that of the TLS estimates but, on the other hand, TLS is less robust than LS and has larger variances. However, when the sample size is increased, the TLS estimates are clearly more accurate than the LS estimates.

When the data significantly violate the assumptions of the model, e.g., when *outliers* are present, the TLS estimates are very *unstable* and their accuracy is strongly affected. LS also encounters stability problems, although less dramatically. In these situations, *robust* procedures, which are quite efficient and rather insensitive to outliers, should be applied.

Finally, Chapter 9 analyzes the significance of TLS in *multicollinearity* problems, i.e., when A_0 is (nearly) rank-deficient. When multicollinearities are present in the data, *biased* estimators, such as principal component (PC) regression, ridge regression (RR), and latent root (LR) regression, are known to perform better in regression models. Multicollinearities also occur in EIV models but, so far, this problem has not been considered. Minimum norm (non)generic TLS is one attempt to deal with this problem.

Algebraic equivalences and interrelations between these biased regression estimators and (non)generic TLS estimation in the presence of exactly defined multicollinearities show that LR and PC estimators eliminate the same multicollinearities from the LS estimator as does the nongeneric TLS estimator from the generic TLS estimator, thereby greatly reducing the estimator variances. In this sense, we could thus say that nongeneric TLS stabilizes the coefficients of the generic TLS solution in EIV models, just as PC and LR stabilize the coefficients of the LS solution in regression models when multicollinearities are present in the data. Furthermore, the equivalence between PC and LR is proved and the difference between LR and TLS is clarified.

10.3. Generalizations and related TLS problems

Every linear parameter estimation problem gives rise to an overdetermined set of linear equations $AX \approx B$. Whenever both A and B are subject to errors, the TLS method can be used for solving this set. Much of the literature concerns the *classical* TLS problem $AX \approx B$, in which all columns of A are subject to errors. This problem, defined in §3.2, is analyzed throughout this book but more general TLS problems, as well as other problems related to classical TLS, have been proposed and are being investigated. In this section, these problems are formulated and briefly discussed. A more thorough discussion of each problem falls outside the scope of this book and the reader is referred to the appropriate references.

If the errors in the measurements A and B are uncorrelated with zero mean and equal variance, then under mild conditions the classical TLS solution $\widehat{X}$ is a strongly consistent estimate of the true solution X_0 of the corresponding unperturbed set $A_0 X = B_0$ (see §8.4). However, in many linear parameter estimation problems, *some* columns of A may be error-free (see §3.5). Moreover, the errors in the remaining data may be *correlated* and *not* equally sized. To maintain consistency of the result when solving these problems, the classical TLS problem formulation can be *generalized* as follows (M^{-T} denotes the transposed inverse of matrix M).

DEFINITION 10.1 (Generalized TLS (GTLS) problem). Given a set of m linear equations in $n \times d$ unknowns X:

$$AX \approx B \qquad A \in \mathcal{R}^{m \times n}, B \in \mathcal{R}^{m \times d}, \text{ and } X \in \mathcal{R}^{n \times d}. \tag{10.1}$$

Partition

$$A = [A_1; A_2] \qquad A_1 \in \mathcal{R}^{m \times n_1}, A_2 \in \mathcal{R}^{m \times n_2}, \text{ and } n = n_1 + n_2 \tag{10.2}$$

$$X = [X_1^T; X_2^T]^T \qquad X_1 \in \mathcal{R}^{n_1 \times d} \text{ and } X_2 \in \mathcal{R}^{n_2 \times d} \tag{10.3}$$

and assume that the columns of A_1 are error-free and that nonsingular error equilibration matrices $D \in R^{m \times m}$ and $C_2 \in R^{(n_2+d)\times(n_2+d)}$ are given such that the errors in $D^{-T}[A_2; B]C_2^{-1}$ are equilibrated, i.e., uncorrelated with zero mean and same variance.

Then, the generalized TLS (GTLS) problem seeks to

$$\underset{[\widehat{A}_2; \widehat{B}] \in \mathcal{R}^{m \times (n_2+d)}}{\text{minimize}} \quad \|D^{-T}[A_2 - \widehat{A}_2; B - \widehat{B}]C_2^{-1}\|_F \tag{10.4}$$

$$\text{subject to} \quad R(\widehat{B}) \subseteq R(\widehat{A}). \tag{10.5}$$

Once a minimizing $[\widehat{A}_2; \widehat{B}]$ is found, then any $X = [X_1^T; X_2^T]^T$ satisfying

$$\widehat{A}X = A_1 X_1 + \widehat{A}_2 X_2 = \widehat{B} \tag{10.6}$$

is called a GTLS *solution* and $[\Delta\widehat{A}_2; \Delta\widehat{B}] = [A_2 - \widehat{A}_2; B - \widehat{B}]$ the corresponding generalized TLS correction.

Whenever the solution is not unique, the solution with *minimal* Frobenius norm, denoted by $\widehat{X} = [\widehat{X}_1^T; \widehat{X}_2^T]^T$, is singled out.

By varying n_1 from n to zero, this formulation can handle any LS ($n_1 = n, C_2 = I, D = I$) and generalized LS problem ($n_1 = n, C_2 = I$), as well as every TLS ($n_1 = 0, C_2 = I, D = I$), mixed LS-TLS ($C_2 = I, D = I$), and GTLS problem. The error equilibration matrices D and C_2 are, up to a factor of proportionality, the square root of the error covariance matrices $\mathcal{C}_2 = \mathcal{E}(\Delta^T \mathcal{D}^{-1} \Delta)$ and $\mathcal{D} = \mathcal{E}(\Delta \mathcal{C}_2^{-1} \Delta^T)$, respectively, where $\Delta_{m \times (n_2+d)}$ represents the errors in the noisy data $[A_2; B]$. Often, only $\mathcal{C}_2$ and $\mathcal{D}$ are known: in these cases, the matrices C_2 and D are simply obtained from their Cholesky decomposition, i.e., $\mathcal{C}_2 = C_2^T C_2$ and $\mathcal{D} = D^T D$, C_2 and D upper triangular.

To solve the GTLS problem and compute the GTLS solution, a computationally efficient generalized TLS algorithm, based on the generalized SVD, has been developed and is outlined in [179], [189].

If $D = I_m$ then C_2 is up to a factor of proportionality given by the square root of the error covariance matrix $\mathcal{C}_2 = \mathcal{E}([\Delta A_2; \Delta B]^T [\Delta A_2; \Delta B])$, which defines the correlations between the errors $[\Delta A_2; \Delta B]$ in $[A_2; B]$. Because of its close relation to the general EIV model, given by (8.7), this special GTLS problem is of utmost importance in EIV regression. Indeed, under mild conditions the GTLS solution of this problem is shown to be a consistent estimate of the true parameters X_1^0, X_2^0 of this general EIV model. Consistency conditions for the one-dimensional GTLS solution ($d = 1$) are investigated by Gallo [55], [54], while Fuller [53, §4.1] proves consistency of the more general multidimensional GTLS solution. An algorithm that computes this GTLS solution by means of the GSVD is outlined in [188]. The efficiency of this algorithm can be further improved by using the Speiser and Van Loan algorithm based on a CS decomposition [153], [196] instead of computing the GSVD. Some further improvements in numerical performance are discussed in [215].

We now turn to a further generalization of the GTLS problem, as shown below [194].

DEFINITION 10.2 (Restricted TLS problem). Consider the following set of m linear equations in $n \times d$ unknowns X:

$$(10.7) \quad AX \approx B \qquad A \in \mathcal{R}^{m \times n}, B \in \mathcal{R}^{m \times d}, \text{ and } X \in \mathcal{R}^{n \times d},$$

where the data matrix $[A; B] = [A_0; B_0] + E^\star$. A_0, B_0 are the error-free data and $E^\star$ represents a "restricted" perturbation matrix of the form

$$(10.8) \quad E^\star = DEC \qquad D \in \mathcal{R}^{m \times p},\ C \in \mathcal{R}^{q \times (n+d)},\text{ and } E \in \mathcal{R}^{p \times q}.$$

D and C are known matrices while E is unknown and arbitrary but bounded. Then, the restricted TLS (RTLS) problem seeks to

$$(10.9) \qquad \underset{\widehat{E} \in \mathcal{R}^{p \times q}}{\text{minimize}} \ \|\widehat{E}\|_F$$

(10.10) subject to $R(B - \Delta\hat{B}) \subseteq R(A - \Delta\hat{A})$ with $[\Delta\hat{A}; \Delta\hat{B}] = D\hat{E}C$.

Once a minimizing $\hat{E}$ is found, then any X satisfying

$$(10.11) \qquad (A - \Delta\hat{A})X = B - \Delta\hat{B}$$

is called an RTLS *solution* and $[\Delta\hat{A}; \Delta\hat{B}] = D\hat{E}C$ the corresponding restricted TLS correction.

Whenever the RTLS solution is *not unique*, a *weighted minimum norm* solution, denoted by $\hat{X}$, is singled out in the sense that $\|F_1\hat{X}F_2\|_F$ is minimized with $F_1 \in \mathcal{R}^{n\times n}$ and $F_2 \in \mathcal{R}^{d\times d}$ appropriately chosen nonsingular weighting matrices.

The name "restricted TLS" originates in its applications. We try to find a matrix $\hat{E}$ of minimal (unitarily invariant) norm that sufficiently reduces the rank of $[A; B] - D\hat{E}C$ with given $[A; B]$, D and C, such that the approximate set $([A; B] - D\hat{E}C)[{X \atop -I}] = 0$ is solvable. Hence, we attempt to reduce the rank of $[A; B]$ by *restricting* the modifications to the column space of D and the row space of C.

Observe that this RTLS formulation is more general than the GTLS formulation given above. Indeed, D and C may be rectangular, even rank-deficient, and error-free columns in A, if any, need not be stored in the first columns of A. Using an appropriately chosen D and C, even columns in B may be error free and equality constraints can be imposed (see below). Finally, if the RTLS solution is not unique, then not only the solution $\hat{X}$ with minimal $\|\hat{X}\|_F$ can be singled out but any other $\hat{X}$ with an appropriately "weighted" minimal norm $\|F_1\hat{X}F_2\|_F$.

It is easy to see that the RTLS formulation can handle any LS and generalized LS problem, as well as every TLS and generalized TLS problem.

- If $D = I_m$ and $C = I_{n+d}$ (respectively, $C = [0_{d\times n}; I_d]$), then we have the *ordinary* TLS (respectively, LS) problem formulation, as formulated in §§2.2.1, 2.3.1, and 3.2.

- The special case $D = I_m$ and $C = [0_{q\times n_1}; I_q]$ with $q = n + d - n_1$ defines the *mixed* LS-TLS problem, which assumes the first n_1 columns of A to be error free, as formulated in §3.5.

- By taking $C = [0_{(n_2+d)\times n_1}; (C_2)_{(n_2+d)\times(n_2+d)}]$ with C_2, D square and nonsingular, the RTLS problem reduces to the GTLS problem, defined above. Indeed, $\hat{E}^* = D\hat{E}C = [0; D\hat{E}C_2] = [0; \Delta\hat{A}_2; \Delta\hat{B}]$. Since D and C_2 are nonsingular, we have $\hat{E} = D^{-1}[\Delta\hat{A}_2; \Delta\hat{B}]C_2^{-1}$, which shows the correspondence between the RTLS problem and the more restrictive GTLS problem.

- If rank$(D)=p$ and $C = [0_{d\times n};\ I_d]$, one obtains the *generalized* LS (GLS) problem formulated as follows [116]:

$$\min \|E\|_F \quad \text{such that} \quad B = AX + DE.$$

  D represents, up to a factor of proportionality σ, the square root of the symmetric nonnegative definite covariance matrix $\mathcal{E}(E_2^\star E_2^{\star T}) = \sigma^2 DD^T$, where $E_2^\star$ represents the errors in the right-hand side matrix B. Many estimation problems are led to the solution of a GLS problem. For instance, in regression analysis, the one-dimensional ($d = 1$) GLS solution provides the best linear unbiased estimate of x in all cases of the general Gauss–Markov model, given by $b = Ax + e^\star$, where $e^\star$ is a zero-mean random vector with covariance matrix $\sigma^2 DD^T$ [116],[18, §23]. Whenever the coefficient matrix A is ill-conditioned, this GLS solution is very sensitive to perturbations. To stabilize the solution and decrease its variance, Zha and Hansen [214] suggested adding Tikhonov regularization [18, §26]. See [214] for more details.

- The RTLS formulation can also handle *any variant* of the GTLS and GLS problem, allowing for error-free columns in A_2 and B as well, e.g., the RTLS problem:
(10.12)
$AX \approx B,\ E^\star = [DEC_1; 0_{m\times d}]$ with $C = [C_1; 0_{q\times d}]$ and D, C_1 known,

  defines a GTLS problem, which assumes B to be error free. If $C_1 = [0_{1\times(i-1)}, 1, 0_{1\times(n-i)}]$, $q = d = 1$, then (10.12) defines a GLS problem in which A and B are error free except for the ith column of A.

- Furthermore, the RTLS formulation also allows us to solve any *(under)determined* set of linear equations (i.e., $m \leq n$), defined as follows:

$$\min_{\widehat{X}} \|F_1 \widehat{X} F_2\|_F \quad \text{such that} \quad A\widehat{X} = B.$$

  This set always has one exact solution $\widehat{X}$. Hence, $E^\star$ must be set to zero in the RTLS formulation, e.g., by taking $D = 0$ or $C = 0$.

- Observe that *equality constraints* can be imposed, given by the error-free rows of $[A; B]$, e.g., if the first m_1 rows of $[A; B]$ represent equality constraints, then $D = [{0_{m_1\times p} \atop D_2}]$. By adding the assumption that n_1 columns of $[A; B]$ are also error free, a whole class of RTLS problems $AX \approx B$ can be defined, characterized by the fact that only one submatrix of $[A; B]$ is perturbed and may be changed by the RTLS algorithm, e.g., $D = [{D_1 \atop 0_{m_1\times p}}]$ and $C = [C_1; 0_{q\times n_1}]$. These structured perturbation problems have been treated by Demmel [35], [36] and arise in the design of control systems based on H^∞ optimization and in stability analysis of various problems in linear algebra. More generally, assume that the noisy elements of $[A; B]$ lie at the intersection of the rows $r_1, \cdots, r_s$ and the columns $k_1, \cdots, k_z$ of $[A; B]$ (i.e., an $s \times z$ noisy submatrix), then C is a $z \times (n+d)$ matrix defined by $c_{ij} = 1$ if $j = k_i$, $1 \leq i \leq z$ otherwise $c_{ij} = 0$. D is an $m \times s$ matrix defined by $d_{ij} = 1$ if $i = r_j$, $1 \leq j \leq s$ otherwise $d_{ij} = 0$.

In practice, engineering design problems are very often formulated as optimization problems using LS (or TLS) approaches. As pointed out in [11], many of these problems involve equality constraints that represent some physical laws, e.g., in inverse kinematics of redundant and parallel manipulators, robot trajectory planning, mechanical systems design, Kirchoff's current and voltage laws, etc. As shown above, constrained linear LS problems are easily transformed into RTLS problems, but constrained nonlinear LS problems can also be solved with RTLS provided the solution method alters the problem in each iteration step to a constrained linear LS problem of the form:

$$\text{(10.13)} \quad \min_x (f - Ex)^T W (f - Ex) \qquad \text{or} \qquad \min_x \|W^{1/2}(f - Ex)\|_2$$

$$\text{(10.14)} \qquad \text{subject to} \qquad Gx = h.$$

W is a positive definite weight matrix and $W^{1/2}$ its square root. By defining $A = [{G \atop E}]^{m_1}$, $B = [{h \atop f}]$, $D = [{0 \atop W^{1/2}}]^{m_1}$, and $C = [0, \cdots, 0, 1]$ in (10.7) and (10.8), (10.13) and (10.14) can be reformulated as an RTLS problem.

- If $F_1 \neq I_m$ or $F_2 \neq I_d$, the corresponding "*weighted*" problems are considered, e.g., the GLS problem given in [195] ($d = 1$):

  $$\min_{e,x} \|e\|_2^2 \text{ such that } b = Ax + De \text{ with } \|F_1 x\|_2 \text{ minimal,}$$

  is a special RTLS problem with $C = [0_{1\times n}; 1]$, $F_2 = 1$ and F_1, D as defined above. This problem is encountered in many important applications, e.g., aircraft-wing flutter analysis [95].

- Finally, observe that the RTLS problem is *not always solvable.* Two kinds of unsolvable RTLS problems must be distinguished. First, the RTLS problem is unsolvable if no matrix $\widehat{E}$ exists such that $\text{rank}([A; B] - D\widehat{E}C) \leq n$, e.g., the following RTLS problem described in [36]:

  $$[A; B] = \begin{bmatrix} 0 & 1 \\ 1 & 0 \end{bmatrix} \quad \text{and} \quad [\Delta\widehat{A}; \Delta\widehat{B}] = D\widehat{E}C$$

  $$\text{given by} \begin{bmatrix} 0 \\ 1 \end{bmatrix} \widehat{e}\,[0\ 1] = \begin{bmatrix} 0 & 0 \\ 0 & \widehat{e} \end{bmatrix}.$$

Since the corrections $[\Delta\widehat{A}; \Delta\widehat{B}]$, applied to the data $[A; B]$, may only affect the perturbed entries in $[A; B]$ (specified by D and C), the rank of $[A; B]$ cannot be sufficiently reduced in these problems to obtain a solvable set $(A - \Delta\widehat{A})X = B - \Delta\widehat{B}$. Second, condition (10.9) may be satisfied but not condition (10.10), i.e., there exists an $\widehat{E}$ with minimal $\|\widehat{E}\|_F$ such that $\text{rank}([A; B] - D\widehat{E}C) \leq n$ but $R(B - \Delta\widehat{B}) \not\subseteq R(A - \Delta\widehat{A})$. In this case, many $\mathcal{O}(\varepsilon)$ perturbations $\widehat{E}_\varepsilon$ of $\widehat{E}$ can make $R(B - \Delta\widetilde{B}) \subseteq R(A - \Delta\widetilde{A})$, where $[\Delta\widetilde{A}; \Delta\widetilde{B}] = D\widehat{E}_\varepsilon C$, but there is no smallest value of $\|\widehat{E}_\varepsilon\|_F$ for which (10.10) is satisfied, e.g.,

$$[A; B] = \begin{bmatrix} 1 & 0 & 1 \\ 0 & 0 & 1 \\ 0 & 0 & 1 \end{bmatrix}, \quad C = I, \quad D = I \text{ and } \widehat{E}_\varepsilon = \begin{bmatrix} 0 & 0 & 0 \\ 0 & \varepsilon & 0 \\ 0 & \varepsilon & 0 \end{bmatrix}.$$

This kind of unsolvability only occurs when A is (nearly) rank-deficient or when the set of equations $AX \approx B$ is highly conflicting. By imposing additional constraints on the solution space, i.e.,

$$\forall \begin{bmatrix} w \\ 0 \end{bmatrix} \in N(D\widetilde{E}C),$$

$$\forall \widetilde{E} \in \{E \mid \|E\|_F \text{ minimal and } \text{rank}([A; B] - DEC) \leq n\}$$

$$(10.15) \qquad [\Delta\widehat{A}; \Delta\widehat{B}] \begin{bmatrix} w \\ 0 \end{bmatrix} \begin{matrix} n \\ d \end{matrix} = 0 \Leftrightarrow D\widehat{E}C \begin{bmatrix} w \\ 0 \end{bmatrix} = 0$$

these RTLS problems can be made solvable and are referred to as *nongeneric* RTLS problems (see also §3.4).

Whenever the elements of E are independently and identically distributed with zero mean and equal variance, consistency of the solution of RTLS problems, where $C = [\,0\,; C_2]$, C_2 and D square and nonsingular, can be proven under mild conditions (see [194], [53, §4.1]). More general consistency results are not yet proved and need to be further analyzed.

To solve the RTLS problem, the restricted singular value decomposition (RSVD), introduced by Zha [212], [213], can readily be applied to (10.9) and (10.10) to yield the solution of the RTLS problem. In essence, the RSVD applies to a given triplet of (possibly complex) matrices T, D, C of compatible dimensions and provides a factorization of the matrix T, relative to the matrices D and C. It can be considered as the ordinary SVD of the matrix T, but with different (possibly nonnegative definite) inner products applied in its column and in its row space. The properties and structure of the RSVD are investigated in detail in [40], [213], as well as its connection to generalized eigenvalue problems, canonical correlation analysis, and other generalizations of the SVD. Additionally, in [40], many applications are discussed. Just as the SVD (respectively, generalized SVD) is a valuable tool for the solution and analysis of the TLS (respectively, generalized TLS) problem, so the RSVD plays the same role of the more general RTLS problem, as pointed out in [194].

Based on this RSVD, a powerful RTLS algorithm is outlined in [194] and can be used in practice to compute the RTLS solution. Its main difference with respect to the classical TLS Algorithm 3.1 is the use of the RSVD instead of the classical SVD. RSVD allows $[A; B]$, D, and C to be rank-deficient and avoids to invert or multiply the matrices explicitly. This guarantees the better numerical performance of the RTLS algorithm with respect to the explicit transformation procedures. Some further improvements in numerical performance are discussed in [215]. Moreover, by first performing orthogonal transformations, the RTLS algorithm only needs to compute the implicit 3-product SVD of a smaller submatrix, hereby improving its computational efficiency.

The main value of the RTLS approach is the fact that only *one* algorithm is needed to solve a wide variety of problems. Of course, there are much faster direct ways of solving special RTLS problems, see, e.g., [69]. Nevertheless, if the problem is not too large and since computing power is generally so cheap, it will often make sense to use the RTLS algorithm at least when code for it becomes part of widely available and reliable subroutine packages. This is because the added information on structure and sensitivity that it provides, the ease of switching from one problem to another and analyzing the impact of any restriction on the sensitivity of the solution, can be very helpful in understanding the problem and finding the best problem formulation.

Although RTLS problems in their most general form are not yet fully understood and investigated, it is the authors' belief that the RTLS problem and the RTLS algorithm will become an important tool in the analysis and numerical solution of numerous problems.

Let us now consider the basic TLS problem, defined in §2.3.1, subject to the *linear equality constraints*:

$$N^T x = d \qquad \operatorname{rank}(N_{n \times n_1}) = n_1 \leq n. \tag{10.16}$$

Although this problem can be formulated as a special RTLS problem and solved with the RTLS algorithm, it is worthwhile to present here a simpler derivation of its solution, due to Golub. We formulate the problem as follows:

$$\min_x \frac{\|Ax - b\|_2^2}{\|x\|_2^2 + 1} \quad \text{subject to} \quad N^T x = d. \tag{10.17}$$

Substituting $[x^T; -1]^T$ by z, $[N^T; d]$ by M^T, and $[A; b]$ by C, (10.17) yields:

$$\min_z \frac{\|Cz\|_2^2}{\|z\|_2^2} \quad \text{subject to} \quad M^T z = 0. \tag{10.18}$$

Now taking the QR factorization of M, i.e.,

$$QM = \begin{bmatrix} R \\ 0 \end{bmatrix} \quad R_{n_1 \times n_1} \text{ upper triangular, } Q_{(n+1)\times(n+1)} \text{ orthonormal,}$$

we obtain

$$M^T z = [R^T; 0] Q z = 0. \tag{10.19}$$

Partitioning Qz as $[{}^g_h]^{n_1}$ and since R is nonsingular, it follows from (10.19) that

$$R^T g = 0 \Rightarrow g = 0. \tag{10.20}$$

Now

$$\frac{\|CQ^TQz\|_2^2}{\|Qz\|_2^2} = \frac{\|CQ^T \begin{bmatrix} g \\ h \end{bmatrix}\|_2^2}{\|g\|_2^2 + \|h\|_2^2} = \frac{\|[\widetilde{C}_1; \widetilde{C}_2] \begin{bmatrix} g \\ h \end{bmatrix}\|_2^2}{\|h\|_2^2} = \frac{\|\widetilde{C}_2 h\|_2^2}{\|h\|_2^2}, \tag{10.21}$$

which solves the problem. Indeed, since (10.21) must be minimized, h is the lowest right singular vector of $\widetilde{C}_2$ corresponding to its smallest singular value. The solution $\widehat{x}$ to the given problem then follows immediately by back transformation, i.e.,

$$\widehat{x} = -\frac{1}{\widehat{z}_{n+1}} \begin{bmatrix} \widehat{z}_1 \\ \vdots \\ \widehat{z}_n \end{bmatrix} \quad \text{where } \widehat{z} = \begin{bmatrix} \widehat{z}_1 \\ \vdots \\ \widehat{z}_{n+1} \end{bmatrix} = Q^T \begin{bmatrix} g \\ h \end{bmatrix} = Q^T \begin{bmatrix} 0 \\ h \end{bmatrix}. \tag{10.22}$$

An example of an equality-constrained TLS problem is discussed in [124] and consists in estimating the granulometry of minerals in a separator. The equality constraints represent mass balance equations of the system in equilibrium.

TLS problems with both nonnegativity constraints

$$x \geq 0$$

and inequality constraints

$$x \geq a \qquad a \neq 0$$

on the solution vector are investigated in [39]. It is shown that the Karush–Kuhn–Tucker conditions for the corresponding optimization problem are a generalized complementarity problem. An algorithm for solving this problem can be used as a preprocessor. The final solution is then found by solving a certain constrained eigenvalue problem when nonnegativity constraints are considered. With equality constraints, the solution scheme is more complicated. The constrained eigenvalue problem now becomes a constrained quadratic eigenvalue problem. The approach is illustrated with several numerical examples.

Another constrained TLS problem has been formulated by Arun [13]. The problem addressed here and arising in signal processing applications is the determination of the TLS solution $\widehat{X}$ of $AX \approx B$, subject to the constraint that $\widehat{X}$ be unitary. As proved by Arun, the solution to this *unitarily constrained* TLS *problem* is the same as the solution to the orthogonal Procrustes problem [69, p. 582].

Abatzoglou and Mendel [3] considered yet another constrained TLS problem, which extends the one-dimensional classical TLS problem $Ax \approx b$ to the case where the *errors* $\Delta A, \Delta b$ in the data A, b are *algebraically related.* In fact, from a statistical point of view, TLS operates under the assumption that the errors in A and b are independently and identically distributed with zero mean. If there is correlation among the errors, a noise-whitening transformation can be applied or a GTLS problem can be solved and the error norm is appropriately modified. However, if there is a linear dependence among the error entries in $[\Delta A; \Delta b]$, then the TLS solution may no longer yield optimal statistical estimators. This happens, for instance, in system identification when we try to estimate the impulse response of a system from its input and output by discrete deconvolution [177], [181]. Suppose that $y(t)$ is the system output and $u(t)$ is its input; then, we want to estimate the impulse response function $h(t)$ from the approximate equations:

(10.23)

$$\begin{bmatrix} y(t-m) \\ y(t+1-m) \\ \vdots \\ y(t) \end{bmatrix} = UH$$

$$= \begin{bmatrix} u(t-m) & \cdots & u(t-n-m) \\ u(t+1-m) & & u(t+1-n-m) \\ \vdots & & \vdots \\ u(t) & \cdots & u(t-n) \end{bmatrix} \begin{bmatrix} h(0) \\ h(1) \\ \vdots \\ h(n) \end{bmatrix}.$$

The errors in U are obviously Toeplitz and this information is not used in the classical TLS problem. Fogel [51] also recognized this problem and proposed to solve a Toeplitz-structured problem.

Another important example, where the errors in the data matrix have a block Toeplitz structure and are hence linearly dependent, is forward-backward linear prediction [3], used to estimate the frequencies of sinusoids from measurements contaminated by white noise.

Other applications occur in frequency estimation [125], estimation of the angular location of emitters [2] and superresolution harmonic analysis [4].

To get more accurate estimates of $\widehat{x}$, Abatzoglou and Mendel extended the classical TLS method to incorporate the algebraic dependence of the errors in $[A; b]$ and called their extension "*constrained* TLS."

DEFINITION 10.3 (Constrained TLS (CTLS) problem). Let

$$C = [A_{m\times n}; b_{m\times 1}], \Delta C = [\Delta A; \Delta b] = [\Delta c_1, \cdots, \Delta c_{n+1}] \tag{10.24}$$

and express each error column Δc_j as $\Delta c_j = F_j v$ where F_j is a matrix of an appropriate size and v is a zero-mean white noise vector of minimal dimensionality. Then, the CTLS solution $\widehat{x}$ is obtained from the following constrained minimization problem involving v and x:

$$\underset{v,\, x}{\text{minimize}} \ \|v\|_2^2 \ \text{ subject to } \ \{C + [F_1 v; \cdots; F_{n+1} v]\} \begin{bmatrix} x \\ -1 \end{bmatrix} = 0. \tag{10.25}$$

This is a *quadratic minimization* problem that is subject to a quadratic constraint equation. In contrast to the classical TLS, GTLS, and RTLS problem, the solution of this problem can no longer be solely computed by means of linear algebra techniques. A closed-form solution exists for special cases [4]. As shown in [3], $\widehat{x}$ can be obtained as the solution of an unconstrained minimization problem. A suboptimal algorithm, as outlined in [5] and [2], is applied to minimize the above functional and rather successful results have been obtained so far. Also, a complex version of the Newton method for finding the minimum of a real function of several complex variables has been derived [4] and applied to find the CTLS solution. A perturbation analysis of the CTLS solution, valid for small noise levels, has been performed and from it the root mean squared error of the CTLS solution is obtained in closed form. It is also shown that the CTLS problem is equivalent to a constrained parameter maximum likelihood problem.

Finally, the CTLS technique is applied to frequency estimation of sinusoids and direction-of-arrival estimation of wavefronts impinging on a linear uniform array. In both applications the root mean squared error of the CTLS solution is very close to the Cramér–Rao bound, up to a threshold signal-to-noise ratio. Simulation results show that the CTLS method is more accurate for frequency estimation than the modified forward-backward linear prediction technique of Tufts and Kumaresan. Also, comparisons are drawn among the CTLS, MUSIC, and root MUSIC techniques for angle-of-arrival estimation. Another class of problems we consider here is obtained by formulating the classical TLS problem in some other norm. These problems, called *total approximation problems*, have been investigated by Späth, Watson, and Osborne [201], [111], [151] for the one-dimensional case ($d = 1$) and can be formulated as follows.

DEFINITION 10.4 (Total approximation problem).

$$\text{Find } \widehat{x} \in \mathcal{R}^n \text{ to minimize } \|[\Delta A; \Delta b]\| \text{ subject to } (A+\Delta A)\widehat{x} = b+\Delta b \tag{10.26}$$

where the norm $||\cdot||$ is an appropriate matrix norm.

For example, if the errors in A and b are independent and normally distributed, then the most likely explanation of the data is achieved by minimizing the Frobenius norm of $[\Delta A; \Delta b]$, i.e., the classical TLS problem. However, if the data give rise to gross errors (wild points), then a more robust estimator may be the l_1 norm of the perturbation matrix regarded as an extended vector. For a situation between these two extremes, some other l_p norm, $1 < p < 2$, may be statistically more appropriate.

If some of the columns of A are known to be error-free, then an additional constraint is that the corresponding columns of ΔA are zero. Without loss of generality, it will be assumed that this is true for the first n_1 columns. If the matrix norm in (10.26) is the l_p norm defined by taking the usual vector l_p norm on the elements of the matrix regarded as an extended vector in $\mathcal{R}^{m\times(n+1)}$ then (10.26) becomes the *total l_p approximation problem.* An effective way in which this problem may be tackled is to exploit its connection with the following constrained vector norm approximation problem. Let $C = [A; b]$, then the problem referred in (10.26) is

$$\text{find } v \in \mathcal{R}^{n+1} \text{ to minimize } ||Cv||_p \text{ subject to } ||v_2||_q = 1, \tag{10.27}$$

where the subscripts on the norms indicate the usual l_p and l_q vector norms, where p and q are dual in the sense that

$$\frac{1}{p} + \frac{1}{q} = 1 \tag{10.28}$$

and where the vector $v_2 \in \mathcal{R}^{n+1-n_1}$ is obtained from v by deleting the first n_1 components.

Both (10.26) and (10.27) are nonconvex problems (the feasible region in (10.27) is the outside of the unit ball), and so it may only be possible to find points satisfying first-order necessary conditions for local solutions (stationary points). The precise relationship between (10.26) and (10.27) is explored in [200].

The problem (10.27) is also meaningful in errors-in-variables problems when p and q are *not* connected by the relationship (10.28) (or indeed when the norms are replaced by arbitrary norms on $\mathcal{R}^m$ and $\mathcal{R}^{n+1-n_1}$, respectively). For example, the so-called *orthogonal l_p approximation problem* corresponds to the choice $q = 2$ in (10.27) (see [149], [198], [151]). In particular, the classical TLS problem corresponds with the orthogonal l_2 approximation problem, i.e., $p = q = 2$ in (10.27). It is shown in [199] that v solving (10.27) with arbitrary norms $||\cdot||_A$ on $\mathcal{R}^m$ and $||\cdot||_B$ on $\mathcal{R}^{n+1-n_1}$ corresponds to a solution of the total approximation problem (10.26) with matrix norm defined on the $m \times (n+1)$ matrix M (with first n_1 columns zero) by

$$||M|| = \max_{||d_2||_B=1} ||Md||_A. \tag{10.29}$$

The vector $d_2 \in \mathcal{R}^{n+1-n_1}$ is obtained from d by deleting the first n_1 components. Osborne and Watson [111] extended these results to a more general class of matrix norms, called separable. They show that the minimizing perturbation matrix of (10.26) has rank one whenever the norm is separable and that this matrix can be computed when the solution to a constrained vector minimization problem of the form (10.27) is known. Attention is focused on the total l_1 approximation problem. In particular, it is proved that in the absence of degeneracy the minimizing perturbation matrix has only one nonzero column and a finite descent algorithm for finding a local minimizing solution is developed.

Watson [201] also analyzed the solution of (10.27) for the values $1 < p, q < \infty$ by a class of algorithms essentially proposed by Späth [149] for the orthogonal l_p problem. Numerical experiments reported by Späth suggest that these algorithms possess certain global convergence properties and this view is reinforced by similar experiments for the total l_p approximation problem [200]. In [201] a theoretical basis for these properties is established and local, as well as some global convergence results are proven. More general properties, characterizing the best approximation to a matrix C in an arbitrary norm by a matrix of any lower rank, are proven in [216], [202].

Finally, let us consider *nonlinear* TLS problems which arise from the fitting of observations $(a_i, b_i), i = 1, \cdots, m$ to a nonlinear model:

$$f(\alpha, x) = \beta \tag{10.30}$$

In nonlinear regression the measurements a_i of the independent variable α are assumed to be exact and only the observations b_i of β are subject to random errors. In nonlinear TLS problems the measurements of the independent variable α also contain errors. Assume that a_i and b_i are subject to errors Δa_i and Δb_i, respectively, so that

$$f(a_i - \Delta a_i, x) = b_i - \Delta b_i.$$

If the errors Δa_i and Δb_i are independent random variables with zero mean and equal variance, then it seems reasonable to choose the parameters x so that the sum of squares of the orthogonal distances $\sqrt{\Delta a_i^2 + \Delta b_i^2}$ from the observations (a_i, b_i) to the curve (10.30) is minimized. Hence the parameters x should be chosen as the solution to

$$\underset{x,\ \Delta a,\ \Delta b}{\text{minimize}} \sum_{i=1}^{m} (\Delta a_i^2 + \Delta b_i^2).$$

$$\text{subject to} \quad f(a_i - \Delta a_i, x) = b_i - \Delta b_i \quad i = 1, \cdots, m.$$

Eliminating Δb_i using the constraints we arrive at the *orthogonal distance problem*

$$\underset{x,\ \Delta a}{\text{minimize}} \sum_{i=1}^{m} \{(b_i - f(a_i - \Delta a_i, x))^2 + \Delta a_i^2\}.$$

If, more generally, $\alpha \in \mathcal{R}^n$ and $\beta \in \mathcal{R}^d$ in (10.30) are vectors, we have the problem

$$\underset{x,\ \Delta a}{\text{minimize}} \sum_{i=1}^{m} \{\|b_i - f(a_i - \Delta a_i, x)\|_2^2 + \|\Delta a_i\|_2^2\}.$$

In particular, when (10.30) reduces to $\alpha^T x = \beta$, $\beta \in \mathcal{R}$, and $x, \alpha \in \mathcal{R}^n$, the orthogonal distance problem is a classical TLS problem. Algorithms for the nonlinear case, based on stabilized Gauss–Newton methods, have been given by Schwetlick and Tiller [141], [142], Boggs, Byrd, and Schnabel [20], and Golub and LeVeque [65] (see also [53, §3.2]).

10.4. Suggestions for future research

Although the entire book is devoted to TLS, the analysis and results presented in it do not claim to have solved all problems associated with the TLS technique and explored all possible applications. Further research concerning the following topics is suggested:

- Experiments show that the TLS technique is sensitive to the *scaling* of the data. To attain the optimal TLS solution properties (see Chapter 8), the data must be properly scaled such that the errors are equilibrated, i.e., independently and identically distributed with zero mean and common covariance matrix of the form $\sigma_\nu^2 I$. Some scaling recommendations are given in §3.6.1. Not only knowledge of the error distribution is required for computing the appropriate scaling factors, but it is not always possible to equilibrate the error matrix appropriately by row and column scaling, e.g., in the case of a nonlinear error distribution. Although in the latter case the optimal properties of the TLS solution are not obtained, they can be fairly well approximated and still motivate the use of TLS, as we could observe in some practical applications. Nevertheless, a more fundamental study of the scaling problem resulting in an adequate *scaling strategy*, would certainly enhance the applicability of TLS and convince most critics.

- If the optimal TLS solution properties cannot be attained by an appropriate scaling of the data, the classical TLS problem can be reformulated, extended, or generalized to improve the accuracy and statistical properties of the classical TLS solution. *Generalizations* of the classical TLS problem, which are currently under investigation, have been presented in §10.2. Promising results have been obtained so far but the study of these problems must be continued to improve the efficiency and reliability of the proposed algorithms and analyze more thoroughly the specific properties of their solutions and practical use.

- In this book, the *nongeneric* TLS solution has been proposed and shown to stabilize the generic TLS solution in the presence of multicollinearities in the data A, B, i.e., when A is (nearly) rank-deficient. This solution can be efficiently computed (see §3.4) and interesting connections with the biased regression estimators have been deduced in Chapter 9. Nevertheless, a more thorough statistical and numerical analysis is needed in order to reveal more clearly the merits of the nongeneric TLS estimator in multicollinearity problems.

- The TLS technique requires a reliable *rank determinator* for computing the correct rank of the data matrix $[A; B]$ (Step 2 of Algorithm 3.1) and for checking (non)singularity of the matrix Γ (Step 3 of Algorithm 3.1). If the rank is unknown, the SVD is the most reliable tool to treat rank deficiency. It allows us to compute the rank from the number of singular values beyond a certain error-dependent bound. The derivation of this bound in the TLS algorithms, presented in this book, is based on the maximal minimal signal-to-noise ratio (see §3.6.1) but other strategies are possible. A deeper insight into the effects of *noise* and its *distribution* onto the SVD of a matrix is certainly required. It enables the design of more refined rank determinators for computing the correct rank of the data $[A; B]$ and the multiplicity of its singular values. Moreover, it allows us to derive reliable and robust rules for checking (non)singularity of matrices in function of the characteristics of the perturbations on the data. This test is required in the last step of the TLS algorithm before computing the TLS solution from the right singular vector(s) of the data.

- Furthermore, the presented TLS algorithms are not well suited for *parallel implementation* and do not yield good *on-line computation schemes*, e.g., in recursive form. These schemes enable us to update the solution at a low computational cost as soon as new data are coming in. The development of efficient on-line schemes for the TLS computations, suitable for parallel implementation, would certainly enhance the use of TLS in signal processing, automatic control, etc. A first scheme for tracking slowly varying TLS problems is already under study [107], [108] but further research is needed.

- Finally, the use of TLS in many applications arising in very diverse domains (see the Introduction) proves the advantages in reliability, accuracy, efficiency, etc. of the TLS method analyzed in this book, and confirms its practical applicability. Nevertheless, a lot of *unexplored* problems still remain where TLS could be successfully applied to improve the quality of the results. It is hoped that this book will aid and *stimulate* the reader to *apply* TLS in his or her own applications and problems.

References

[1] J. O. AASEN, *On the reduction of a symmetric matrix to tridiagonal form,* BIT, 11 (1971), pp. 233–242.

[2] T. J. ABATZOGLOU, G. A. HARADA, AND M. SHINE, *Total least squares techniques for high resolution direction finding,* Proc. IEEE MILCOM88, San Diego, CA, 1988, pp. 405–409.

[3] T. J. ABATZOGLOU AND J. M. MENDEL, *Constrained total least squares,* Proc. IEEE Internat. Conf. on Acoustics, Speech and Signal Processing, Dallas, TX, April 1987, pp. 1485–1488.

[4] T. J. ABATZOGLOU, J. M. MENDEL, AND G. A. HARADA, *The constrained total least squares technique and its application to harmonic superresolution,* IEEE Trans. Signal Processing, SP-39 (1991), pp. 1070–1087.

[5] T. J. ABATZOGLOU AND V. SOON, *Constrained total least squares applied to frequency estimation of sinusoids,* 4th IEEE ASSP Workshop on Spectrum Analysis and Modeling, Minneapolis, MN, 1988, pp. 250–252.

[6] R. J. ADCOCK, *A problem in least squares,* The Analyst, 5 (1878), pp. 53–54.

[7] L. AMMANN and J. VAN NESS, *A routine for converting regression algorithms into corresponding orthogonal regression algorithms,* ACM Trans. Math. Software, 14 (1988), pp. 76–87.

[8] L. P. AMMANN AND J. W. VAN NESS, *Standard and robust orthogonal regression,* Comm. Statist. B–Simulation Comput., 18 (1989), pp.145–162.

[9] T. W. ANDERSON, *Estimation of linear functional relationships: approximate distributions and connections with simultaneous equations in econometrics,* J. Roy. Statist. Soc. Ser. B, 38 (1976), pp. 1–20.

[10] ——, *The* 1982 *Wald memorial lectures: Estimating linear statistical relationships,* Ann. Statist., 12 (1984), pp. 1–45.

[11] J. ANGELES, K. ANDERSON, AND C. GOSSELIN, *An orthogonal-decomposition algorithm for constrained least-squares optimization,* in Proc. 13th ASME Design and Automat. Conf., Boston, 1987, pp. 215–220.

[12] M. AOKI AND P. C. YUE, *On a priori estimates of some identification methods,* IEEE Trans. Automat. Control, AC-15 (1970), pp. 541–548.

[13] K. S. ARUN, *A unitarily constrained total least squares problem in signal processing,* SIAM J. Matrix Anal. Appl., 13 (1992), pp. 729–745.

Illinois, Urbana Champaign, IL, 1080; SIAM J. Matrix Anal. Appl., to appear.

[14] V. BARWELL AND A. GEORGE, *A comparison of algorithms for solving symmetric indefinite systems of linear equations*, ACM Trans. Math. Software, 2 (1976), pp. 242–251.

[15] F. L. BAUER, *Optimally scaled matrices*, Numer. Math, 5 (1963), pp. 73–87.

[16] S. BEGHELLI, R. P. GUIDORZI, AND U. SOVERINI, *Structural identification of multivariable systems by the eigenvector method*, Internat. J. Control, 46 (1987), pp. 671–678.

[17] C. H. BISCHOF AND P. C. HANSEN, *Structure-preserving and rank-revealing QR-factorizations*, SIAM J. Sci. Stat. Comput., 12 (1991), pp. 1332–1350.

[18] Å. BJÖRCK, *Least squares methods*, in Handbook of Numerical Analysis, Vol. I: Finite Difference Methods; Solution of Equations in $\mathcal{R}^n$, P. G. Ciarlet and J. L. Lions, eds., North-Holland, Amsterdam, 1990.

[19] A. M. BLOCH, *Total least squares estimation in infinite dimensions and completely integrable Hamiltonian systems*, in Theory and Applications of Nonlinear Control Systems, Proc. 7th Internat. Symp. on Mathematical Theory of Networks and Systems (MTNS), Royal Inst. Tech., Stockholm, Sweden, 1985, C. I. Byrnes and A. Lindquist, eds., Elsevier, New York, 1986, pp. 241–246.

[20] P. T. BOGGS, R. H. BYRD, AND R. B. SCHNABEL, *A stable and efficient algorithm for nonlinear orthogonal distance regression*, SIAM J. Sci. Statist. Comput., 8 (1987), pp. 1052–1078.

[21] H. BOWDLER, R. S. MARTIN, C. REINSCH, AND J. H. WILKINSON, *The* QR *and* QL *algorithms for symmetric matrices*, Numer. Math., 11 (1968), pp. 293–306.

[22] J. R. BUNCH, *Analysis of the diagonal pivoting method*, SIAM J. Numer. Anal., 8 (1971), pp. 656–680.

[23] J. R. BUNCH AND L. KAUFMAN, *Some stable methods for calculating inertia and solving symmetric linear systems*, Math. Comput., 31 (1977), pp. 163–179.

[24] J. R. BUNCH AND B. N. PARLETT, *Direct methods for solving symmetric indefinite systems of linear equations*, SIAM J. Numer. Anal., 8 (1971), pp. 639–655.

[25] R. J. CAROLL, P. GALLO, AND L. J. GLESER, *Comparison of least squares and errors-in-variables regression, with special reference to randomized analysis of covariance*, J. Amer. Statist. Assoc., 80 (1985), pp. 929–932.

[26] T. F. CHAN, *An improved algorithm for computing the singular value decomposition*, ACM Trans. Math. Software, 8 (1982), pp. 72–83.

[27] ——, *Rank revealing* QR *factorizations*, Linear Algebra Appl., 88/89 (1987), pp. 67–82.

[28] T. F. CHAN and P. C. HANSEN, *Computing truncated* SVD *least squares solutions by rank revealing* QR *factorizations*, SIAM J. Sci. Statist. Comput., 11 (1990), pp. 519–530.

[29] ——, *Some applications of the rank revealing QR factorization*, SIAM J. Sci. Statist. Comput., 13 (1992), pp. 727–741.

[30] C. L. CHENG AND J. W. VAN NESS, *Robust errors-in-variables regr*
Report 179, Programs in Math. Sciences, University of Texas, Dallas, TX, 1987.

[31] K. L. KUNG, *A Course in Probability Theory*, Harcourt, Brace and World, New

[32] P. Comon and G. H. Golub, *Tracking a few extreme singular values and vectors in signal processing*, Proc. IEEE, 78 (1990), pp. 1327–1343.

[33] J. K. Cullum and R. A. Willoughby, *Lanczos algorithms for large symmetric eigenvalue computations,* Vol. I, *Theory*, Progr. Sci. Comput., Vol. 3, Birkhäuser, Boston, 1985.

[34] C. Davis and W. M. Kahan, *The rotation of eigenvectors by a perturbation* III, SIAM J. Numer. Anal., 7 (1970), pp. 1–46.

[35] J. W. Demmel, *The smallest perturbation of a submatrix which lowers the rank and constrained total least squares problems*, SIAM J. Numer. Anal., 24 (1987), pp. 199–206.

[36] ——, *On structured singular values*, Proc. 27th Conf. on Decision and Control, Austin, TX, 1988, Vol. 3, pp. 2138–2143; also appeared in Recent Advances in Robust Control, P. Dorato and R. K. Yedavalli, eds., IEEE Press, 1990, pp. 397–402.

[37] J. Demmel and W. Kahan, *Accurate singular values of bidiagonal matrices*, SIAM J. Sci. Statist. Comput., 11 (1990), pp. 873–912.

[38] B. De Moor, *Total linear least squares with inequality constraints.* ESAT-SISTA Report 1990-02, Department of Electrical Engrg., Katholieke Universiteit Leuven, Leuven, Belgium, 1990.

[39] ——, *Total linear least squares with inequality constraints*, ESAT-SISTA Report 1990-02, Dept. of Electrical Engrg., Katholieke Universiteit Leuven, Leuven, Belgium, 1990 SIAM J. Matrix Anal. Appl., to appear.

[40] B. L. R. De Moor and G. H. Golub, *The restricted singular value decomposition: properties and applications*, Tech. Report NA-89-03, Computer Sciences Dept., Stanford University, Stanford, CA, 1989; SIAM J. Matrix Anal. Appl., 12 (1991), pp. 401–425.

[41] B. De Moor, J. Staar, and J. Vandewalle, *Oriented energy and oriented signal-to-signal ratio concepts in the analysis of vector sequences and time series*, in SVD and Signal Processing: Algorithms, Applications and Architectures, E. F. Deprettere, ed., Elsevier, 1988, pp. 209–232.

[42] B. De Moor and J. Vandewalle, *An adaptive singular value decomposition algorithm based on generalized Chebyshev recursion*, in Proc. Conf. Mathematics in Signal Processing, University of Bath, Bath, UK, 1985, T. S. Durrani, J. B. Abbiss, J. E. Hudson, R. W. Madan, J. G. McWhirter, and T. A. Moore, eds., Clarendon Press, Oxford, UK, 1987, pp. 607–635.

[43] ——, *A geometrical approach to the maximal corank problem in the analysis of linear relations*, Proc. 25th IEEE Conf. Decision and Control, Athens, Greece, 1986, pp. 1990–1995.

[44] ——, *A unifying theorem for linear and total linear least squares*, IEEE Trans. Automat. Control, AC-35 (1990), pp. 563–566.

[45] J. J. Dongarra, J. R. Bunch, C. B. Moler and G. W. Stewart, LINPACK *user's guide*, Society for Industrial and Applied Mathematics, Philadelphia, PA, 1979.

[46] J. J. Dongarra and E. Grosse, *Distribution of mathematical software via electronic mail*, Comm. ACM, 30 (1987), pp. 403–407.

[47] N. R. Draper and H. Smith, *Applied regression analysis*, 2nd ed., John Wiley, New York, 1981.

[48] G. ECKART AND G. YOUNG, *The approximation of one matrix by another of lower rank*, Psychometrica, 1 (1936), pp. 211–218.

[49] K. V. FERNANDO AND H. NICHOLSON, *Identification of linear systems with input and output noise: the Koopmans-Levin method*, IEE Proc. D, 132 (1985), pp. 30–36.

[50] G. W. FISHER, *Matrix analysis of metamorphic mineral assemblages and reactions*, Contrib. Mineralogy and Petrology, 102 (1989), pp. 69–77.

[51] E. FOGEL, *Total least squares for Toeplitz structures*, in Proc. 21st IEEE Conf. Decision and Control, Orlando, FL, 1982, Vol. 3, pp. 1003–1004.

[52] J. FRANCIER, *Iteratieve on-line algoritmen voor het oplossen van traagvariërende totale kleinste kwadratenproblemen*, Master's thesis, Dept. of Electrical Engrg., Katholieke Universiteit Leuven, Leuven, Belgium, 1990.

[53] W. A. FULLER, *Measurement Error Models*. John Wiley, New York, 1987.

[54] P. P. GALLO, *Properties of estimators in errors-in-variables models*, Ph.D. thesis, Institute of Statistics Mimeoseries # 1511, University of North Carolina, Chapel Hill, NC, 1982.

[55] ——, *Consistency of regression estimates when some variables are subject to error*, Comm. Statist. Theory Methods, 11 (1982), pp. 973–983.

[56] F. R. GANTMACHER, *The Theory of Matrices*, Chelsea, New York, 1959.

[57] C. F. GAUSS, *Theoria combinationis observationum erroribus minimis obnoxiae*, Comment. Soc. Reg. Sci. Gotten. Recent., 5 (1823), pp. 33–90.

[58] L. J. GLESER, *Calculation and simulation in errors-in-variables regression problems*, Mimeograph Ser. 78-5, Statistics Dept., Purdue University, West Lafayette, IN, 1978.

[59] ——, *Estimation in a multivariate "errors in variables" regression model: Large sample results*, Ann. Statist., 9 (1981), pp. 24–44.

[60] L. J. GLESER AND G. S. WATSON, *Estimation of a linear transformation*, Biometrika, 60 (1973), pp. 525–534.

[61] G. H. GOLUB, *Some modified matrix eigenvalue problems*, SIAM Rev., 15 (1973), pp. 318–344.

[62] G. H. GOLUB, A. HOFFMAN, AND G. W. STEWART, *A generalization of the Eckart–Young–Mirsky matrix approximation theorem*, Linear Algebra Appl., 88/89 (1987), pp. 322–327.

[63] G. H. GOLUB AND W. KAHAN, *Calculating the singular values and pseudoinverse of a matrix*, SIAM J. Numer. Anal. Ser. B, 2 (1965), pp. 205–224.

[64] G. H. GOLUB AND M.D. KENT, *Estimates of eigenvalues for iterative methods*, Numer. Anal. Project, Manuscript NA-87-02, Computer Science Dept., Stanford University, Stanford, CA, 1987.

[65] G. H. GOLUB AND R. J. LEVEQUE, *Extensions and uses of the variable projection algorithm for solving nonlinear least squares problems*, in Proc. 1979 Army Numerical Analysis and Computers Conf., White Sands Missile Range, White Sands, NM, 1979, pp. 1–12.

[66] G. H. GOLUB, F. T. LUK, AND M. L. OVERTON, *A block Lanczos method for computing the singular values and corresponding singular vectors of a matrix*, ACM Trans. Math. Software, 7 (1981), pp. 149–169.

[67] G. H. GOLUB AND C. REINSCH, *Singular value decomposition and least squares solutions*, Numer. Math., 14 (1970), pp. 403–420.

[68] G. H. GOLUB AND C. F. VAN LOAN, *An analysis of the total least squares

problem, SIAM J. Numer. Anal., 17 (1980), pp. 883–893.

[69] G. H. GOLUB AND C. F. VAN LOAN, *Matrix Computations*, 2nd ed., Johns Hopkins University Press, Baltimore, MD, 1989.

[70] R. P. GUIDORZI, *Canonical structures in the identification of multivariable systems*, Automatica, 11 (1975), pp. 361–374.

[71] R. P. GUIDORZI, M. P. LOSITO AND T. MURATORI, *The range error test in the structural identification of linear multivariable systems*, IEEE Trans. Automat. Control, 27 (1982), pp. 1044–1054.

[72] R. F. GUNST AND R. L. MASON, *Biased estimation in regression: an evaluation using the mean squared error*, J. Amer. Statist. Assoc. 72 (1977), pp. 616–628.

[73] ——, *Regression Analysis and Its Application: A Data-Oriented Approach*, Marcel Dekker, New York, 1980.

[74] R. F. GUNST, J. T. WEBSTER, AND R. L. MASON, *A comparison of least squares and latent root regression estimators*, Technometrics, 18 (1976), pp. 75–83.

[75] H. R. HALL AND R. J. BERNHARD, *Total least squares solutions to acoustic boundary element models*, in Proc. Internat. Symposium on Numerical Techniques in Acoustic Radiation, ASME Winter Annual Meeting, San Francisco, CA, 1989, R. J. Bernhard and R. F. Keltie, eds., American Society of Mechanical Engineers, New York, 1989, NCA, Vol. 6, pp. 145–152.

[76] F. R. HAMPEL, E. M. RONCHETTI, P. J. ROUSSEEUW, AND W. A. STAHEL, *Robust Statistics*, John Wiley, New York, 1986.

[77] P. C. HANSEN, SVD-*theory and applications*, Report No. 84-05, Technical High School, Lyngby, Denmark, 1984.

[78] D. M. HAWKINS, *On the investigation of alternative regressions by principal component analysis*, Appl. Statist., 22 (1973), pp. 275–286.

[79] R. R. HOCKING, *The analysis and selection of variables in linear regression*, Biometrics, 32 (1976), pp. 1–49.

[80] ——, *Developments in linear regression methodology* 1959–1982, Technometrics, 25 (1983), pp. 219–230.

[81] A. E. HOERL AND R. W. KENNARD, *Ridge regression* —1980. *Advances, algorithms and applications*, Amer. J. Math. Management Sci., 1 (1981), pp. 5–83.

[82] P. J. HUBER, *Robust Statistics*, John Wiley, New York, 1981.

[83] Y. S. HUNG AND A. G. J. MACFARLANE, *Multivariate feedback: a quasi-classical approach*, in Lecture Notes in Control and Inform. Sci., 40, Springer-Verlag, New York, 1982.

[84] L. S. JENNINGS AND M. R. OSBORNE, *A direct error analysis for least squares*, Numer. Math., 22 (1974), pp. 322–332.

[85] J. H. JUSTICE AND A. A. VASSILIOU, *Diffraction tomography for geophysical monitoring of hydrocarbon reservoirs*, in Proc. IEEE, 78 (1990), pp. 711–722.

[86] R. E. KALMAN, *System identification from noisy data*, in Dynamical Systems II, A. R. Bednarek and L. Cesari, eds., Academic Press, New York, 1982, pp. 135–164.

[87] G. KELLY, *The influence function in the errors in variables problem*, Ann. Statist., 12 (1984), pp. 87–100.

[88] M. G. KENDALL AND A. STUART, *The Advanced Theory of Statistics*, 4th ed., Vol. 2, Griffin, London, 1979.

[89] R. W. KENNY, D. M. ACKERY, J. S. FLEMING, B. A. GODDARD, AND R. W. GRANT, *Deconvolution analysis of the scintillation camera renogram*, British J. Radiol., 48 (1975), pp. 481–486.

[90] R. H. KETELLAPPER, *The relevance of large-sample properties of estimators for the errors-in-variables model: A Monte Carlo study*, Comm. Statist., Ser. B, 11 (1982), pp. 625–634.

[91] ——, *On estimating parameters in a simple linear errors-in-variables model*, Technometrics, 25 (1983), pp. 43–47.

[92] T. C. KOOPMANS, *Linear Regression Analysis of Economic Time Series*, De Erven F. Bohn, N.V. Haarlem, the Netherlands, 1937.

[93] S. KOUROUKLIS AND C. C. PAIGE, *A constrained least squares approach to the general Gauss–Markov linear model*, J. Amer. Statist. Assoc., 76 (1981), pp. 620–625.

[94] W. S. KRASKER AND R. E. WELSCH, *Efficient bounded-influence regression estimation*, J. Amer. Statist. Assoc., 77 (1982), pp. 595–604.

[95] W. E. LARIMORE AND F. T. LUK, *System identification and control using* SVD*s of systolic arrays*, Proc. SPIE, High Speed Comput., Vol. 880, 1988, pp. 37–48.

[96] C. L. LAWSON AND R. J. HANSEN, *Solving Least Squares Problems*, Prentice–Hall, Englewood Cliffs, NJ, 1974.

[97] J. LEURIDAN, D. DE VIS, H. VAN DER AUWERAER, AND F. LEMBREGTS, *A comparison of some frequency response function measurement techniques*, Proc. 4th Int. Modal Anal. Conf., Los Angeles, CA, 1986, pp. 908–918.

[98] M. J. LEVIN, *Estimation of a system pulse transfer function in the presence of noise*, IEEE Trans. Automat. Control AC-9 (1964), pp. 229–235.

[99] J. LEWIS, *Algorithms for sparse matrix eigenvalue problems*, Tech. Report STAN-CS-77-595, Dept. of Computer Science, Stanford University, Stanford, CA, 1977.

[100] D. LUDWIG, C. J. WALTERS, AND J. COOKE, *Comparison of two models and two estimation methods for catch and effort data*, Natur. Re. Modeling, 2 (1988), pp. 457–498.

[101] A. MADANSKY, *The fitting of straight lines when both variables are subject to error*, J. Amer. Statist. Assoc., 54 (1959), pp. 173–205.

[102] J. MANDEL, *Use of the singular value decomposition in regression analysis*, Amer. Statist., 36 (1982), pp. 15–24.

[103] L. MIRSKY, *Symmetric gauge functions and unitarily invariant norms*, Quart. J. Math. Oxford, 11 (1960), pp. 50–59.

[104] C. MOLER, J. LITTLE, S. BANGERT, AND S. KLEIMAN, PRO-MATLAB *for* VAX/VMS *Computers*, Version 3.34-VMS, MathWorks, Sherborn, MA, 1988.

[105] M. MOONEN, B. DE MOOR, L. VANDENBERGHE, AND J. VANDEWALLE, *On- and off-line identification of linear state-space models*, Internat. J. Control, 49 (1989), pp. 219–232.

[106] M. MOONEN AND J. VANDEWALLE, *A* QSVD *approach to on- and off-line state-space identification*, Internat. J. Control, 51 (1990), pp. 1133–1146.

[107] M. MOONEN, P. VAN DOOREN, AND J. VANDEWALLE, *Updating singular value decompositions. A parallel implementation*, in Proc. SPIE, Advanced Algorithms and Architectures for Signal Processing, IV, San Diego, CA, 1989, Vol. 1152, paper 1152–8.

[108] ——, *An SVD updating algorithm for subspace tracking*, SIAM J. Matrix Anal. Appl., 13 (1992), pp. 1015–1038.

1989-13(a), ESAT Lab., Dept. of Electrical Engrg., Katholieke Universiteit Leuven, Leuven, Belgium, 1989; SIAM J. Matrix Anal. Appl., to appear.

[109] ——, *A systolic array for SVD updating*, SIAM J. Matrix Anal. Appl., 14 (1993), pp. 353–371.

[110] P. A. P. Moran, *Estimating structural and functional relationships*, J. Multivariate Anal., 1 (1971), pp. 232–255.

[111] M. R. Osborne and G.A. Watson, *An analysis of the total approximation problem in separable norms, and an algorithm for the total l_1 problem*, SIAM J. Sci. Statist. Comput., 6 (1985), pp. 410–424.

[112] B. Ottersten, M. Viberg, and T. Kailath, *Asymptotic analysis of the total least squares ESPRIT algorithm*, in Proc. 33rd SPIE Internat. Tech. Symp., Advanced Algorithms and Architectures for Signal Processing, IV, San Diego, CA, 1989.

[113] ——, *Performance analysis of the total least squares ESPRIT algorithm*, IEEE Trans. on Signal Processing, 1991, SP-39, May 1991.

[114] C. C. Paige, *Error analysis of the Lanczos algorithm for tridiagonalizing a symmetric matrix*, J. Inst. Math. Appl. 18 (1976), pp. 341–349.

[115] ——, *Accuracy and effectiveness of the Lanczos algorithm for the symmetric eigenproblem*, Linear Algebra Appl., 34 (1980), pp. 235–258.

[116] ——, *The general linear model and the generalized singular value decomposition*, Linear Algebra Appl., 70 (1985), pp. 269–284.

[117] C. C. Paige and M. A. Saunders, *Towards a generalized singular value decomposition*, SIAM J. Numer. Anal. 18, (1981), pp. 398–405.

[118] B. N. Parlett, *The Rayleigh quotient iteration and some generalizations for nonormal matrices*, Math. Comput., 28 (1974), pp. 679–693.

[119] ——, *The symmetric eigenvalue problem*, Prentice–Hall, Englewood Cliffs, NJ, 1980.

[120] W. M. Patefield, *Multivariate linear relationships: maximum likelihood estimation and regression bounds*, J. Roy. Statist. Soc. Ser. B, 43 (1981), pp. 342–352.

[121] K. Pearson, *On lines and planes of closest fit to points in space*, Philos. Mag., 2 (1901), pp. 559–572.

[122] G. Peters and J. H. Wilkinson, *Inverse iteration, ill-conditioned equations and Newton's method*, SIAM Rev., 21 (1979), pp. 339–360.

[123] V. F. Pisarenko, *The retrieval of harmonics from a covariance function*, Geophys. J. Roy. Astron. Soc., 33 (1973), pp. 347- -366.

[124] J. Ragot and M. Aubrun, *Application de la régression orthogonale sous contrainte linéaire à un problème d'équilibrage de bilan-matière*, Rev. Statist. Appl., Vol. XXX, No. 2 (1982), pp. 45–56.

[125] M. A. Rahman and K. B. Yu, *Total least squares approach for frequency estimation using linear prediction*, IEEE Trans. Acoust. Speech Signal Process., ASSP-35 (1987), pp. 1440- 1454.

[126] B. D. Rao and K. V. S. Hari, *Performance analysis of* ESPRIT, *minimum-norm method and TAM in determining the direction of arrival of plane waves in noise*, IEEE Trans. Acoust. Speech Signal Process., ASSP-37 (1989), pp. 1990–1994.

[127] P. M. Reilly and H. Patino-leal, *A Bayesian study of the errors-in-*

variables models, Technometrics, 23 (1981), pp. 221–231.

[128] T. J. Rivlin, *The Chebyshev Polynomials*, John Wiley, New York, 1976.

[129] R. Rost and J. Leuridan, *A comparison of least squares and total least squares for multiple input estimation of frequency response functions*, ASME Paper No. 85-DET-105 presented at the ASME Design Engineering Division Conference and Exhibit on Mechanical Vibration and Noise, Cincinnati, OH, 1985.

[130] R. H. Roy, ESPRIT—*Estimation of signal parameters via rotational invariance techniques*, Ph.D. thesis, Dept. of Electrical Engrg., Stanford University, Stanford, CA, 1987.

[131] R. Roy and T. Kailath, *Total least-squares* ESPRIT, Proc. 21st Ann. Asilomar Conf. on Signals, Systems and Computers, Pacific Grove, CA, 1987, pp. 297–301.

[132] ——, ESPRIT—*Estimation of signal parameters via rotational invariance techniques*, Optical Engrg., 29 (1990), pp. 296–313.

[133] ——, ESPRIT—*Estimation of signal parameters via rotational invariance techniques*, IEEE Trans. Acoust. Speech Signal Process., ASSP-37 (1989), pp. 984–995.

[134] R. Roy, A. Paulraj, and T. Kailath, ESPRIT—*A subspace rotation approach to estimation of parameters of cisoids in noise*, IEEE Trans. Acoust. Speech Signal Process., ASSP-34 (1986), pp. 1340–1342.

[135] H. Rutishauser, *Computational aspects of F. L. Bauer's simultaneous iteration method*, Numer. Math., 13 (1969), pp. 4–13.

[136] ——, *Simultaneous iteration method for symmetric matrices*, Numer. Math., 16 (1970), pp. 205–223.

[137] Y. Saad, *Chebyshev acceleration techniques for solving nonsymmetric eigenvalue problems*, Math. Comput., 42 (1984), pp. 567–588.

[138] P. Schmidt, *Econometrics*, Marcel Dekker, New York, 1976.

[139] R. O. Schmidt, *Multiple emitter location and signal parameter estimation*, IEEE Trans. Antennas Propagat., AP-34 (1986), pp. 276–281.

[140] H. Schneeweiss, *Consistent estimation of a regression with errors in the variables*, Metrika, 23 (1976), pp. 101–115.

[141] H. Schwetlick and V. Tiller, *Numerical methods for estimating parameters in nonlinear models with errors in the variables.* Technometrics, 27 (1985), pp. 17–24.

[142] ——, *Nonstandard scaling matrices for trust region Gauss–Newton methods*, SIAM J. Sci. Statist. Comput., 10 (1989), pp. 654–670.

[143] D. S. Scott, *Analysis of the symmetric Lanczos process*, UCB-ERL Tech. Report M78/40, University of California, Berkeley, CA, 1978.

[144] M.T. Silvia and E. C. Tacker, *Regularization of Marchenko's integral equation by total least squares*, J. Acoust. Soc. Amer., 72 (1982), pp. 1202–1207.

[145] B. T. Smith, J. M. Boyle, Y. Ikebe, V. C. Klema, and C. B. Moler, *Matrix Eigensystem Routines*: EISPACK *Guide*, 2nd ed., Springer–Verlag, New York, 1970.

[146] F. W. Smith and W. B. Hilton, *Monte Carlo evaluation of methods for pulse transfer function estimation*, IEEE Trans. Automat. Control, AC–12 (1967), pp. 568–576.

[147] T. Söderström, *Identification of stochastic linear systems in presence of input*

noise, Automatica, 17 (1981), pp. 713–725.

[148] T. SÖDERSTRÖM AND P. STOICA, *Instrumental variable methods for system identification*, Springer-Verlag (LNICS series), Berlin, 1983.

[149] H. SPÄTH, *On discrete linear orthogonal L_p-approximation*, Z. Angew. Math. Mech., 62 (1982), pp. 354–355.

[150] ——, *Orthogonal least squares fitting with linear manifolds*, Numer. Math., 48 (1986), pp. 441–445.

[151] H. SPÄTH AND G. A. WATSON, *On orthogonal linear l_1 approximation*, Numer. Math., 51 (1987), pp. 531–543.

[152] J. M. SPEISER, *An overview of matrix-based signal processing*, in Proc. 21st Asilomar Conf. Signals, Systems and Computers, Pacific Grove, CA, 1987, Vol. 1, pp. 284–289.

[153] J. SPEISER AND C. VAN LOAN, *Signal processing computations using the generalized singular value decomposition*, in Proc. SPIE, Real Time Signal Processing VII, Vol. 495, paper 495–32, 1984, pp. 47–55.

[154] P. SPRENT, *A generalized least-squares approach to linear functional relationships*, J. Roy. Statist. Soc. Ser. B28, 1966, pp. 278–297.

[155] ——, *Models in Regression and Related Topics*, Methuen, London, 1969.

[156] J. STAAR, *Concepts for reliable modelling of linear systems with application to on-line identification of multivariable state space descriptions*, Ph.D. thesis, Dept. of Electrical Engrg., Katholieke Universiteit Leuven, Leuven, Belgium, June 1982.

[157] C. M. STEIN, *Multiple regression contributions to probability and statistics*, in Essays in Honor of Harold Hotelling, I. Olkin, ed., Stanford University Press, Stanford, CA, 1960, pp. 424-443.

[158] G. W. STEWART, *Error and perturbation bounds for subspaces associated with certain eigenvalue problems*, SIAM Rev., 15 (1973), pp. 727–764.

[159] ——, *On the perturbation of pseudo-inverses, projections and linear least squares problems*, SIAM Rev., 19 (1977), pp. 634–662.

[160] ——, *On the invariance of perturbed null vectors under column scaling*, Numer. Math., 44 (1984), pp. 61–65.

[161] ——, *Rank degeneracy*, SIAM J. Sci. Statist. Comput., 5 (1984), pp. 403–413.

[162] ——, *Collinearity and least squares regression*, Statist. Sci., 2 (1987), pp. 68–100.

[163] ——, *Perturbation theory and least squares with errors in the variables*, Tech. Report UMIACS-TR-89-97, CS-TR 2326, Dept. of Computer Sciences, University of Maryland, College Park, MD, 1989; also appeared in Contemporary Mathematics 112: Statistical Analysis of Measurement Error Models and Applications, P. J. Brown and W. A. Fuller, eds., Amer. Math. Soc., Providence, RI, pp. 171–181.

[164] ——, *An updating algorithm for subspace tracking*. IEEE Trans. Signal Process., 40 (1992), pp. 1535–1541.

[165] G. W. STEWART AND J. G. SUN, *Matrix perturbation theory*, Academic Press, New York, 1990.

[166] A. STOIAN, P. STOICA, AND S. VAN HUFFEL, *Comparative performance study of least-squares and total least-squares Yule-Walker estimates of autoregressive parameters*, Scientific Bulletin, Electrical Eng. Series, Polytechnic Institute of Bucharest, Vol. 52, No. 2, 1990, pp. 81–89.

of Bucharest, Roumania, 1990; Sci. Bull. of the Bucharest Polytechnic Inst., to appear.

[167] P. STOICA, B. FRIEDLANDER, AND T. SÖDERSTRÖM, *A high-order Yule-Walker method for estimation of the* AR *parameters of an* ARMA *model*, Syst. Control Lett., 11 (1989), pp. 99–105.

[168] P. STOICA AND T. SÖDERSTRÖM, *Bias correction in least squares identification*, Internat. J. Control, 35 (1982), pp. 449–457.

[169] R. C. THOMPSON, *Principal submatrices* IX: *interlacing inequalities for singular values of submatrices*, Linear Algebra Appl., 5 (1972), pp. 1–12.

[170] C. F. TIRENDI AND J. F. MARTIN, *Quantitative analysis of* NMR *spectra by linear prediction and total least squares*, J. Magnetic Resonance, 85 (1989), pp. 162–169.

[171] D. W. TUFTS AND R. KUMARESAN, *Singular value decomposition and improved frequency estimation using linear prediction*, IEEE Trans. Acoust. Speech Signal Process., ASSP-30 (1982), pp. 671–675.

[172] M. E. VALENTINUZZI AND E. M. MONTALDO VOLACHEC, *Discrete deconvolution*, Med. Biol. Engrg., 13 (1975), pp. 123–125.

[173] A. VAN DER SLUIS, *Condition, equilibration, and pivoting in linear algebraic systems*, Numer. Math., 15 (1970), pp. 74– 86.

[174] ——, *Stability of the solution of linear least squares problems*, Numer. Math., 23 (1975), pp. 241–254.

[175] A. VAN DER SLUIS AND G. W. VELTKAMP, *Restoring rank and consistency by orthogonal projection*, Linear Algebra Appl. , 28 (1979), pp. 257–278.

[176] S. VAN HUFFEL, *Documented Fortran 77 programs of the extended classical total least squares algorithm, the partial singular value decomposition algorithm and the partial total least squares algorithm*, Internal. Report ESAT-KUL 88/1, ESAT Lab., Dept. of Electrical Engrg., Katholieke Universiteit Leuven, Leuven, Belgium, 1988.

[177] ——, *Analysis of the total least squares problem and its use in parameter estimation*, Ph.D. thesis, Dept. of Electrical Engrg., Katholieke Universiteit Leuven, Leuven, Belgium, 1987.

[178] ——, *The extended classical total least squares algorithm*, J. Comput. Appl. Math., 25 (1989), pp. 111–119.

[179] ——, *The generalized total least squares problem: formulation, algorithm and properties*, in Numerical Linear Algebra, Digital Signal Processing and Parallel Algorithms, G. Golub and P. Van Dooren, eds., NATO ASI Series F, Springer-Verlag, Berlin, 1990, pp. 651- -660.

[180] ——, *Iterative algorithms for computing the singular subspace of a matrix associated with its smallest singular values*, Linear Algebra Appl., 154-156 (1991), pp. 675–709.

[181] S. VAN HUFFEL AND J. VANDEWALLE, *The use of total linear least squares techniques for identification and parameter estimation*, in Proc. 7th IFAC/IFORS Symposium on Identification and System Parameter Estimation, York, UK, 1985, Vol. 2, pp. 1167–1172.

[182] ——, *Subset selection using the total least squares approach in collinearity problems with errors in the variables*, Linear Algebra Appl., 88/89 (1987), pp. 695–714.

[183] ——, *Algebraic relationships between classical regression and total least-squares estimation*, Linear Algebra Appl., 93 (1987), pp. 149–162.

[184] ——, *The partial total least squares algorithm*, J. Comput. Appl. Math., 21 (1988), pp. 333–341.

[185] ——, *Analysis and solution of the nongeneric total least squares problem*, SIAM J. Matrix Anal. Appl., 9 (1988), pp. 360–372.

[186] ——, *Iterative speed improvement for solving slowly varying total least squares problems*, Mech. Syst. Sign. Proc., 2 (1988), pp. 327–348.

[187] ——, *Algebraic connections between the least squares and total least squares problems*, Numer. Math., 55 (1989), pp. 431–449.

[188] ——, *Analysis and properties of the generalized total least squares problem $AX \approx B$ when some or all columns of A are subject to errors*, SIAM J. Matrix Anal. Appl., 10 (1989), pp. 294–315.

[189] ——, *Reliable and efficient techniques based on total least squares for computing consistent estimators in models with errors in the variables*, in Mathematics in Signal Processing II, J. G. McWhirter, ed., Clarendon Press, Oxford, UK, 1990, pp.593–603.

[190] ——, *On the accuracy of total least squares and least squares techniques in the presence of errors on all data*, Automatica, 25 (1989), pp. 765–769.

[191] ——, *Comparison of total least squares and instrumental variable methods for parameter estimation of transfer function models*, Internat. J. Control, 50 (1989), pp. 1039–1056.

[192] S. Van Huffel, J. Vandewalle, M. Ch. De Roo, and J. L. Willems, *Reliable and efficient deconvolution technique based on total linear least squares for calculating the renal retention function*, Med. Biol. Engrg. Comput., 25 (1987), pp. 26–33.

[193] S. Van Huffel, J. Vandewalle, and A. Haegemans, *An efficient and reliable algorithm for computing the singular subspace of a matrix, associated with its smallest singular values*, J. Comput. Appl. Math., 19 (1987), pp. 313–330.

[194] S. Van Huffel and H. Zha, *The restricted total least squares problem: formulation, algorithm and properties*, SIAM J. Matrix Anal. Appl., 12 (1991), pp. 292–309.

[195] C. F. Van Loan, *Generalizing the singular value decomposition*, SIAM J. Numer. Anal., 13 (1976), pp. 76–83.

[196] ——, *Computing the CS and generalized singular value decomposition*, Numer. Math., 46 (1985), pp. 479–491.

[197] R. K. Ward, *Comparison and diagnosis of errors for six parameter estimation methods*, Internat. J. Sys. Sci., 15 (1984), pp. 745–758.

[198] G. A. Watson, *Numerical methods for linear orthogonal L_p approximation*, IMA J. Numer. Anal., 2 (1982), pp. 275–287.

[199] ——, *The total approximation problem*, in approximation theory IV, C. K. Chui, L. L. Schumacher, and J. D. Ward, eds., Academic Press, New York, 1983, pp. 723–728.

[200] ——, *The numerical solution of total l_p approximation problems*, Proc. Numerical Analysis, Dundee, UK, 1983, Lecture Notes in Math., Springer-Verlag, Berlin, 1984, Vol. 1066, pp. 221–237.

[201] ——, *On a class of algorithms for total approximation*, J. Approx. Theory., 45

(1085), pp. 219 231.

[202] ——, *The smallest perturbation of a submatrix that lowers the rank of the matrix*, IMA J. Numer. Anal., 8 (1988), pp. 295–303.

[203] J. T. WEBSTER, R. F. GUNST, AND R. L. MASON, *Latent root regression analysis*, Technometrics, 16 (1974), pp. 513–522.

[204] P. A. WEDIN, *Perturbation bounds in connection with the singular value decomposition*, BIT, 12 (1972), pp. 99–111.

[205] ——, *Perturbation theory for pseudo-inverses*, BIT, 13 (1973), pp. 217–232.

[206] M. WEI, *The perturbation of consistent least squares problems*, Linear Algebra Appl., 112 (1989), pp. 231–245.

[207] ——, *The analysis for the total least squares problem with more than one solution*, SIAM J. Matrix. Anal. Appl., 13 (1992), pp. 746–764.

[208] J. H. WILKINSON, *The algebraic eigenvalue problem*, Clarendon Press, Oxford, UK, 1965.

[209] S. WOLD, A. RUHE, H. WOLD, AND W. J. DUNN, *The collinearity problem in linear regression, the partial least squares* (PLS) *approach to generalized inverses*, SIAM J. Sci. Statist. Comput., 5 (1984), pp. 735–743.

[210] D. YORK, *Least squares fitting of a straight line*, Canad. J. Phys., 44 (1966), pp. 1079–1086.

[211] R. H. ZAMAR, *Robust estimation in the errors-in-variables model*, Biometrika, 76 (1989), pp. 149–160.

[212] H. ZHA, *A numerical algorithm for computing the restricted singular value decomposition of matrix triplets*, Linear Algebra Appl., 168 (1992), pp. 1–25.

[213] ——, *The restricted singular value decomposition of matrix triplets*, SIAM J. Matrix Anal. Appl., 12 (1991), pp. 172–194.

[214] H. ZHA AND P. C. HANSEN, *Regularization and the general Gauss–Markov linear model*, Math. Comp., 55, (1990), pp. 613–624.

[215] H. ZHA, *Implicit QR factorization of a product of three matrices*, BIT, 31 (1991), pp. 375–379.

[216] K. ZIETAK, *Properties of the approximations of a matrix which lower its rank*, IMA J. Numer. Anal., 9 (1989), pp. 545–554.

[217] M. D. ZOLTOWSKI, *Signal processing applications of the method of total least squares*, in Proc. 21st Asilomar Conf. on Signals, Systems and Computers, Pacific Grove, CA, 1987, pp. 290–296.

[218] ——, *Generalized minimum norm and constrained total least squares with applications to array processing*, in Proc. SPIE, Advanced Algorithms and Architectures for Signal Processing, III, San Diego, CA, 1988, Vol. 975, pp. 78–85.

[219] M. D. ZOLTOWSKI AND T. S. LEE, *Frequency and power estimation of complex sinusoids via minimum norm* ESPRIT *and projection operator based total least squares*, Tech. Report, School of Electrical Engrg., Purdue University, West Lafayette, IN, 1989.

[220] M. D. ZOLTOWSKI AND D. STAVRINIDES, *Sensor array signal processing via a Procrustes rotations based eigenanalysis of the* ESPRIT *data pencil*, IEEE Trans. Acoust. Speech Signal Process., ASSP-37 (1989), pp. 832–861.

Index